AA001112

# 2006 Asian Optical Fiber Communication & Optoelectronic Exposition & Conference

# 24 - 27 October 2006

# Shanghai, China

**Copyright © 2004 by the Institute of Electrical and Electronics Engineers, Inc.**
**All Rights Reserved**

*Copyright and Reprint Permissions:* Abstracting is permitted with credit to the source. Libraries are permitted to photocopy beyond the limit of U.S. copyright law for private use of patrons those articles in this volume that carry a code at the bottom of the first page, provided the per-copy fee indicated in the code is paid through Copyright Clearance Center, 222 Rosewood Drive, Danvers, MA 01923.

For other copying, reprint or republication permission, write to IEEE Copyrights Manager, IEEE Service Center, 445 Hoes Lane, Piscataway, NJ 08854. All rights reserved.

IEEE Catalog Number:        06EX1532

ISBN 13:                    978-0-9789217-0-5

Additional Copies of This Publication Are Available From:

IEEE Service Center
445 Hoes Lane
Piscataway, NJ 08854
Phone:        (800) 701-4333
              (732) 981-1393
Fax:          (732) 981-9667
E-mail:       customer-service@ieee.org

# 2006 Asia Optical Fiber Communication & Optoelectronic Exposition & Conference

Shanghai, China
24-27 October 2006

IEEE Catalog Number:   CFP0639B-POD
ISBN:                  978-0-97892-170-5

AOE 2006 General Chair
for Europe

**Professor Dr.-Ing. Hans-Joachim Grallert**

**Heinrich-Hertz-Institut, Berlin, Germany**
Hans-Joachim Grallert is currently the managing director of the Fraunhofer Institute for Telecommunications Heinrich-Hertz-Institute (HHI) and professor with the Technical University Berlin at the faculty "Electrical Engineering and Computer Science". After graduation from RWTH Aachen in communications engineering he became assistant and senior assistant at the universities of Aachen and Duisburg, Germany. Thereafter he joined Siemens AG, Munich and held several positions in the department for development of transmission systems. In this area he was responsible for the development of multiplexers, line systems, crossconnects and TMN-Systems due to commercial transmission systems. As head of this department, member of the board and Senior Vice President of Siemens Optical Networks he left the company and took over the position as Managing Director of Marconi Communications Ondata Inc. and as Vice President in Marconi Optical Networks. In April 2005 he accepted a call to the Heinrich-Hertz-Institute in Berlin. Professor Hans-Joachim Grallert holds 31 patents and inventions in optical communications and digital signal processing. He is author of more than 40 publications and was furthermore chairman of several national and international conferences e.g. ECOC 2000 in Munich.

AOE 2006 General Chair
for US & Canada

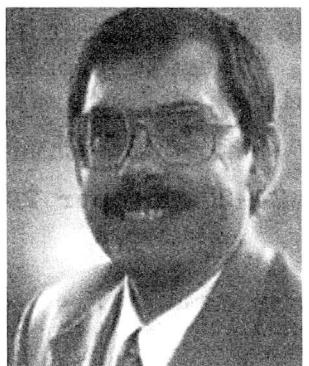

**Dr. JACEK CHROSTOWSKI, P.Eng.**
is currently President and CTO of Scheelite Technologies LLC, Reno, USA . He is also as Adjunct Professor at University of Ottawa, Canada . Previously, he was Vice-President, technology strategy at EMCORE Co. From 1998 to 2003 he worked at Cisco Systems in a number of roles, including Technology Deputy, Optics in the Office of CTO and a member of the University Research Board. Previous to that he had a distinguished carrier at the National Research Council of Canada. He co-authored over 100 scientific papers, large number of conference presentations, 6 US patents in optoelectronics and co-edited 3 books. He has served on many program committees of international conferences in fiber optic networks and IEEE, he currently serves on a number of industrial and venture advisory boards, including EXFO E. O. Inc., Signal Lake Management LLC, iN3Parters Inc. and Syntune AB. He has been a reviewer/auditor in many Government research programs, including US NSF, European Commission and Canadian NSERC. In 2002 he served on the 10 year technology roadmap committee in optical networking for the US Congress. He is currently technical editor of IEEE ComSoc Magazine and has been a co-editor of the special September 2003 ComSoc Magazine issue "After the optical bubble - the reality check". He received IEEE Millennium Medal for his activities at IEEE and Outstanding Achievement Award from NRC in 1994 for his work on erbium-doped optical amplifiers.

AOE 2006 General Chair
for Asia

**Dr. Mou-Tion Lee
Corning Incorporated**
Dr. Mou-Tion Lee is currently Manager, International Engineering for Corning Optical Fiber, Corning Incorporated, New York, USA. He has twenty-five years of experience in the fiber-optics industry and was with ITT and Philips U.S. prior to joining Corning in 1984. Dr. Lee has held various technical and managerial positions in R&D, engineering and international operations during his career. Since 1995, he has been involved with Corning's international business for Greater China and Asia Pacific regions including four years of residence in China (1999- 2002). He is well known in the telecommunications industry, especially in Greater China where he has conducted numerous presentations and was on the board of the China Optinet Conference during its formation in 2000. Dr. Lee is actively engaged in the China Communication Standards Association (CCSA), a China Standards Body. His work with the CCSA is concentrated in the areas of standards review and discussion and promotion of the latest ITU-T and IEEE standards to China industry.

# Welcome to the 2006 Asia Optical Fiber Communications & Optoelectronic
## Exposition & Conference (AOE)
## Shanghai, China
## 24-27 October 2006

On behalf of the AOE Program Committee, we are pleased to welcome you to AOE 2006 at the Shanghai New International Expo Center!

AOE 2006 provides a unique opportunity to reach service providers in optical equipment manufacturing as well as technical researchers in the field by combining an exposition with a technical program of scholarly papers. Networking opportunities abound, as you visit the 120 exhibits covering 100,000 square feet and papers selected from a filed of more than 100 submissions, it including plenary talks, panel discussions, workshops and poster sessions over four days. Finally, make certain to attend the reception where you will have yet another opportunity to interact with conference and exhibition participants!

We are particularly pleased to welcome our two technical co-sponsors for AOE 2006: the Optical Society of America (OSA) and the Institute of Electrical and Electronics Engineers / Lasers and Electro-Optics Society (IEEE / LEOS).

Thank you for your attendance and we hope you enjoy the experience!

## AOE 2006 Conference General Chairs

**General Chair for Asia:**
Dr. Mou-Tion Lee, Corning Incorporated

**General Chair for Europe:**
Prof. Hans-Joachim Grallert, Heinrich-Hertz-Institut, Berlin, Germany

**General Chair for US & Canada:**
Dr. Jacek Chrostowski, Scheelite Technologies
Professor at University of Ottawa, Ottawa, Canada
CTO of Scheelite Technologies LLC, Reno, USA

**Steering committee:**
Prof. Alan Willner
IEEE/LEOS President, USA
Dept. of Electrical Engineering - Systems
Viterbi School of Engineering
University of Southern California, USA
Prof. Yi Xin Chen
Bookham Technology
Shanghai Jiao Tong University, China
Mr. Ake Lidstrom
Optical Transport Networking, Ericsson
WebCast Intl. Sweden

**Chairmen of the sessions:**
Dr. Yanjun Li, Chinese Academy of Sciences
Prof. Wei-Ping Huang, McMaster University, Canada
Ed Y. Zhang, Chief Technology Officer, Hometown Innovation Automation Inc.
**Program Committee:**
Dr. Sudhir Dixit, Nokia, Finland
Dr. David McDonald, GM/VP Engineering Intune Technologies, Ireland
Prof. Wei-Ping Huang, McMaster University, Canada
Prof. Chennupati Jagadish, Federation Fellow, Department of Electronic Materials
Engineering, Research School of Physical Sciences and Engineering
Australian National University, Canberra, ACT 0200, Australia
Dr. Minjun Li, Science & Technology, Corning Incorporated, USA
Dr. Kexing Liu, Ciena Corporation, Canada
Dr. Ray T. Chen, Microelectronic Research Center, Department of Electrical and Computer
Engineering, The University of Texas at Austin, Austin. Omega Optics, Austin, TX, USA
Dr. Gregory W. Schinn, Chief Technology Officer (Optical Layer)
Directeur de la technologie (Couche optique), EXFO Electro-Optical Engineering, Inc.
Canada
Dr. Wojtek J. Bock,University of Quebec, Canada
Prof. Alan Willner, IEEE/LEOS President, USA
Dept. of Electrical Engineering – Systems, Viterbi School of Engineering
University of Southern California, USA
Prof. Jintong Lin, Beijing University of Posts and Telecoms, China
Dr. Ronghui Qu, Shanghai Institute of Optics and Fine Mechanics, China
Prof. Weisheng Hu,Shanghai Jiao Tong University, China
Prof. Liangzhen Du,Shenzhen Photon Technology, China
Dr. Jiping Hua, Optical & Electroinc Cable Association of China
Prof. Rujian Lin, Shanghai University, China
Dr. John Feng, Avanex USA
Dr. Bo Cai, JDS Uniphase
Dr. Zaiyun Fan,Shanghai Electric Cable Research Institute, China
Dr. Zhengguo Tang, No:23 Institute Reaserch. China
Prof. Jim Yang,Dalian University of Technology, China
Dr. Wei Jin,Ploy University, Hong Kong
Dr. Guoqing Tang, Shanghai Optoelectronics Association, China
Dr. Lanxing Shao, Shanghai Laser Society, China
Dr. Chun Qiang Li, Communication Industry Association, China
Dr. Bolin Du, Shanghai Communication Society, China
Dr. Yongzhi Liu, University of Electronic Sciences and Technology of China
Dr. Yuliang Liu, Beijing Institute of Semiconductors, China

**AOE-Expo/AOE-TV**
**2006 Copyright Agreement**

This publication agreement and assignment of copyright relates to all the material published concerning the conference during AOE-Expo 2006 such as, but not limited to, Technical Digest and webpage and other electronic forms. In submitting a paper to AOE-Expo 2006 the authors assign to AOE-Expo/AOE-TV copyright ownership to the submitted work, this assignment to be effective as of the day of registration of the material at AOE-Expo 2006. AOE-Expo 2006 shall have the right to register copyright to the article in its name as claimant, whether separately or as part of another medium in which such work is included. Also, AOE-Expo 2006 shall have the right to grant reprint and web-publishing permission to third parties and to receive reasonable royalties in such instances.

Authors reserve the following rights:

· The right to use all or portions of the submitted article in future works of their own
· All proprietary rights, such as patent rights

This agreement must be confirmed by the author. Ticking the "accept" box in the registration indicates acceptance of this agreement. Any questions related to this agreement should be directed to:

AOE-Expo 2006

Wen Global Solutions
Tel: 86.755.2583.4722 Fax: 86.755.2583.4922
Website: http://www.wgs-china.com
Email: info@wgs-china.com
6503, Shun Hing Square 65th Floor
Di Wang Commercial Centre
5002 Shennan Road East
Shenzhen, Guangdong 518008
China

The copyright of the paper must be assigned to the AOE-Expo 2006 by means of the Publication Agreement and Assignment of Copyright.

1. If the copyright of the paper is owned by the author, the agreement should be accepted as described above, the author asserts to AOE-Expo 2006 that he/she/they has/have the moral right to be identified as the author of the work.

2. In the USA, copyright protection is not available for work prepared by an officer or employee of the US government as part of that person's official duties, or on behalf of the US government. Similar provisions may apply to work done by government employees in other countries. By approving as described above, authors should affirm the warranties.

3. If the copyright in the paper is owned by the organization that employs the author, AOE-Expo 2006 accepts that it has been authorized by the representative of the copyright owner when approval is accepted as described above.

The organization in assigning the copyright in the paper gives no warranty expressed or implied that the paper is free from defamatory matter, nor that the paper does not infringe the rights of any third party.

The organization warrants that it:

a) has advised the author's of his/her/their responsibilities to avoid defamation or infringement of third party rights, and

b) will notify AOE-Expo 2006 of any adverse claim that comes to its knowledge prior to publication.

4. No other intellectual property rights, such as patents, that may be referred to or described in the paper shall transfer to the AOE-Expo 2006 save in relation to copyright and database rights.

5. The party assigning to AOE-Expo 2006 copyright and database rights in the paper shall be entitled, without payment to the AOE-Expo 2006, to:

a) reproduce figures or extracts from the paper with proper acknowledgment to the AOE-Expo 2006;

b) reuse all or portions of the paper in other works with proper acknowledgment to the AOE-Expo 2006; and

c) make and have made copies of the paper as published by the AOE-Expo 2006 for his/her/their/its own purposes but not for sale, provided that any reference to the AOE-Expo 2006 does not imply endorsement by the AOE-Expo 2006 of any organization, its products or its business.

6. If AOE-Expo 2006 agrees to publish the paper, but does not publish the paper within two years of agreeing to do so, then the author or organization that assigned copyright and database rights in the paper to the AOE-Expo 2006 shall have the right to require AOE-Expo 2006 to assign back the copyrights and the database rights in the paper. If for any reason AOE-Expo 2006 fails to publish the paper then the only right and remedy against AOE-Expo 2006 of the party that submitted the paper and/or assigned copyright and database rights in the paper to the AOE-Expo 2006 shall be to require that the copyright and any database rights in the paper are assigned back.

By approving the copyright agreement the author's asserts to AOE-Expo 2006 that he/she/they has/have the moral right to be identified as the authors of the paper. All authors (or an authorized agent for joint authors) of the paper accept and approve, by clicking in the box while registering the paper and simultaneously confirming:

I/We, the named authors of the paper, confirm that the work is original, that it contains nothing defamatory, and that it may be submitted to the AOE-Expo 2006 with the intent for publication. I/We agree that the AOE-Expo 2006 shall be entitled to be fully indemnified by me/us against all claims:

(a) that the paper is defamatory, or
(b) that the paper infringes the copyright of any third party, or
(c) arising from any negligent or willful miss-statements.

2006 Asia Optical Fiber Communication & Optoelectronic Exposition and Conference

IEEE Catalog Number 06EX1532
ISBN 978-0-9789217-0-5

**Asia Optical Fiber Communication & Optoelectronic Exposition & Conference**

 www.aoe-expo.com

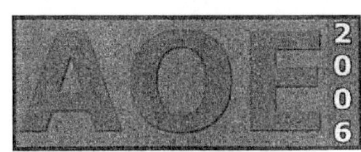

# Conference Schedule

| October 24, Tuesday | October 25, Wednesday | October 26, Thursday | | October 27, Friday | |
|---|---|---|---|---|---|
| Registration Opening | Plenary Session Keynote speeches | Plenary Session Keynote speeches Invited Speakers | | Technical Session A-2 | Technical Session B-2 |
| Workshop Panel Discussion | Plenary Session Special Invited Speakers | Technical Session A-1 | Technical Session B-1 | Technical Session A-3 | Technical Session B-3 |

| | Tuesday October 24 | Wednesday October 25 | Thursday October 26 | Friday October 27 |
|---|---|---|---|---|
| Registration | 9:00am-5:00pm | 8:30am-5:00pm | 8:30am-5:00pm | 8:30am-5:00pm |
| Panel Discussion | 2:00pm-5:00pm | | | |
| Plenary Session Keynote Speech | | 9:00am-5: 30pm | 9:00am – 10:00am | |
| Plenary Session Invited Speech | | | 9:00am-5: 30pm | |
| Technical Session | | | | 9:00am-5: 30pm |
| Exposition | | 9:00am-5: 00pm | 9:00am-5: 00pm | 9:00am-4: 00pm |
| Conference Reception | | 5:30pm-7: 30pm | | |

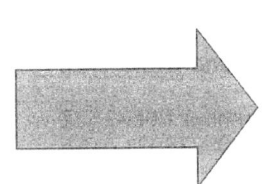

## AOE 2006 Key to Presenters

### LEGEND

Tu = Tuesday

W = Wednesday

Th = Thursday

F = Friday

PD = Panel Discussion

PK = Plenary Session Keynote Speech

PI = Plenary Session Invited Speech

TS = Technical Session

**Asia Optical Fiber Communication & Optoelectronic Exposition & Conference**

 www.aoe-expo.com

# Table of Contents

**TUPD Panel Discussion Chairman** .................................**22**
Panel Discussion. October 24. Tuesday. 2:00pm - 5:00pm
Innovation in Optoelectronics/Photonics Industry-Past, Present and
Future


0
Edward Zhang, System Design Frontier Journal

**Chairs speech**
October 25, Wednesday, 9:00am – 9:30am

**WPK1 - Plenary Session Keynote speech** ....................**23**
October 25, Wednesday. 9:30am – 10:30am
China's Next Generation Internet (CNGI) Project Introduction
Mr. Wu Hequan, Vice President / A Member of Academia Sinica,
Professor
Chinese Academy of Engineering
中国工程院: 邬贺铨, 副院长, 国家院士, 教授

**WPK2 - Plenary Session Keynote speech** ....................**24**
October 25, Wednesday. 10:30am – 11:30am
The Future of Asian Optical Communications
Mr. Clark Kinlin, CEO
Corning China
康宁公司: 康克明, 大中华区首席执行官

**WPK3 - Plenary Session Keynote speech**....................**28**
October 25, Wednesday.11:30am – 12:30am
Testing Transceivers fast, economic and reliable
Dr. Chen Li
PRC/HK EMG Marketing Manager
Agilent Technologies

**WPI 1 - Special Invited Speaker**...........................**46**
October 25, Wednesday. 1:30pm – 2:00pm
Telecommunication Market in China
Dr. Xie Yi, Vice President
China Academy of Telecom. Research, MII
信息产业部电信科学研究院: 谢毅 副院长, 教授

**Asia Optical Fiber Communication & Optoelectronic Exposition & Conference**

www.aoe-expo.com

### WPI 2 - Special Invited Speaker...................................47
October 25, Wednesday. 2:00pm – 2:30pm
Service-driven Advanced Optical Network
Dr. Ji Yuefeng, School of Telecomm.Engineering
Beijing University of Post and Telecomm, China
邮电大学电信工程学院: 纪越峰 副院长, 教授

### WPI 3 - Special Invited Speaker...................................52
October 25, Wednesday. 2:00pm – 3:00pm
Technology Development of Flat Panel Display (FPD)
Prof. Li Dejie, Dept. of Electronic Engineering, Professor
Tsinghua University, China
清华大学电子工程系: 李德杰 教授

### WPI 4 - Special Invited Speaker...................................56
October 25, Wednesday. 3:30pm – 4:00pm
Broadband Access Services Requirements for Optical Transport Network
Mr. Zhang Chengliang, Deputy Chief Engineer
China Telecom Corporation Limited, Beijing Research Institute
中国电信北京设计院: 张成良 副总工, 教授

### WPI 5 - Special Invited Speaker...................................57
October 25, Wednesday. 4:00pm – 4:30pm
Optical Access in China: Application Models and Market Prospects
Dr. Tang Xionyan, Vice President
China Netcom Research Institute
中国网通研究院: 唐雄燕 副院长, 教授

### WPI 6 - Special Invited Speaker...................................58
October 25, Wednesday. 5:00pm – 5:30pm
Future high bit rate transmission and Ultra Long Haul network applications
David Z. Chen
PMTS-Technology
Verizon, USA
美国电信公司

### THPK - Plenary Session Keynote speech...........................59
October 26, Thursday. 9:00am – 10:00am
Physical-Layer Challenges in Stable, High-Capacity Optical
Communication Networks
Prof. Alan Willner
IEEE/LEOS President, USA
Dept. of Electrical Engineering - Systems

**Asia Optical Fiber Communication & Optoelectronic Exposition & Conference**

 www.aoe-expo.com

Viterbi School of Engineering
University of Southern California, USA

**THPI 1 - Special Invited Speaker...................................62**
October 26, Thursday.10:00am – 10:30am
Recent Research Activities on Specialty Fibers
Dr. Ming-Jun Li
Corning Incorporated, New York, USA

**THPI 2 - Special Invited Speaker...................................63**
October 26, Thursday. 10:30am – 11:00am
FTTH Network Installation Implementation Innovations
Michael Kunigonis, Jr.
Product Line Manager Access
Corning Cable Systems, New York, USA

**THPI 3 - Special Invited Speaker...................................64**
October 26, Thursday. 11:00am – 11:30am
Characterization of a transmission system on RX, TX, Channel and the
Clocking system
Dr. Haiyang Hu
Director of Open Lab and Solution
Agilent Technologies
**THPI 4**
October 26, Thursday. 11:30am – 12:00
Enabling fast, reliable and cost-effective sensing through highly
integrated optical measurement systems.
Dr. Haiyang Hu
Director of Open Lab and Solution
Agilent Technologies

**THPI 5 - Invited Speaker...................................65**
October 26, Thursday. 12:00 – 12:30pm
Low-cost Radio-over-Fiber Network for 2.45GHz In-Building Signal
Distribution
Michael Ong Ong1, M. L. Yee1, B. Luo1 and M. Fujise2
1Institute for Infocomm Research, Singapore Science Park II, Singapore
2Wireless Communications Laboratory, Singapore Science Park II,
Singapore

**THTSA1-1 - Invited Speaker...................................68**
October 26, Thursday. 1:40pm – 2:10pm
OBS based transparent optical networking policy
Professor Anshi Xu
Peking University, China

**THTSA1-2 - Invited Speaker...................................72**
October 26, Thursday. 2:10pm – 2:40pm
High Performance Burst-Mode Upstream Transmission for Next
Generation PONs

**Asia Optical Fiber Communication & Optoelectronic Exposition & Conference**

 www.aoe-expo.com

Prof. Xing-Zhi Qiu
Ghent University, INTEC/IMEC, Sint-Pietersnieuwstraat 41, 9000 Gent, Belgium

**THTSA1-3** ...............................................................................**75**
October 26, Thursday. 2:40pm – 3:00pm
Restoration Mechanism and Maximum Dual Failure Restorability of Span-restorable Optical Networks_
Dr. Ling Zhou
Dept. Information Technology and Electrical Engineering
Swiss Federal Institute of Technology Zuerich
Dept. Electronics/Metrology and Reliability Center
Swiss Federal Laboratories for Materials Testing and Research (EMPA)
Ueberlandstrasse 129, CH-8600 Duebendorf. Switzerland

**THTSA1-4**...............................................................................**78**
October 26, Thursday. 3:00pm – 3:20pm
Performance Limitations of an Optical Heterodyne CPFSK System Impaired by Polarization Mode Dispersion in a Single Mode Fiber
Dr. Md Saiful Islam
(1): Institute of Information and Communication Technology, Bangladesh University of Engineering and Technology, Dhaka-1000, Bangladesh
(2): Department of Electrical and Electronic Engineering
Bangladesh University of Engineering and Technology, Dhaka-1000, Bangladesh

**THTSA1-5**...............................................................................**81**
October 26, Thursday. 3:40pm – 4:00pm
Ultrafast Bit and Byte addressing of All-Optical Memory based on Microring Resonators for Next-Generation Optical Networks
Dr. Seng-Tiong Ho
Northwestern University
2145 Sheridan Rd, Dept of ECE, Evanston, IL 60208, USA

**THTSA1-6**...............................................................................**84**
October 26, Thursday. 4:00pm – 4:20pm
Implementation of Token-based Optical Burst Switching Ring Network Node and Testbed Using Fixed Transmitter and Tuneable Receiver
Dr. Hongxiang Wang
Beijing University of Posts & Telecommunications, China

**THTSA1-7**...............................................................................**87**
October 26, Thursday. 4:20pm – 4:40pm
G-TEP: A GMPLS Testing and Emulation Platform
Dr. Weiqiang Sun
State Key Lab of Advanced Optical Communication System and Networks
Shanghai Jiao Tong University, China

**THTSA1-8**...............................................................................**90**
October 26, Thursday. 4:40pm – 5:00pm
Phase noise due to ASK amplitude un-equality and Kerr nonlinearity in

**Asia Optical Fiber Communication & Optoelectronic Exposition & Conference**

www.aoe-expo.com

optical ASK-DPSK, ASK-DQPSK systems
Dr. Aiying Yang
Beijing Institute of Technology, China

**THTSB1-1 – Invited Speaker**................................................**93**
October 26, Thursday. 1:40pm – 2:10pm
Building Blocks for Electronic- Photonic Integrated Chips
Dr. Jurgen Michel
Massachusetts Institute of Technology, USA

**THTSB1-2 - Invited Speaker**................................................**94**
October 26, Thursday. 2:10pm – 2:40pm
CAD for Photonics Devices and Circuit
Professor Chenglin Xu
Shandong University, China

**THTSB1-3**................................................**100**
October 26, Thursday. 2:40pm – 3:00pm
Recent development of ion-exchanged glass waveguide technology
Dr. Yinlei Hao
Department of Information Science & Electronics Engineering
Zhejiang University, Hangzhou, 310027, China

**THTSB1-4**................................................**110**
October 26, Thursday. 3:00pm –3:20pm
Guided Modes in a Slab Waveguide with an Anisotropic Dispersive
Plasmonic Core
Transmission Property of the Nonmagnetic Media with a Hyperbolic
Dispersion Relation
Slow Propagation of Light in a Dielectric-Metal-Dielectric Waveguide
Dr. Guoan Zheng
Zhejiang University, HangZhou, China

**THTSB1-5**................................................**119**
October 26, Thursday. 3:40pm –4:00pm
Performance and Reliability Issues of SC/APC Plug Style Attenuators
Dr. Andrzej Tymecki
Telekomunikacja Polska, Poland

**THTSB1-6**................................................**123**
October 26, Thursday. 4:00pm –4:20pm
BER performance analysis of PIN photodiode in 10Gbps fiber optical
communication

**Asia Optical Fiber Communication & Optoelectronic Exposition & Conference**

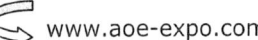 www.aoe-expo.com

Dr. Shang-Bin Li
Amertron-global, China

**THTSB1-7**.................................................................**126**
October 26, Thursday. 4:20pm –4:40pm
Optimization Quasielastic Phenomenological Simulation For Medium
Phenomenal Isotropic Liquid Crystal
Prof. Chia -Fu,Chang (1) Wi-Ci,Chen (2) Zou-ni,Win (3)
Kun Shan University Of Technology, Taiwan

**THTSB1-8**.................................................................**127**
 October 26, Thursday. 4:40pm –5:00pm
A Critical Study of the Soot Deposition Rate in ACVD Process
Mr. Gopal Mishra
Sterlite Optical Technologies Ltd, India

**FTSA2-1 -  Invited Speaker**................................................**130**
October 27, Friday. 8:40am – 9:10am
Upgrades of Submarine Systems using Higher Bitrates and Advanced
Modulation Formats
Dr. Jorg Schwartz
Jorg Schwartz (1), Ronald Freund (2), Lutz Molle (2), Christoph Caspar
(2), Steve Webb (1) and Stuart Barnes (1)
1: Azea Networks Ltd., Bates House, Harold Wood, Romford RM3 0SD,
United Kingdom,
2: Fraunhofer-Institute for Telecommunications, Heinrich-Hertz-Institut,
Einsteinufer 37, 10587 Berlin, Germany

**FTSA2-2 -  Invited Speaker**................................................**133**
October 27, Friday. 9:10am – 9:40am
Prospective Optical Communication Industry in China
Feng Wang, Infostone Communication Consultant
Yixin Chen, Shanghai Jiao Tong University

**FTSA2-3**.................................................................**134**
October 27, Friday. 9:40am – 10:00am
Calculation on Finesse of Fabry-Perot Interferometer in Millimeter-Wave
Radio-over-Fiber System
Dr. Minglei Xiu
School of Communication and Information Engineering
Shanghai University, China

**FTSA2-4**.................................................................**138**
October 27, Friday. 10:00am – 10:20am

**Asia Optical Fiber Communication & Optoelectronic Exposition & Conference**

www.aoe-expo.com

Research on the light source of 60GHz millimeter wave ROF system
Miss Yinghua Pan
Shanghai University, China

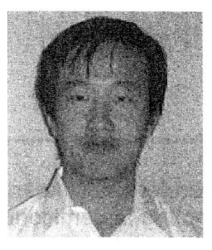

**FTSA2-5**..................................................141
October 27, Friday. 10:40am – 11:00am
Research of Wireless Indoor Channel in 60GHz Radio over Fiber System
Mr. Zhang qi
Shanghai University, China

**FTSA2-6**..................................................146
October 27, Friday. 11:00am – 11:20am
Model of PPM Receiver used in Deep Space Communication Systems
Free Space Optics
Prof. Moncef B Tayahi
Advanced Photonics Research laboratory, USA

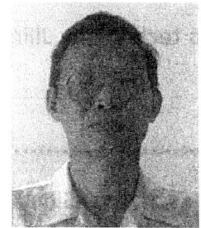

**FTSA2-7**..................................................153
October 27, Friday. 11:20am – 11:40am
All-wave Non-Zero Dispersion-Flattened Single Mode Fibers
Prof. Yeheng Wang
Shanghai Transmission Lines Research Institute, China

**FTSA2-8**..................................................157
October 27, Friday. 11:40am – 12:00
Design & Usage of Non-zero Dispersion Wideband Transport (NZDWT)
Fiber
Mr. Saurav Dutta
Sterlite Optical Tecnologies Limited, India

**FTSB2-1 - Invited Speaker**..............................160
October 27, Friday. 8:40am – 9:10am
1. Single-Frequency Operation of a Widely Tunable SOA-Based Fiber
Ring Laser
2, Tunable Fiber Ring Lasers and their Applications to Fiber-Optic Test
and Measurement
Dr. Hongxin Chen
EXFO Electro-Optical Engineering Inc. Canada

**FTSB2-2 - Invited Speaker**..............................163

**Asia Optical Fiber Communication & Optoelectronic Exposition & Conference**

 www.aoe-expo.com

October 27, Friday. 9:10am – 9:40am
80-micron Interaction Length Silicon Nano-Photonic CrystalWaveguide Modulator
Prof. Ray T. Chen
1 Microelectronic Research Center, Department of Electrical and Computer Engineering,
The University of Texas at Austin, Austin, TX 78758, USA
2 Omega Optics, Austin, TX 78758, USA

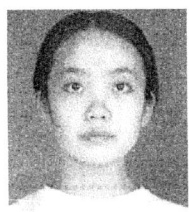

**FTSB2-3**............................................................................**164**
October 27, Friday. 9:40am – 10:00am
Analysis of the influences of polarization-dependent nonlinear gain and nonlinear polarization rotation on optical sampling in semiconductor optical amplifier
Dr. Maotong Liu
Beijing Institute of Technology, China

**FTSB2-4**............................................................................**169**
October 27, Friday. 10:00am –10:20am
Gain leveling in a two-level system for EDFA use Quantum-interference effects
Dr. Xuemei Su
Department of optical information science and technology Jilin University, China

**FTSB2-5**............................................................................**172**
October 27, Friday. 10:40am – 11:00am
Sidelobes Suppression of Periodically Placed Resonators Side-Coupled to Photonic Crystal Waveguide
Mr. Hanhui Li
Key Laboratory of OCLT, Ministry of Education
Beijing University of Posts and Telecommunications, China

**FTSB2-6**............................................................................**175**
October 27, Friday. 11:00am – 11:20am
Photon Tunneling Effect in a One-dimensional Photonic Crystal Containing a Partially Negative Permittivity Uniaxial Media
Dr. Ke Chen
State Key Laboratory for Modern Optical Instrument, Zhejiang Univesity, China

**FTSB2-7**............................................................................**178**
October 27, Friday. 11:20am – 11:40am
High Performance and Low-cost Fiber Grating Laser Module Package Employing a Hyperbolic Fiber Microlens
Professor Huei-Min Yang
I-SHOU University, Taiwan

**Asia Optical Fiber Communication & Optoelectronic Exposition & Conference**

 www.aoe-expo.com

**FTSA3-1 - Invited Speaker.................................180**
October 27, Friday. 1:40pm – 2:10pm
Fiber to the home
FTTP Optical Transceivers:
When, how and if integration can make a real impact on performance-cost ratio?
Prof. Wei-Ping Huang
McMaster University, Canada

**FTSA3-2 - Invited Speaker.................................181**
October 27, Friday. 2:10pm – 2:40pm
Optional Solutions for FTTH Market in China
Dr. River Huang
Shenzhen First Mile Communications Ltd. China

**FTSA3-3.................................182**
October 27, Friday. 2:40pm – 3:00pm
Design and Implement of practical EPON Network Management System
Sun Jie
Shanghai University, China

**FTSA3-4.................................185**
October 27, Friday. 3:00pm – 3:20pm
Design of a Hybrid Optical CDMA/WDMA System with the Position Code
Dr. Po-Hao Chang
National Dong Hwa University, Taiwan

**FTSA3-5.................................188**
October 27, Friday. 3:40pm – 4:00pm
A Novel Implementation of VLAN-Based Multicast
Mr. Min Zhu
Shanghai University, China

**FTSA3-6.................................192**
October 27, Friday. 4:00pm – 4:20pm
1. Simple analysis of the distribution of birefringence in fiber optics links
2. Polarization tracking based on Stokes curve analysis
Dr. Krzysztof Perlicki
Warsaw University of Technology, Institute of Telecommunications
Nowowiejska 15/19, 00-665 Warsaw, Poland

**FTSA3-7.................................196**
October 27, Friday. 4:20pm – 4:40pm
PMD Distribution Measurement by an OTDR with Polarimetry considering

**Asia Optical Fiber Communication & Optoelectronic Exposition & Conference**

 www.aoe-expo.com

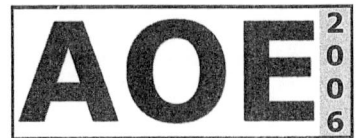

Depolarization of Rayleigh Backscattered waves
Prof. Takeshi Ozeki
Sophia University, Tokyo, Japan

**FTSA3-8**..................................................................................**199**
October 27, Friday. 4:40pm – 5:00pm
Geometry Modulation in Optic Communication
Dr. Mingwu Gao
Zhejiang University, China

**FTSB3-1 -** ................................................................................**200**
October 27, Friday. 2:10pm – 2:40pm
Randomly non-lithographic masking and MOCVD regrowth of GaN micro-hillocks to improve light emitting diode efficiencies
Professor Wei-I Lee
Department of Electrophysics
National Chiao Tung University
Hsinchu, 30010 Taiwan

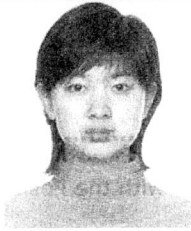

**FTSB3-2**..................................................................................**201**
October 27, Friday. 2:40pm – 3:00pm
High Efficiency Deep-Blue Organic Light-Emitting Devices
Dr. Yingfang Zhang
State Key Laboratory of Integrated Optoelectronics, Jilin University, China

**FTSB3-3**..................................................................................**204**
October 27, Friday. 3:00pm – 3:20pm
2X2 juxtaposed Type Optical Fiber Raman Temperature Sensor
Dr. Lixun Zhang
University of Electronic Science and Technology of China

**FTSB3-4**..................................................................................**209**
October 27, Friday.3:40pm – 4:00pm
Portable Laser Therapeutic Device With Pulse - modulated Regimes Mode
Dr. Michael Titov
Technica-Pro
Moscow Russia

**FTSB3-5**..................................................................................**210**
October 27, Friday. 4:00pm – 4:20pm
Full Progress of Digital Signal Processing in Open Loop-IFOG
M.R.Nasiri Avanaki
Electrooptics Department, MS.c Student of University of Semnan, Iran

**Asia Optical Fiber Communication & Optoelectronic Exposition & Conference**

 www.aoe-expo.com

**FTSB3-6**..................................................................**211**
October 27, Friday. 4:20pm – 4:40pm
A Fast Convergence Self-Adaptive Bit Allocation Algorithm for OFDM Systems
ZHENG Pei-chao1, SONG Han-tao1, LIU Bin1,2
(1. School of Computer Science and Technology, Beijing Institute of Technology, Beijing 100081, China;
2. College of Economics and Management, Hebei University of Science and Technology, Shijiazhuang Hebei 050018, China)

**FTSB3-7**..................................................................**212**
October 27, Friday. 4:40pm – 5:00pm
Dynamic Load Balancing Based on Restricted Multicast Tree in Homogeneous Multiprocessor Systems
Mr. Bin Liu
Beijing Institute of Technology, China

## Innovation in Optoelectronics/Photonics Industry - Past, Present and Future

Optoelectronics/Photonics as one of today's enabling technologies with explosive growth in information exchange and the plethora of data available nowadays, has start to replace the dominating position of electronics as the photonics part of optoelectronic technology surges steadily forward, revealing its vast potential.

Optoelectronics/Photonics, as an industry, its inception could be dated back to 1960s when the laser was invented, followed in the late 1970's as the perfection of optical fibers as an effective means of transmitting information using these intense laser beams. Later on, key technologies moved from the laboratory into the diffusion and application phases, and Optoelectronics/Photonics as an industry becomes mature accordingly. During the past decade, two of optoelectronics/photonics's drivers are the rapid growth of the Internet industry, and the consolidation of the telecommunication industry. In the coming decade, with the projected quick development of 3C (Computer, Communication and Consumer Electronics) convergence and Web 2.0, the future of optoelectronics/photonics deserves to be bright.

This panel, consisting of both industry veterans and academic scholars with versatile business and/or technical experience in optoelectronics / photonics industry, will deliver their visions and insights on how innovation plays its critical role in research, development, and manufacturing of optoelectronics/photonics industry, past, present and future.

A unique and informative panel session will reveal the historical facts and discover new challenges ahead!

**Panel Discussion Chairman**

Ed Y. Zhang
Chief Technology Officer
Hometown Innovation Automation Inc
Editor in Chief, System Design Frontier Journal
Mr. Zhang received his B.S. in electrical engineering from Tsinghua University, China, in 1992, and his M.S. in information and computer science from the University of California at Irvine in 1997. He worked at Beijing IC Design Center as a member of technical staffs before he went to Canada and the United States for further study in 1994. Mr. Zhang worked at Avant! Corporation and later at Magma Design Automation as a member of technical staffs between 1998 and 2001. In early 2002, Mr. Zhang co-founded Hometown Innovation Automation Inc with a mission to empower innovation automation. Currently, Mr. Zhang serves as Hometown Innovation Automation's Chief Technology Officer, as well as Editor-in-chief of System Design Frontier Journal.

# Plenary Session
# Keynote speech
**October 25, Wednesday**
**9:30am – 10:30am**

### Mr. Wu Hequan, Vice President / A Member of Academia Sinica, Professor
### Chinese Academy of Engineering
邬贺铨 中国工程院 副院长, 国家院士, 教授

### China's Next Generation Internet (CNGI) Project Introduction
中国下一代互联网示范工程(CNGI)介绍

### CNGI Project Introduction

**@Survey on Internet of China**
**@New Generation Network and CNGI**
**@Optical Internet Network**

### Survey on Internet of China

Wu Hequan, born on 1943 in Guangzhou, China; graduated from Wuhan Post and Telecommunications Institute in 1964. Wu Hequan was work for the China Academy of Post and Telecommunications (CAPT) of Ministry of Post and Telecommunications since 1964. He was Vice-President and Chief Engineer of China Academy of Telecommunications Technology (CATT) in 1997-2003. He has undertaken to study on optical fiber transmission system and broadband network and manage R&D projects for a long time. He takes charge to study the development strategy on NGN and NGI as well as 3G in recent years. He was elected the academician of Chinese Academy of Engineering (CAE) in 1999 and Vice-President of CAE since June 2002. He was assigned as Vice-Director of China Advisory Committee for State Informatization. He was Vice-Director of an Executive Council of China Institute of Communications and an Adviser of Communication S&T Committee of MII. He currently takes on director of Experts Committee of China's Next Generation Internet (CNGI) project. He was a senior member of IEEE.

邬贺铨，1943年出生于中国广州，1964年毕业于武汉邮电学院。毕业后在邮电部邮电科学研究院（CAPT）从事研究工作，在1997~2003年间他担任电信科学技术研究院（CATT）的副院长兼总工程师。他长期从事光纤通信传输系统和宽带网的研究开发和项目管理工作，近年从事下一代网络和第三代移动通信的发展战略研究。1999年当选为中国工程院院士，2002年任中国工程院副院长。现兼任国家信息化专家咨询委员会副主任、中国通信学会副理事长、信息产业部通信科技委顾问中国下一代互联网示范工程（CNGI）专家委员会主任、IEEE高级会员。

# Plenary Session
# Keynote speech
**October 25, Wednesday**
**10:30am – 11:30am**

**Clark S. Kinlin**
**CEO**
**Corning Greater China**

**The Future of Asian Optical Communications**

Asia has been the leader in broadband deployments and long-term potential for all parts of the telecomm value chain remain. The market is shifting away from copper and toward access deployments, and among access architecture choices we believe the less expensive alternatives to FTTH do not overcome the large difference in capability.

**Clark S. Kinlin** joined Corning in 1981 responsible for sales and marketing roles in the Technical Products Division. Since then, he has led several projects focused on Corning's expanding and commercial entry interests in Japan and Asia as director-Latin America/Asia Pacific and ultimately led the division's North America Sales and Marketing team. From 1995 onwards, Kinlin joined and led several different divisions and international office as head of sales and marketing. In 2000, Kinlin was named president – Corning International Corporation.

Kinlin assumed his current responsibilities in 2003, when he was appointed CEO, Greater China. In this role he leads operations in the region, in support of the division managers. He develops business unit strategies for Greater China with the intent of building long-term sustainable commercial and manufacturing success within the region. Kinlin is a director of the National Bureau of Asian Research and a trustee of Elmira College. Kinlin received his undergraduate degree from Kenyon College and his MBA from Harvard University.

Jan 2006
Corporate Communications Greater China

# The Future of Asian Optical Communications

Asian Optical Fiber Communication & Optoelectronic
Exposition & Conference 2006

Clark Kinlin, CEO Corning China

Emergence of Optical Communications [1]

Optical Communications emerged from the confluence of four technologies which appeared over a very short time period. In 1969 the first internet message was sent over what was then called ARPANET. The following year the first semiconductor laser diode to operate at room temperature was demonstrated and the first low loss optical fiber was invented. Finally in 1971 the first computer chip was brought to market.

The result has been astounding. The Internet has grown so that 1 billion people connected is in sight. And our connection to the Internet, the personal computer, has grown to nearly 800 million units in use. The early PC was slow and cumbersome, but PC's have become much faster, smaller, and easier to use. And in order to connect those millions of PCs, more than 700 million kilometers of optical fiber have been installed.

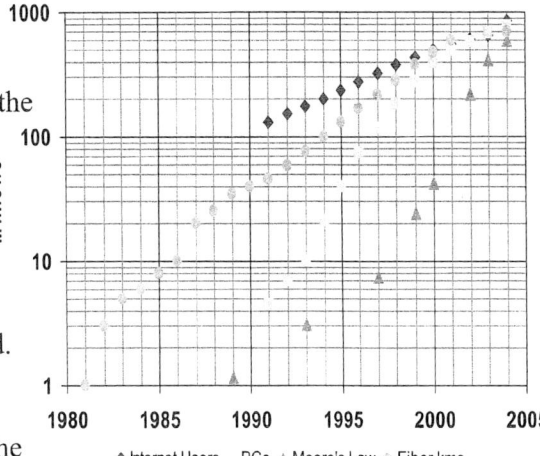

While progress in computer technology has been rapid, in optical fiber communications the rate of increase in transmission bandwidth has also been impressive. It took 120 years to increase the transmission capacity by one-million fold, the time period from about 1850 to 1970. Since then it has taken just 30 years to again increase our capability by one-million fold.

The Future of Optical Communications

Twenty years ago there were 100,000 subscribers, connected by copper lines, for each optical termination. Today we have more than 7 million homes connected directly to optical fiber, and the marketplace, technology, and public policy are aligning for broad fiber to the home deployments on a worldwide scale.

After the three difficult years which followed the year 2000 global telecommunications downturn, capital investment by the world's service providers has resumed growing. And this growth is expected to continue. This trend is driven by a shift from investments in copper technology to optical fiber and from the long-haul segment to access.

The focus on access is the result of the burgeoning demand for residential broadband services and the focus on building the capability to offer triple-play services. Despite six years of rapid growth in broadband, it remains a significant growth opportunity with over

80% of households worldwide remaining to be connected. Worldwide broadband growth has been led by Asia, with more than 90 million subscribers.

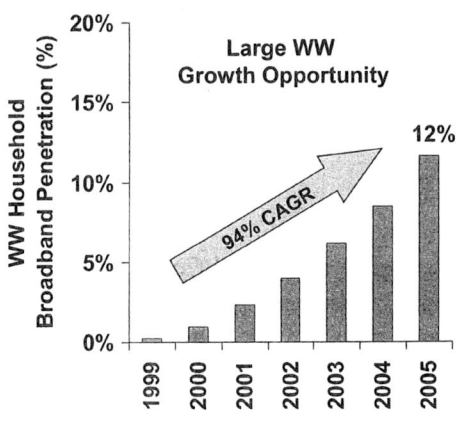

Source: ITU, Corning analysis

Asia is also the worldwide leader in Fiber to the Home (FTTH), with about 80% of all FTTH subscribers residing in Japan. FTTH in Japan resulted from "eJapan", an innovative forward-looking public policy initiative that has distinguished Japan as the birthplace of FTTH on a broad scale. We believe the demand by consumers for bandwidth will continue for years to come, due to advances in video such as HDTV, and new services such as on-line gaming, distance learning, and telecommuting that will dramatically increase the bandwidth needs of each household's internet connection.

Industry Challenges

Despite the positive activity and improving environment, challenges remain. Public policy that encourages facilities-based competition and advanced broadband adoption is critical to fostering the digital economy. In the United States, Public Policy has made substantial progress in recent years and the result has been a resurgence of facilities-based investment and the associated job creation. In Europe we are hopeful that a compromise to ex-ante regulation can be identified. And in China we are encouraged by the recently published 11th 5-year plan which calls for explicit telecomm legislation.

Access Architecture Alternatives

The gold standard among access architectures, and that chosen by NTT, Verizon, and municipalities and utilities worldwide, is Fiber to the Home (FTTH). This architecture eliminates the use of expensive active components in the field, components which drive up the operating cost, by extending optical fiber all the way from the central office to the household, multiple dwelling unit, or small business.

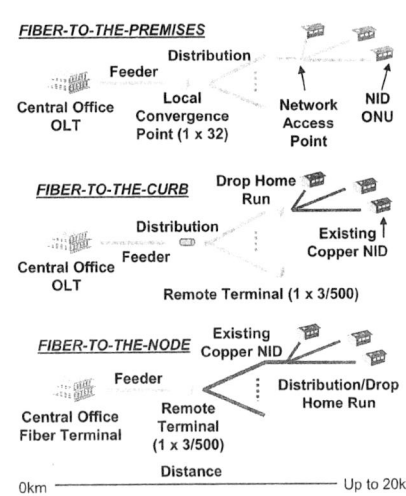

Historically FTTH's drawback has been cost, but that has improved significantly. Due to manufacturing scale effects associated with rapid increases in demand, and innovative new passive products from Corning, the cost to deploy FTTH and the differential between FTTH and less capable alternatives has declined substantially.

22

<u>Summary</u>

Asia has been the leader in broadband deployments and long-term potential for all parts of the telecomm value chain remain. The market is shifting away from copper and toward access deployments, and among access architecture choices we believe the less expensive alternatives to FTTH do not overcome the large difference in capability.

---

[1] Excerpts from "The Future will be Full of Light", 2005 OFC/NFOEC Keynote, Dr. Don Keck.

**Dr. Chen Li**
**Agilent Technologies**
**Testing Transceivers fast, economic and reliable**

**This keynote speech is focused on a total solution approach for testing electrical and optical transceiver parameters in manufacturing and R+D.**

Chen Li is Agilent PRC/HK EMG Marketing Manager.

Chen Li joined Hewlett Packard in 1991 as a Staff Marketing Engineer of Test and Measurement Organization (TMO), focus on telecom market development. From 1991 to 1993, he worked as a Marketing Engineer in China TMO. In 1994, he worked at Hewlett Packard Microwave Instrument Division (MID) Marketing Program Manager in China. From 1996 to 1997, he moved to MID located in Santa Rosa, Calif. USA and worked in Division's Marketing Department.

In Sep. 1997, Chen Li relocated to Beijing and named as Market Development Manager of China TMO Marketing. 1998, he was named as PRC/HK Marketing Manager of China TMO.

In 1999, Agilent Technologies Co., Ltd. split from Hewlett Packard, Chen Li was delegated as Asia Component Manufacture Market Segment Manager and PRC/HK EPSG Marketing Manager.

Chen Li was born in Beijing, China. He holds a Bachelor's degree in electrical engineering (EE) from Tsinghua University in Beijing and a Master's degree in EE from Beijing Institute of Technology.

陈力现任安捷伦科技有限公司电子仪器与系统事业部大中国区市场部总监。
陈力于1991年加入惠普公司，1991年至1993年间任电子测量仪器部市场工程师，负责电信市场开发。1994年，任惠普公司微波仪器工厂中国市场项目经理。1996年初至1997年底，陈力到美国加州惠普公司微波仪器工厂市场部工作，任项目经理。

1997年9月，陈力回国任电子仪器与系统事业部市场部市场开发经理，1998年，升任大中国区市场部总监。

1999年底，惠普公司进行战略重组。仪器部分离成立安捷伦公司。2000年起，
陈力任安捷伦科技有限公司电子仪器与系统事业部大中国区市场部总监。

陈力生于北京。获清华大学电子工程系电子工程学士学位，北京理工大学电子工程系电子工程硕士学位。

# Testing Transceivers fast, economic and reliable

Agilent Technologies

This key note speech is focused on a total solution approach for testing electrical and optical transceiver parameters in manufacturing and R+D.

Agilent Technologies

---

# Outline

Transceiver market overview

Test parameter

•Device test

•Transmitter test

•Receiver test

Summary

Agilent Technologies

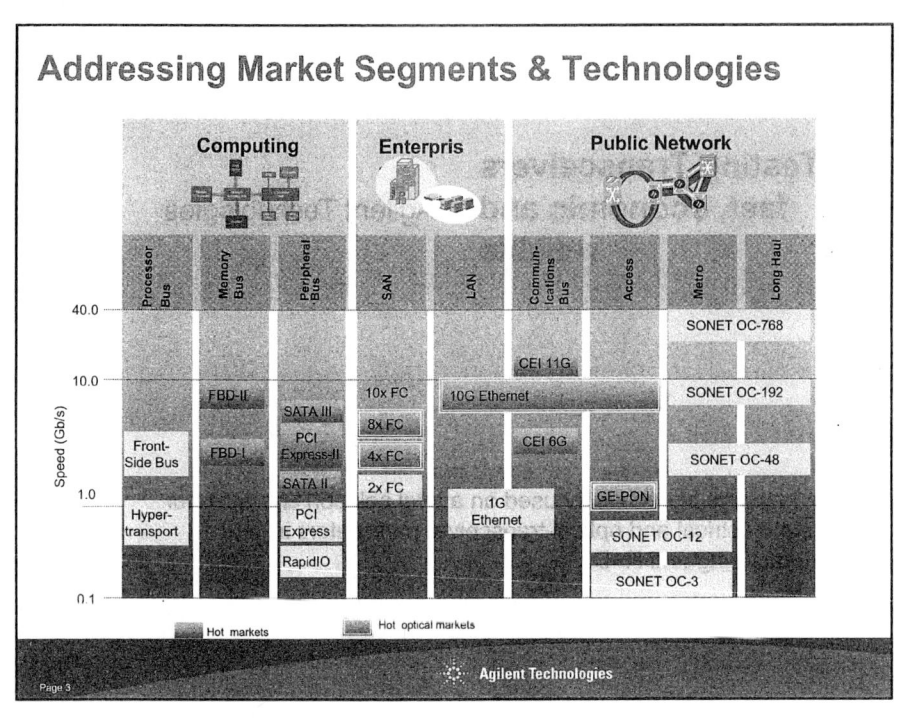

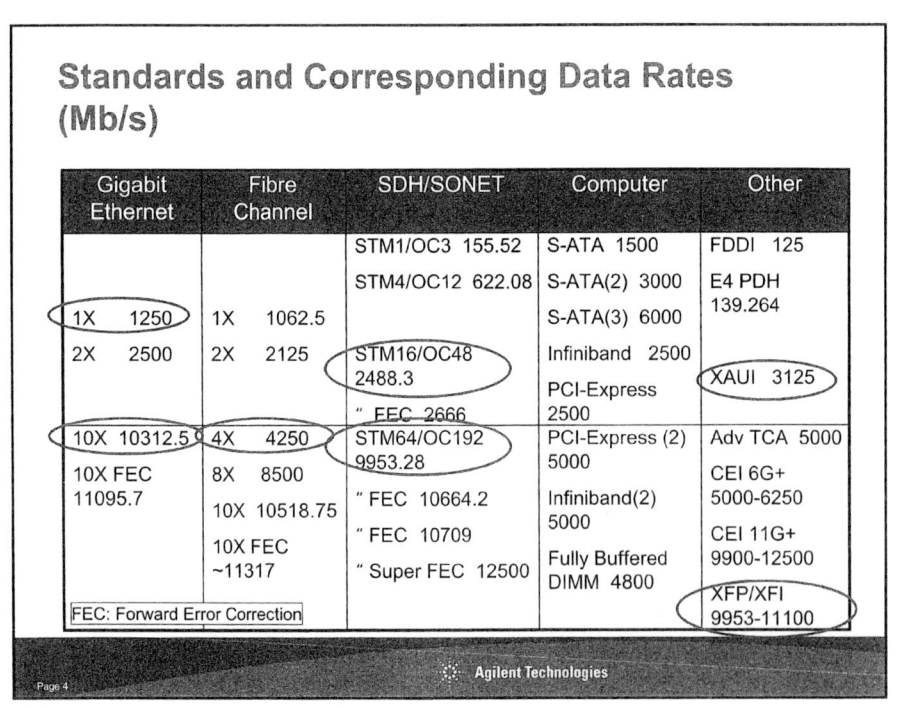

# Low rate optical transceivers example

| MSA | 1x9 pin | GBIC | SFF | SFP |
|---|---|---|---|---|
| Application | SONET/SDH GbE FiberChannel | GbE FiberChannel | SONET/SDH GbE FiberChannel | |
| Size (mm) | 40x26x10 | 66x32x10 | 50x14x10 | 57x14x10 |
| Pluggable ? | No | Yes | No | Yes |

1x9/2x9 pin MMF     GBIC MSA     SFF/SFP MSA

Photo Courtesy AVAGO TECHNOLOGIES

SFF: Small Form Factor
SFP :Small Form-factor pluggable

Agilent Technologies

---

# 10Gb/s Optical Transceiver example

| MSA | 300-pin | XENPAK | X PAK | X2 | XFP |
|---|---|---|---|---|---|
| Optical Interface | IEEE 802.3ae SONET/SDH ITU G.709 | IEEE 802.3ae | IEEE 802.3ae 10G-Fiber Channel SONET/SDH ITU G.709 | | |
| Electrical Interface | SFI-4, XSBI | XAUI | XAUI, SFI-4-P2 | | XFI (10G) |
| Size (inch) | 4.0x3.5x0.53 3.0x2.2x0.53 | 4.8x1.4x0.7 | 2.7x1.4x0.4 | 3.0x1.4x0.4 | 2.2x0.68x0.4 |
| Power Consumption | < 12 W | < 11 W | < 4W | < 4W | <2.4W +SerDes |
| Pluggable ? | No | Yes | Yes | Yes | Yes |

Agilent Technologies

# Typical Test Parameters

|  | Parameter | Test Instrument | Agilent solution |
|---|---|---|---|
| Device | Modulation bandwidth | Lightwave Component Analyzer | N4373A |
|  | Chirp | Chirp Test system | 86146B OSA +86100 DCA |
| Tx | Center Wavelength Side-Mode Suppression | Optical Spectrum Analyzer | 86146B |
|  | Rise-time/fall-time Extinction Ratio Eye Mask | Optical-scope (DCA) + pulse pattern generator | 86100C series |
|  | Optical Output power Return Loss | Optical Powermeter Return loss testset | 8163B series |
| Rx | Optical Receiver Sensitivity Stressed Receiver Sensitivity Dispersion Penalty | BERT & optical attenuator test system | N4900 series |
|  | Jitter Tolerance |  |  |

Agilent Technologies

---

# Test of Modulation Bandwidth

$$\text{E/O measurement} = \frac{P_{mod}^{Out}}{I_{mod}^{In}}$$

$$\text{O/E measurement} = \frac{I_{mod}^{Out}}{P_{mod}^{In}}$$

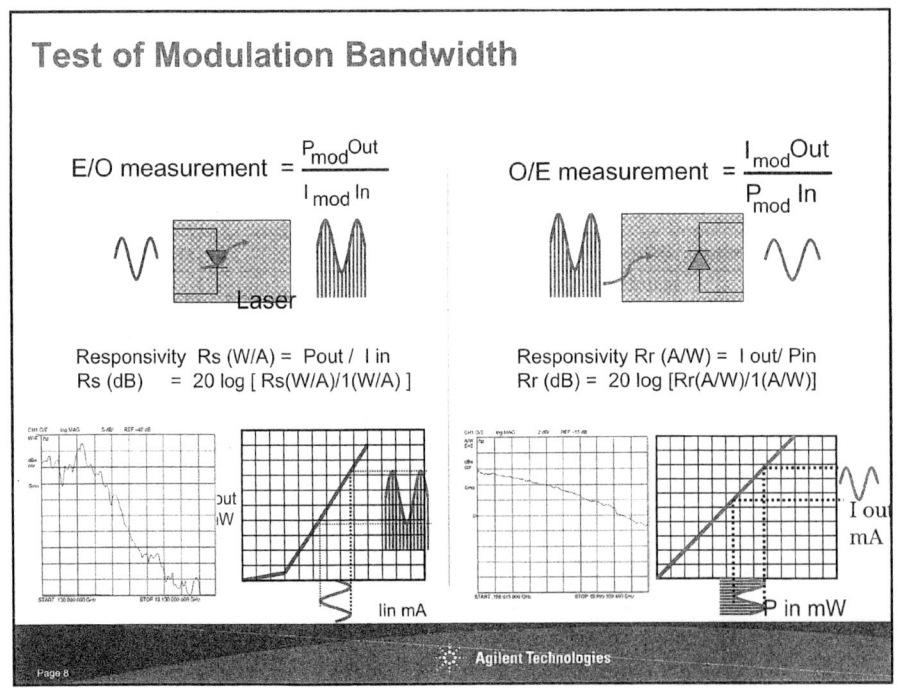

Responsivity Rs (W/A) = Pout / I in
Rs (dB)   = 20 log [ Rs(W/A)/1(W/A) ]

Responsivity Rr (A/W) = I out/ Pin
Rr (dB) = 20 log [Rr(A/W)/1(A/W)]

Agilent Technologies

# Frequency Response – Flatness

## Optical CATV

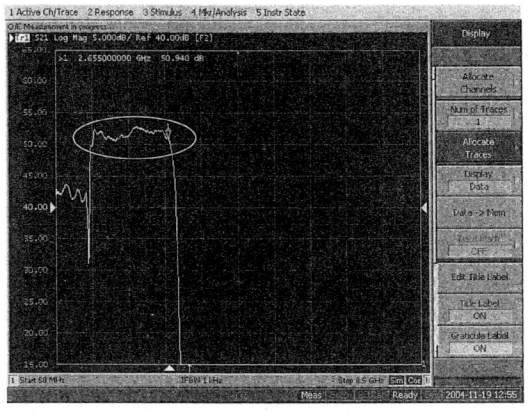

- Signal flatness in the transmission band
- Bandwidth
- Modulation depth (option)

Agilent Technologies

---

# Lightwave Component Analyzer

**Features:**
- NIST traceability for the full **solution**
- Single measurement gives all S-parameters within seconds
- 850nm (MM), 1310nm (SM), 1550nm (SM), 1310 + 1550nm (SM)
- **300kHz** up to **67GHz** in various options
- Single-ended and **Differential** interfaces
- Single component analysis for E/E, E/O, O/E and O/O components; calibrated **wafer characterization**
- Defined calibration procedures for every measurement setup
-

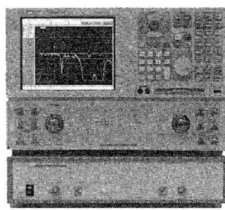

Agilent N4373A

Agilent Technologies

# Electrical Differential Interface

## 4xFC,10GbE : TOSA/ROSA, Optical Transceivers

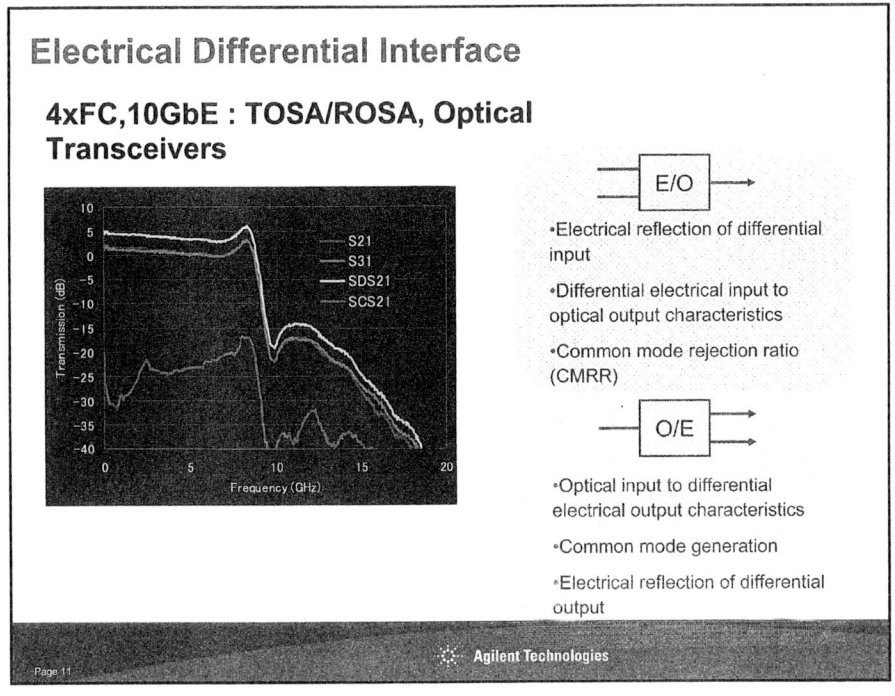

- Electrical reflection of differential input
- Differential electrical input to optical output characteristics
- Common mode rejection ratio (CMRR)

- Optical input to differential electrical output characteristics
- Common mode generation
- Electrical reflection of differential output

Agilent Technologies

---

# LCA with Differential Electrical interface

- **Electrical Return Loss (Differential/Common) [SDD11/SCC11 ]**
- **Electrical to Optical Conversion (Differential/Common) [SSD21/SSC21]**
- **Optical to Electrical Conversion (Differential/Common)**

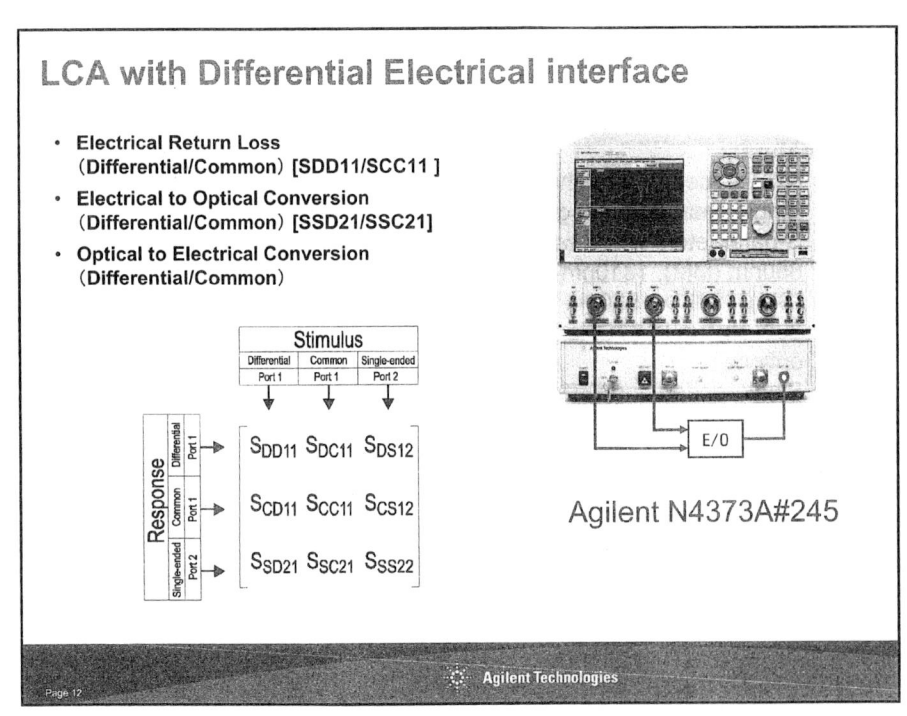

Agilent N4373A#245

Agilent Technologies

# Test of Optical Spectrum

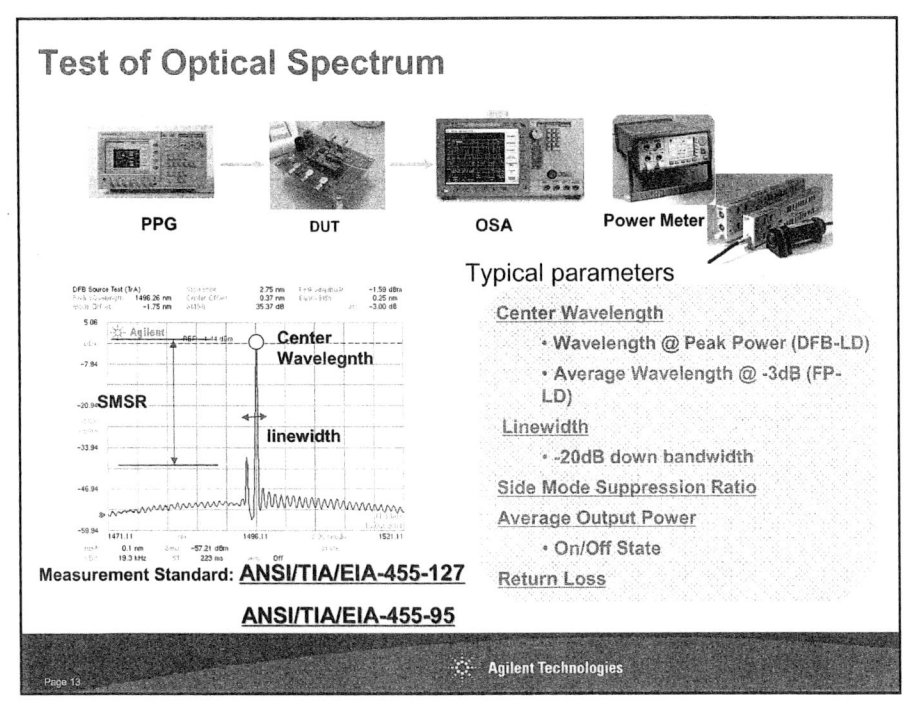

**Typical parameters**

Center Wavelength
- Wavelength @ Peak Power (DFB-LD)
- Average Wavelength @ -3dB (FP-LD)

Linewidth
- -20dB down bandwidth

Side Mode Suppression Ratio

Average Output Power
- On/Off State

Return Loss

**Measurement Standard: <u>ANSI/TIA/EIA-455-127</u>**

**<u>ANSI/TIA/EIA-455-95</u>**

---

# Chirp Measurement

**What is Chirp ?** When a laser is modulated, a variation in both optical intensity and frequency occurs. <u>Chirp is the frequency modulation</u> (usually unintended) caused by the modulation signal.

- Chirp interacts with the dispersion characteristics of the transmission media affecting the distance over which the signal can propagate. More chirp = less propagation distance

- Chirp causes spectral broadening which can interfere with adjacent channels in ultra-dense WDM.

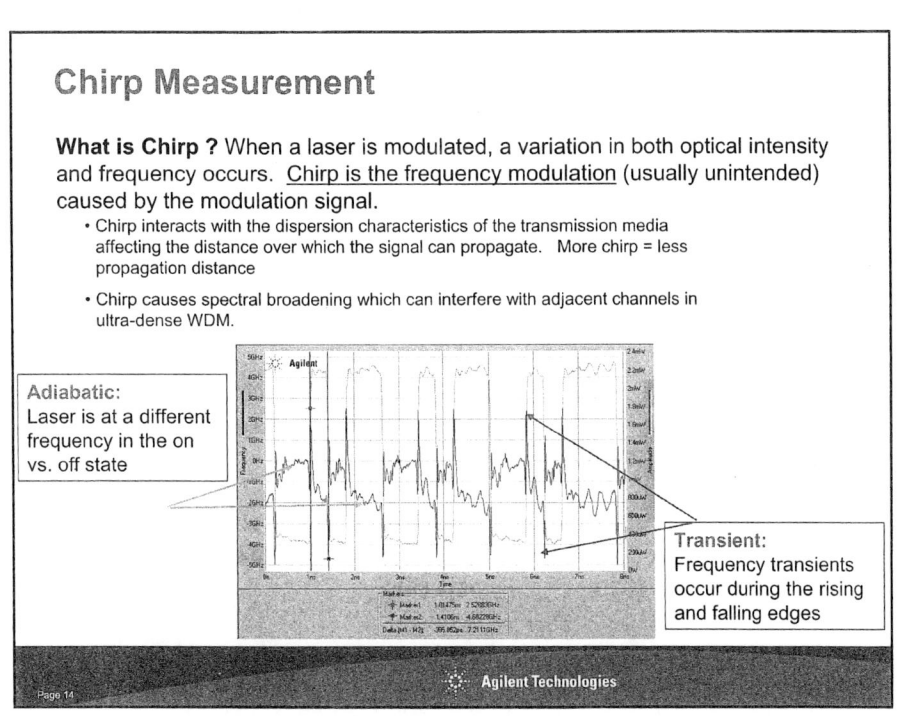

**Adiabatic:**
Laser is at a different frequency in the on vs. off state

**Transient:**
Frequency transients occur during the rising and falling edges

31

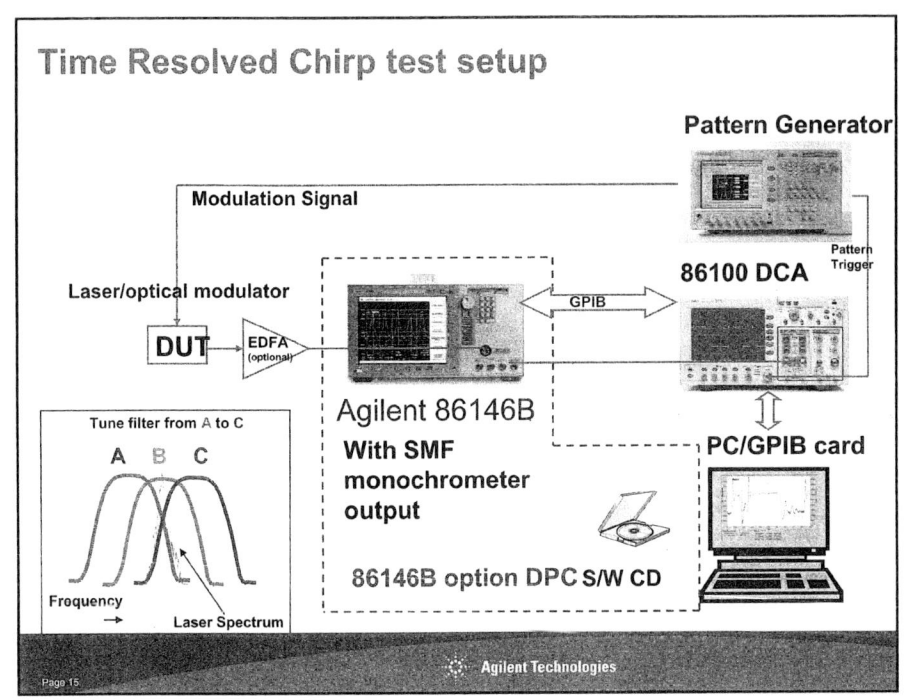

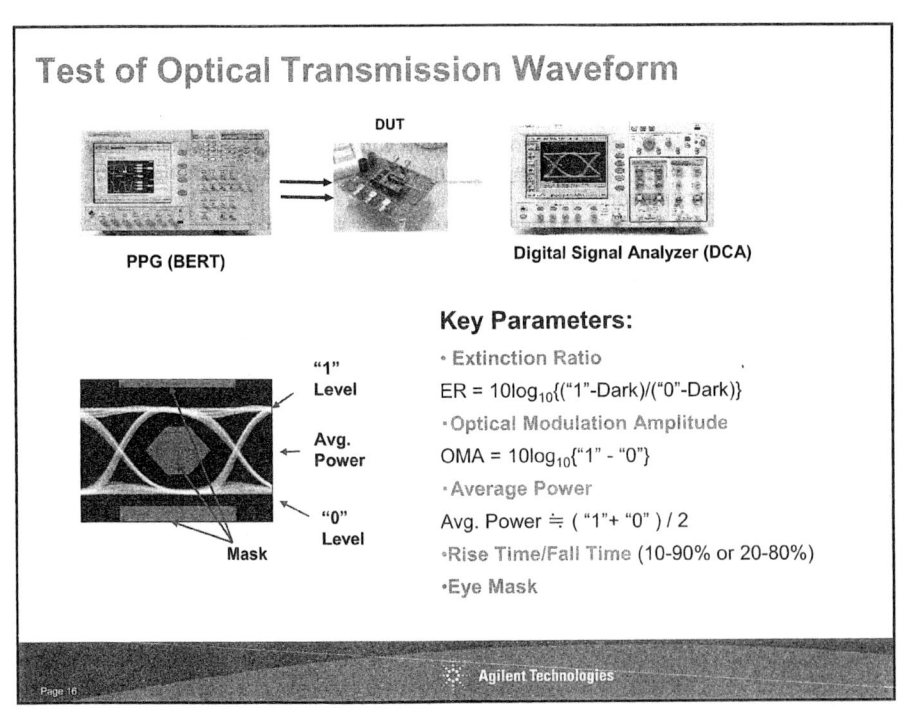

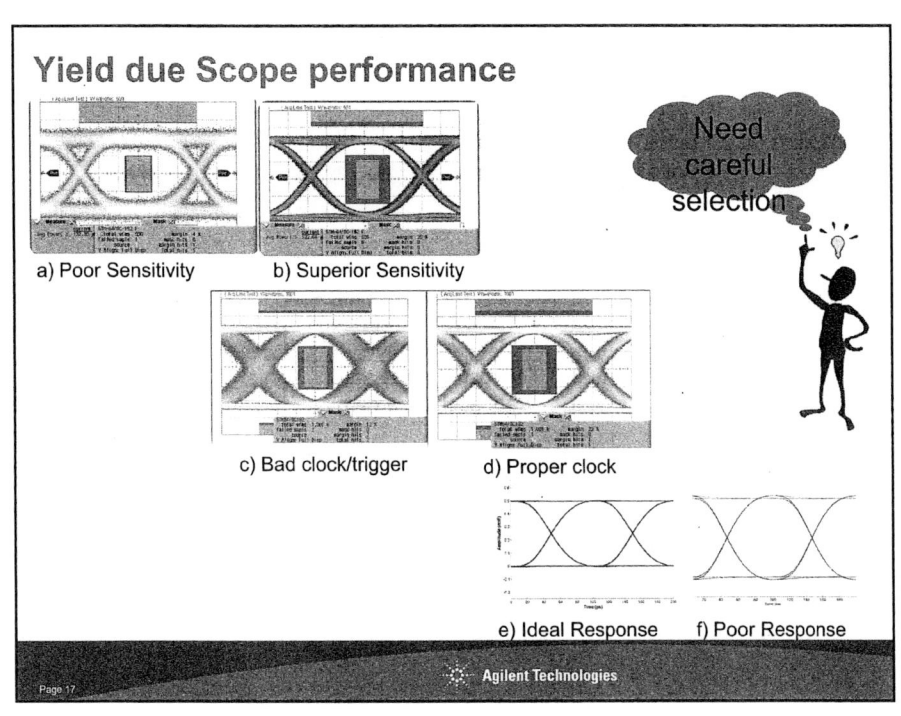

# Yield due Scope performance

a) Poor Sensitivity    b) Superior Sensitivity

Need careful selection

c) Bad clock/trigger    d) Proper clock

e) Ideal Response    f) Poor Response

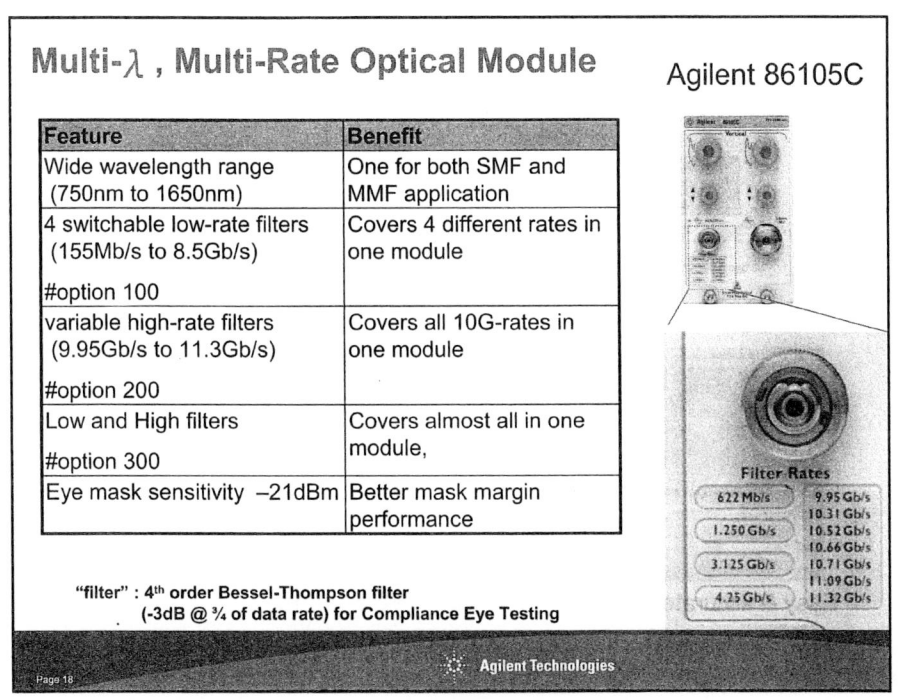

# Multi-λ , Multi-Rate Optical Module

## Agilent 86105C

| Feature | Benefit |
|---|---|
| Wide wavelength range (750nm to 1650nm) | One for both SMF and MMF application |
| 4 switchable low-rate filters (155Mb/s to 8.5Gb/s) #option 100 | Covers 4 different rates in one module |
| variable high-rate filters (9.95Gb/s to 11.3Gb/s) #option 200 | Covers all 10G-rates in one module |
| Low and High filters #option 300 | Covers almost all in one module, |
| Eye mask sensitivity −21dBm | Better mask margin performance |

**"filter" : 4th order Bessel-Thompson filter**
**(-3dB @ ¾ of data rate) for Compliance Eye Testing**

Filter Rates
622 Mb/s    9.95 Gb/s
1.250 Gb/s   10.31 Gb/s
          10.52 Gb/s
          10.66 Gb/s
3.125 Gb/s   10.71 Gb/s
          11.09 Gb/s
4.25 Gb/s   11.32 Gb/s

## Clock Recovery , best for scope

Sampling Oscilloscope (DCA) needs trigger signal or clock.
but Trigger is not available for System validation or Xenpak transceiver.

➡ Requires the Clock Recovery.

| Feature | Benefit |
|---------|---------|
| Wide wavelength range (750nm to 1650nm) | One for both SMF and MMF application |
| 50Mb/s to 13.5Gb/s continuous tuning w/#option 200 | No gap for any data rate such as emerging 8.5Gb/s |
| variable loop bandwidth w/#option 300 | Evaluate "receiver" performance correctly |
| Low jitter < 0.5 ps (RJ-rms) | Allow clean eye |

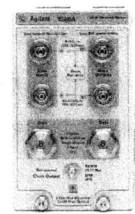

Agilent 83496A

Agilent Technologies

---

## DCA is versatile tool for High Speed Signal Integrity Characterization : Jitter , TDR/S-parameter

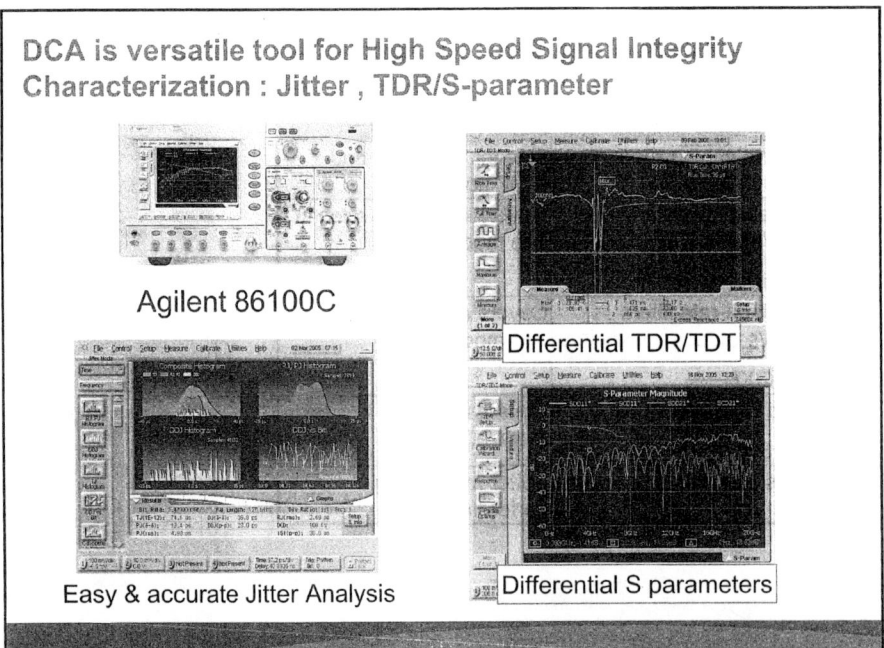

Agilent 86100C

Easy & accurate Jitter Analysis

Differential TDR/TDT

Differential S parameters

Agilent Technologies

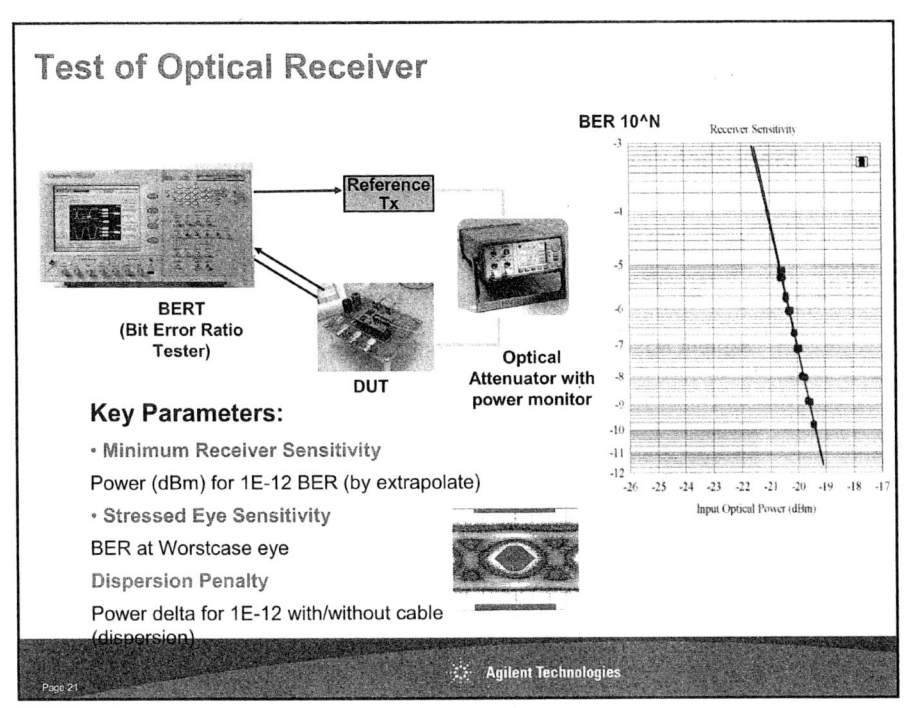

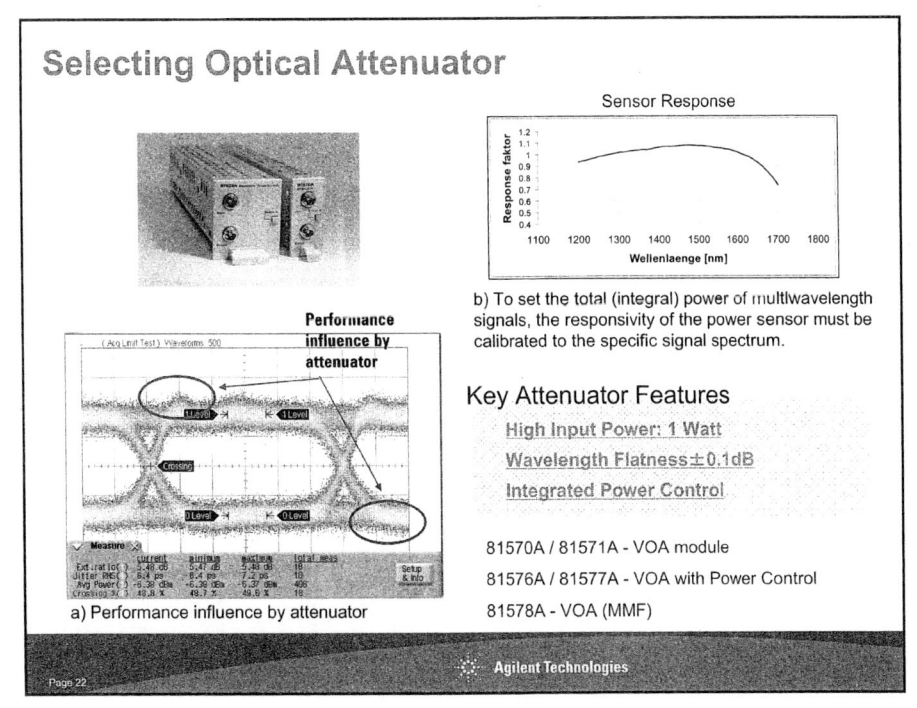

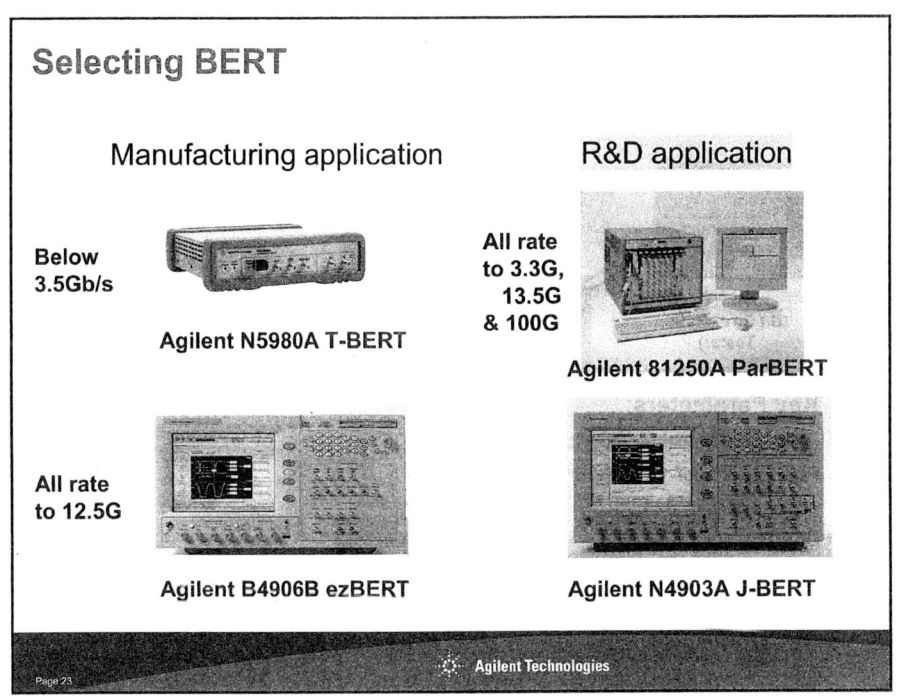

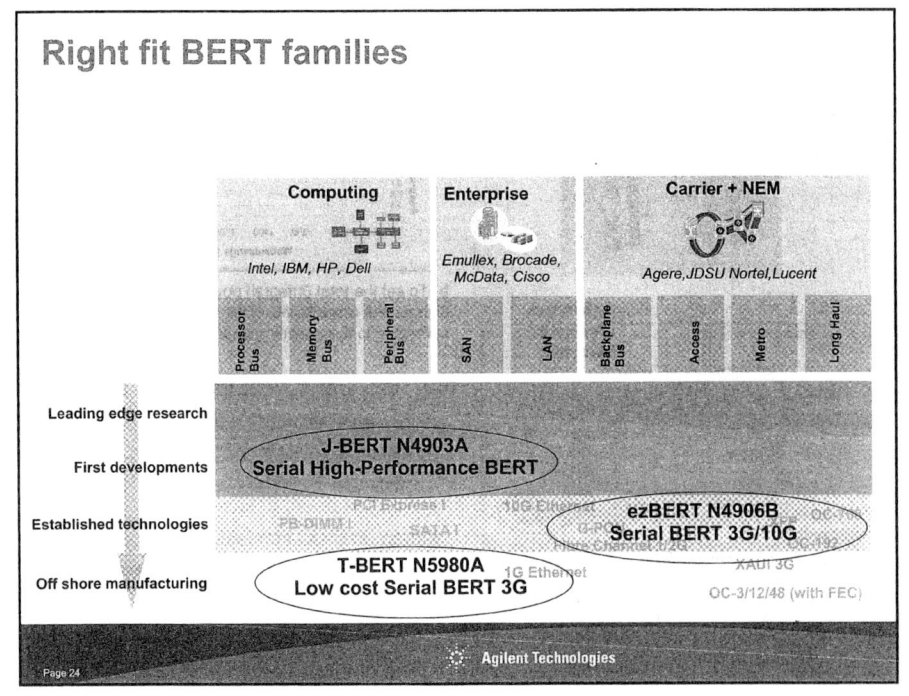

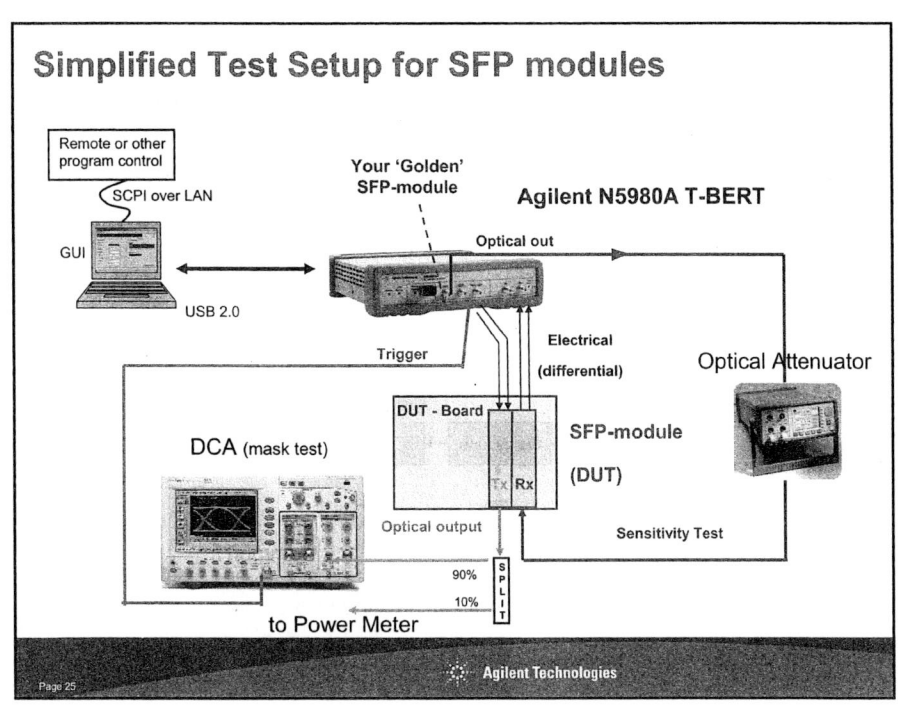

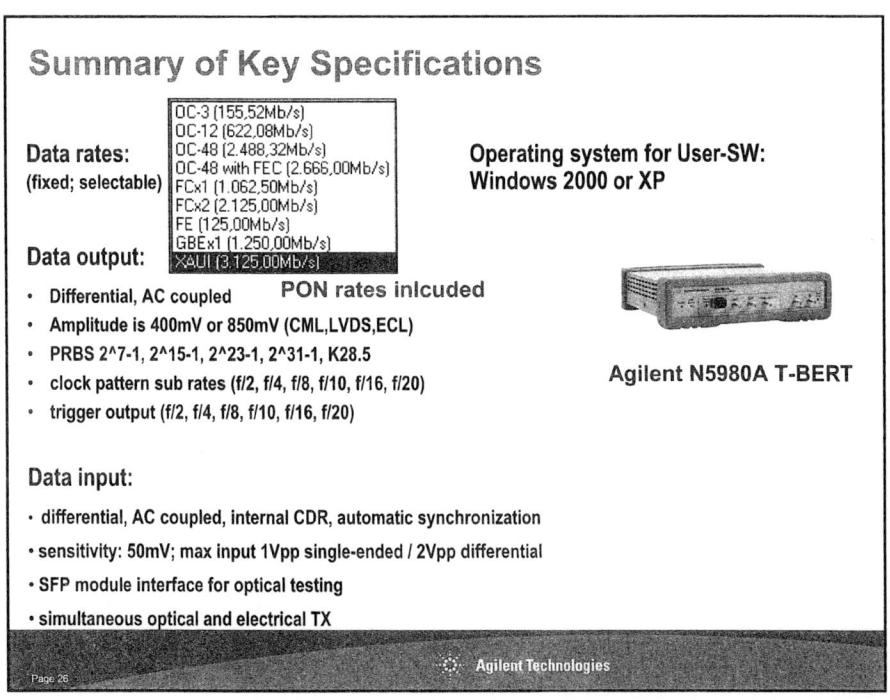

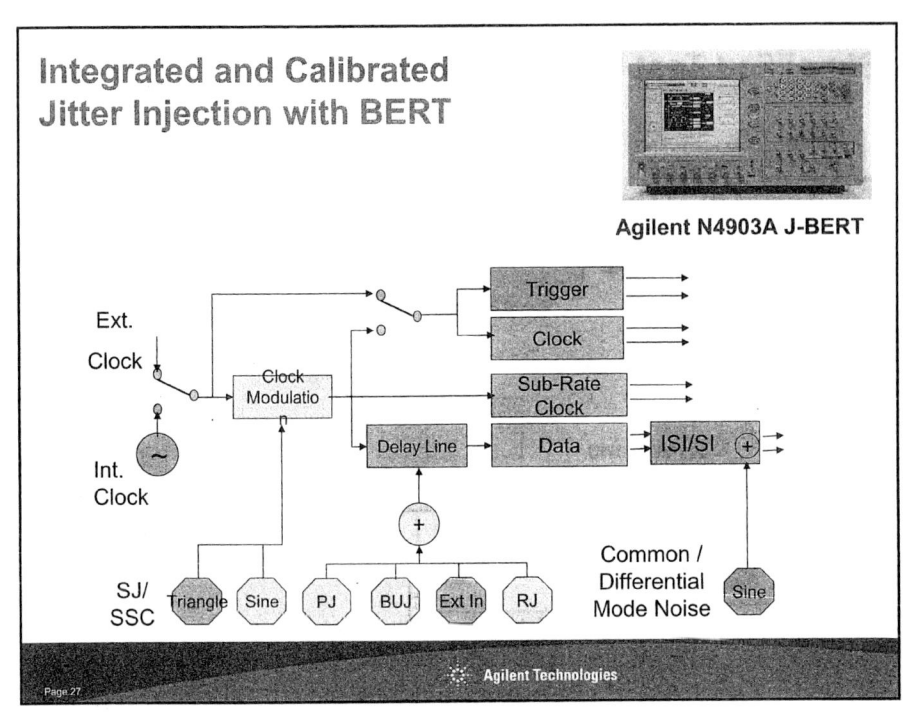

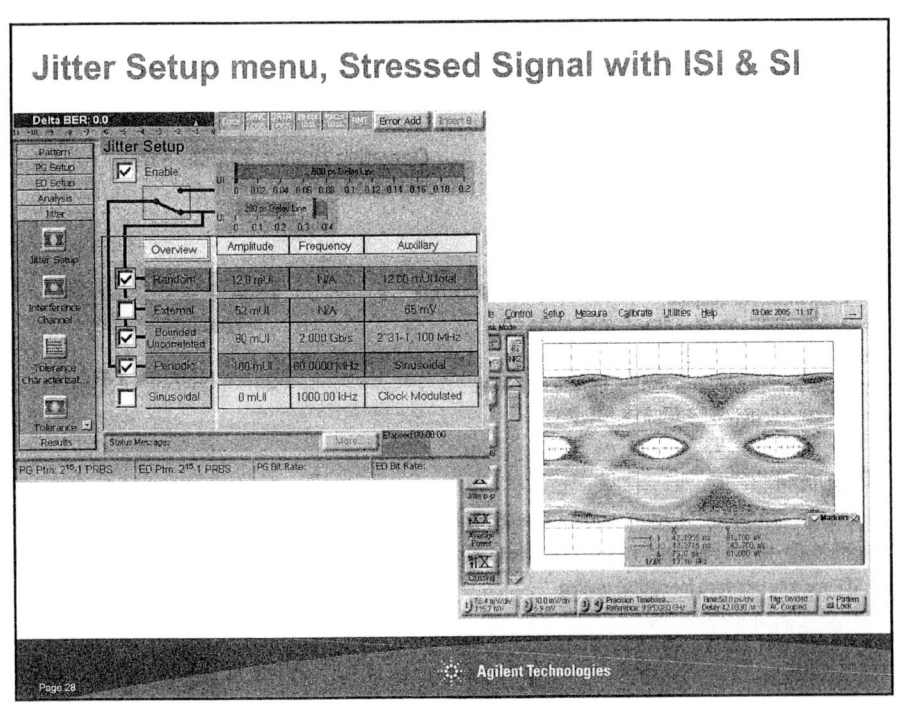

# Result : Jitter Tolerance Curve

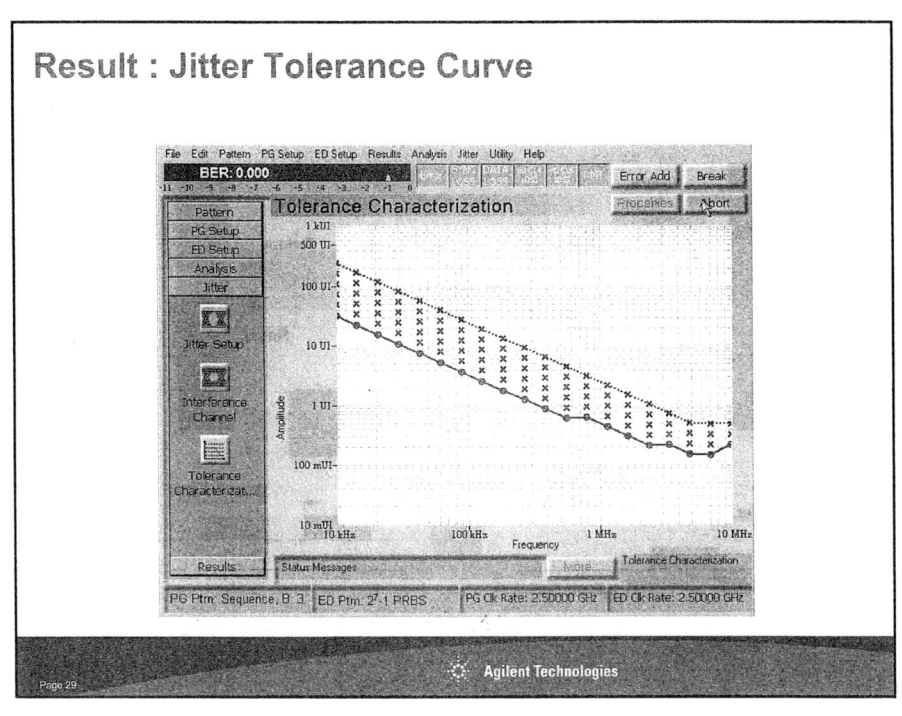

# Additional feature

**Measurement Suite**

- •Output timing
- • Jitter
- • Output level/Q-factor
- • Eye opening
- • Fast eye mask

New •Quick EYE Diagram

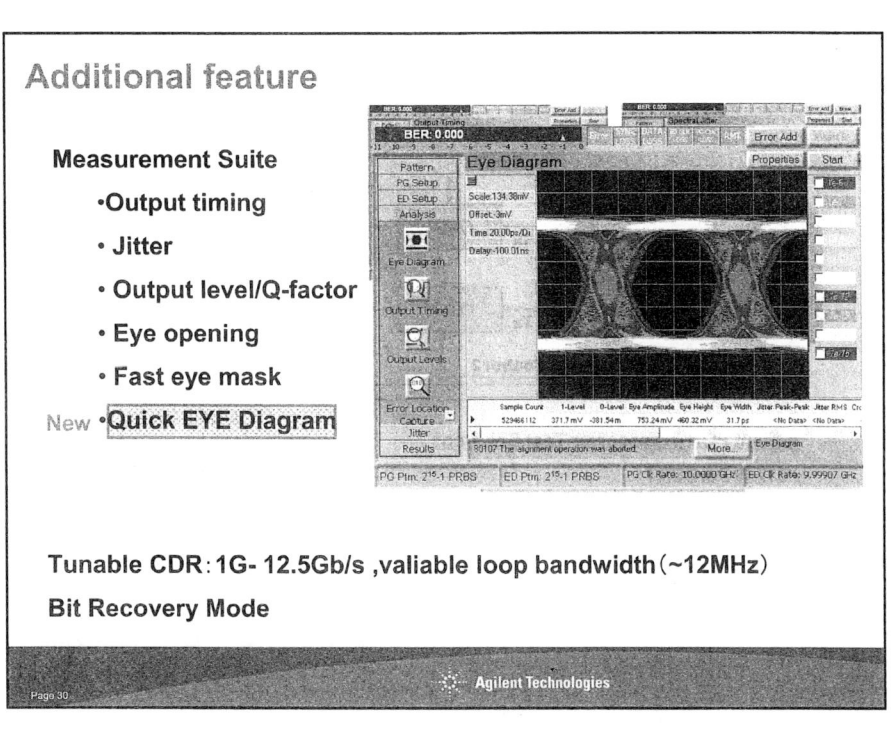

Tunable CDR：1G- 12.5Gb/s ,valiable loop bandwidth（~12MHz）

**Bit Recovery Mode**

# Typical PON Transceiver Measurement Setup
# - > OLT Burst Receiver test is most critical

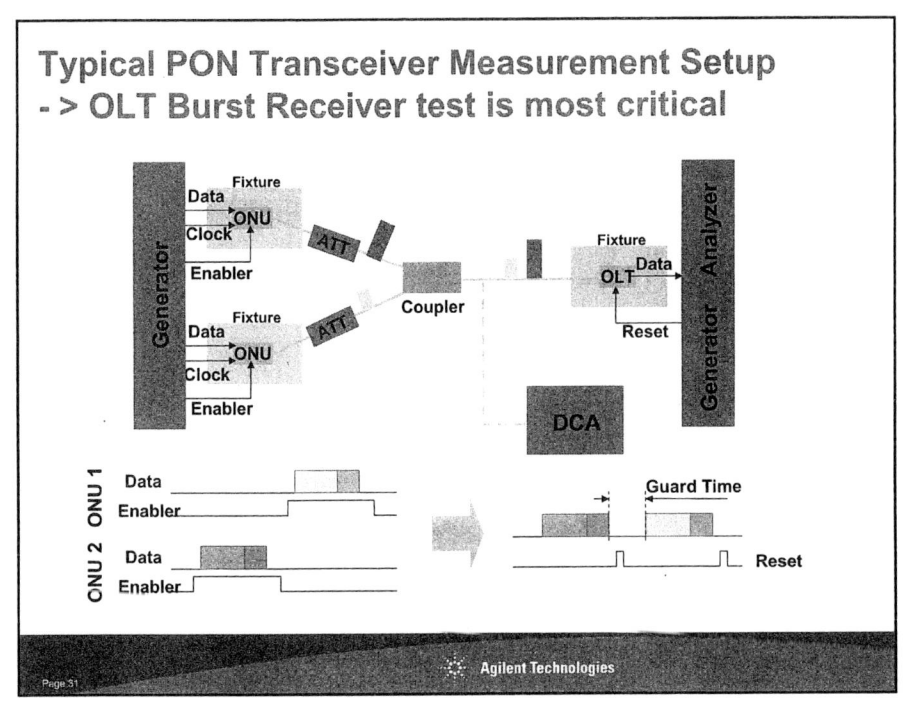

# ParBERT based solution for PON transceiver

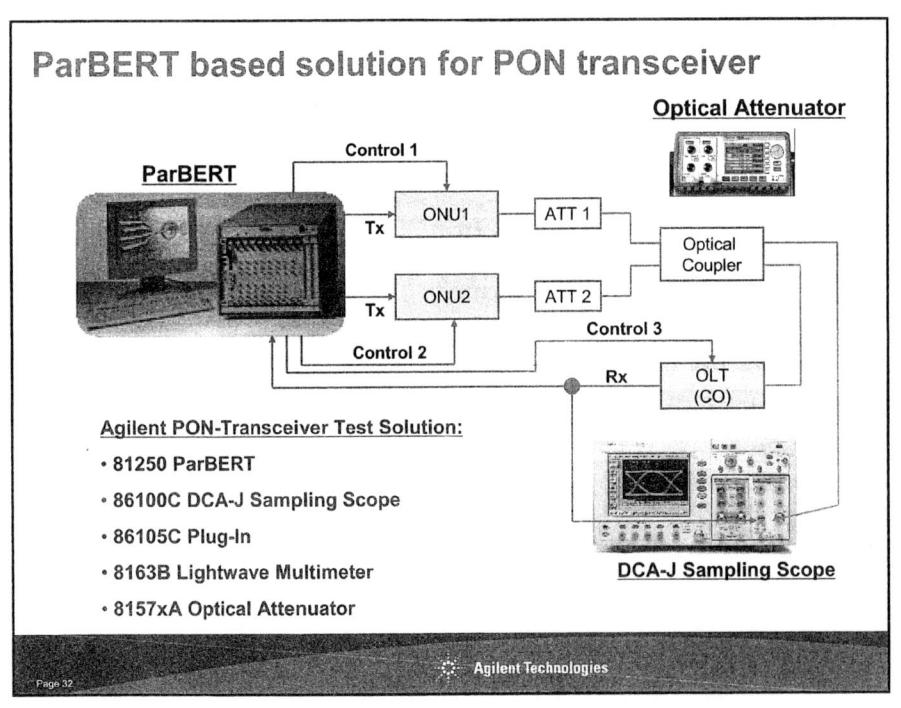

**Agilent PON-Transceiver Test Solution:**

- 81250 ParBERT
- 86100C DCA-J Sampling Scope
- 86105C Plug-In
- 8163B Lightwave Multimeter
- 8157xA Optical Attenuator

40

## Summary

**Agilent has full line up of total solution in optical transceiver testing**

Time to market

Time to 1st production

Time to volume

# Plenary Session
## Special Invited Speakers
**October 25, Wednesday**
**1:30pm – 2:00pm**

**Telecommunication Market in China**
**Dr. Xie Yi, Vice President**
谢毅 副院长, 教授
信息产业部电信科学研究院

## China Academy of Telecommunication Research, Ministry of Information Industry

The latest development of telecommunication market in China will be reviewed in this talk, especially on terminal market aspects and special features for mobile telecommunication. The author will also provide his point of view of China communication market trend and requirements and viable technologies for upcoming years.

Dr. Xie Yi is Vice President, China Academy of Telecommunication Research (CATR), Ministry of Information Industry; and Chairman of Technical Committee 9, China Communication Standards Association. Dr. Xie has been heavily engaged in research area of optical communications for numerous years. He is a Professor level on his professional and well known in industry in China.

谢毅，教授，工学博士，现任信息产业部电信研究院副院长，兼任中国通信标准化学会TC9主席，从事光通信和通信测试领域的研究工作。

# Plenary Session
## Special Invited Speakers
**October 25, Wednesday**
**2:00pm – 2:30pm**

**Service-driven Advanced Optical Network**
**Prof. Yuefeng Ji**
**Beijing University of Posts & Telecommunications, Beijing, 100876, China**
业务驱动的先进光网络，纪越峰 教授，北京邮电大学，北京，100876，中国

### Service types and characteristics
- Service types and characteristics
- Performance requirements of different services
- Optical network evolution driven by new services

### Key technologies of service-driven advanced optical network
- National 863 optical communication R&D program (O-TIME Program)
- Broadband optical access technology (PON)
- Adaptation between IP layer and optical layer (IP over Optical)
- Burst service oriented optical network (OBS network)
- High efficient optical switching network (OPS network)
- Control in advanced optical network (GMPLS)
- Distributed optical network  (Photonic Grid)

### Prospect of next generation optical network
- Impetus: services
- Self-aware optical network: service and resource self-ware
- Self-adaptive optical switching network: multi-granular service self-adaptive
- Self-organized optical network: self-routing，self-linking and self-protection/ restoration

Yuefeng Ji , Ph. D., graduated from Beijing University of Posts & Telecommunications (BUPT)，P. R. China. Now he is a professor and vice dean of the School of Telecommunication Engineering of BUPT. He is also an expert in the Communication Technology Group of the National High-Tech Research and Development Program (National 863 Program) of China, etc.. His research interests are primarily in the areas of optical communication and broadband information networks, with emphasis on key theory, realization of technology, and service applications.

业务的分类及其特征
- 业务类型与特征
- 不同业务对网络性能的需求
- 业务驱动下光网络的演进

业务驱动先进光网络的若干关键技术
- 中国863光通信高技术研究开发计划（O-TIME计划）
- 宽带光接入技术（PON）
- IP层与光层的适配技术（IP over Optical）
- 面向突发业务的光交换网络（OBS）
- 光网络中的控制（GMPLS）
- 分布式光网络（Photonic Grid）

下一代光网络发展方向
- 推动力－多业务
- 自感知光网络－业务网络资源的自感知
- 自适应光交换网络－多粒度业务的自适应
- 自组织光网络－光网络的自选路、自建链与自保护／自恢复

纪越峰，博士，毕业于北京邮电大学。现任北京邮电大学电信工程学院教授/博士生导师/副院长，兼任"十五"国家863计划通信技术主题成员、"十五"国家863 /O-TIME（光时代）计划总负责人等，主要研究领域为光通信与宽带信息网，包括关键理论、实现技术与业务应用等。

# Service-Driven Advanced Optical Network

Yuefeng Ji

Beijing University of Posts & Telecommunications, Beijing, China

jyf@bupt.edu.cn

**Abstract** *with the rapid growth of Internet applications, service-driven technology requirement keeps on increasing. To meet this tide, some advanced optical network technologies are proposed. This paper presents these technologies such as IP adaptation, PON, OBS, Photonics Grid, Control, Plan and new conceptions of next generation optical network – Autonomic Optical Network (AON) and convergence are proposed.*

## 1. Introduction

Much of the almost hyperactivity within the optical communications industry over the past decade can be traced to the perceived and real growth of the traffic on the Internet. According to the seventeenth statistical report of CNNIC, up to Jan, 2006, the number of China's Internet users has already reached 111,000,000 [1]. Compared with investigation of one year ago, the total number of China's Internet user increases 17,000,000 in 2005 and grows at 18.1% per year. The total number of China's Internet users still keeps on increasing. With the rapid development of Internet applications and user requirements in China, optical networks have become the best choice to fulfil the exponentially-growing bandwidth. Therefore, R&D activities on optical technologies have been greatly promoted in China. Service-driven optical network will be the best way for future carrier networks.

This paper mainly analyzes the current service-driven advanced optical network. Firstly, the national project plans, such as the National High Technology R&D Program (863 program) and related application achievements, are presented. Secondly, the key technologies of optical network are discussed. Then, the R&D status of next generation optical network is introduced.

## 2. National R&D Plans about Optical Communications R & D in China

To research and develop new optical communications technologies, government of China has launched several R&D plans as follows:

(1)The National Natural Science Foundation of China (NSFC): promoting and financing the basic and applied research in China [2].

(2)The National Basic Research Program (973 Program): organizing and implementing the national strategic research in China [3].

(3)The National High Technology Research and Development Program of China (863 Program): promoting the development of forefront and foresighted high technology in China [4].

(4)The Key Technologies R&D Programme: the national science & technology(S&T) program to address major S&T issues in national economic construction and social development in China [5].

(5)China Torch Program: a guiding program designed to develop new and high technology industries in China, promoting commercialization, industrialization and internationalization of the technology research [6].

Each plan described above has supported optical communication from different emphasis. Take national 863 Program for instance, with the successful implementation of the Seventh, Eighth and Ninth Five-Year Plan, the State Council continuously carried out the Tenth Five-Year Plan (2001~2005) for 863 Program. In this period, Chinese government had supported the studies of carrying multiple applications on optical networks, such as O-Time program and 3T-net project.

O-Time (**O**ptical **T**echnology for **I**P with **M**ulti-wavelength **E**nvironment) is a part of National 863 program in Tenth Five-Year Plan (as Figure-1 shows). O-Time focuses on the high-speed, intelligent, merged and new generation optical system and optical network technology [7].

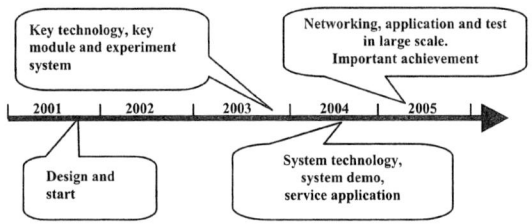

Figure-1 Time Schedule of O-TIME Program

There are 4 main research areas in O-Time program: high-speed optical transmission , broadband optical access, optical switch node, and intelligent optical networking.

With the successful CANONET (China Advanced Info-Optical Network) implemented in 2001, the National 863 program proposes a "High Performance Information Demonstration Network" project (3T-net, as was shown in Figure-2) in Tenth Five-Year Plan. The project objective is to implement Tbps-level router, Tbps-level optical transmission system and Tbps-level switching capacity equipment [8].

The program has independently developed a new generation of network core equipments and application-supported environment involving Tbps-level router, switch, transmission equipment, and constructed high performance broadband information demonstration networks supporting large-scale concurrent stream video service and interactive multimedia service. Proceeding with broadband interactive service, the service application result is mainly to be oriented towards large-scale concurrent DTV/HDTV broadband stream video and Peer to Peer service application. 3T-net has been built to form a large-scale interactive network TV testing network.

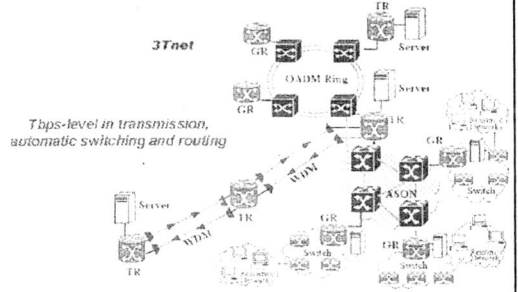

Figure-2  3T-net broadband network

To further the research in optical communications, some key laboratories are built up with government's support in some famous universities, such as Tsinghua University [9], Beijing University of Posts and Telecommunications [10], Peking University [11], Shanghai Jiaotong University [12], and so on.

### 3. Key Technologies of Advanced Optical Network

In order to fit the development of new services, many researching works of new technologies are ongoing.

**(1) IP adaptation to optical**:  As we know, IP is transmitted in upper layer, while optical fibre is used in lower layer. So how to adapt IP with optical in high-efficiency is a key question. A new technology, named Dynamic Adaptation Unit (DAU), is presented to improve network efficiency by BUPT [Figure-3]. The main features are IP over GFP-e over WDM, flexible network management, high efficiency adaptation, high capacity access, multi-application mode, bandwidth on demand.

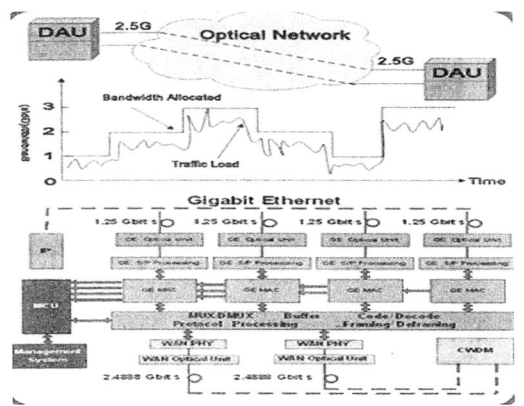

Figure-3  Dynamic Adaptation Unit (DAU)

**(2) FTTH**: In the last few years, China has seen a rapid growth in broadband access. With the increasing requirements for bandwidth by new applications, Fiber To The Home (FTTH) has been drawing attention of both research and industry. Currently, the three major FTTH technologies commonly considered in China are EPON, GPON and Point to Point (P2P). APON will not be deployed due to its complexity and high equipment cost. Those three technologies are complimentary rather than conflicting. EPON is mainly used for FTTH in the area with high population density; GPON will be mainly used for corporations that need a large number of TDM leased line services. P2P can be used for both business cases and residential cases. For business case, P2P will be used when the required bandwidth is more than that a PON can provide. For residential case, P2P is preferred than EPON when the residential homes are scarcely located. Other PON technologies like WDM PON are still under research. Until now, most local vendors focus on GEPON system [13] [14] [15]. GPON system is under consideration by some vendors.

**(3) Optical switching**: OBS is a compromised product of optical circuit switching (OCS) and optical packet switching (OPS) [16]. It adopts the advantages and feasible technology of OCS and OPS, and evades the disadvantages and unpractical ones. In OBS network, data classified by different of services are assembled into bursts instead of packets. Variable bursts make OBS network becomes more adaptive to transmit multi-services. Research on OBS has been carried on in Beijing University of Posts & Telecommunications (BUPT) and other Univ. for several years. It is known two OBS test-bed have been established in BUPT, one for OBS ring network, the other for OBS mesh network [17][18]. Both of the two OBS test-beds are well-designed and well-running (Figure-4, Figure-5). Many experiments about OBS network are completed on these two test-beds [19] [20].

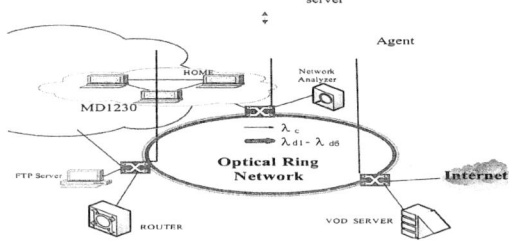

Figure-4 OBS Ring Network Test-bed

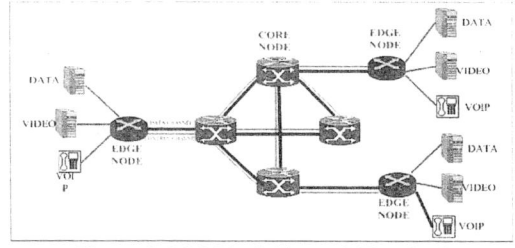

Figure-5 OBS Mesh Network Test-bed

**[4] Photonic Grid**: Grid networks will be new

communication infrastructures for data-intensive applications, data Grids and computational Grids. Figure-6 shows some results finished by BUPT [21]. A middleware-based Architecture is proposed and implemented to support GT4, capacity management and light-path provisioning. Hosts of the Grid network have the capability of throughput request and control. As circuit-based switching is efficient to transmit bulk datasets in data intensive applications, a middleware based on GMPLS is proposed for circuit-based switching networks. The middleware and the optical network interconnect distributed computational communities and satisfy the demands of various transmissions. The measurements and simulations show that the middleware supports traditional transmission functionalities and is efficient to transmit bulk datasets. Today's technologies enable optical networks to provide grid services for distributed computing. High performance end-to-end optical Ethernet light-paths have high speed, stable and low packet-loss bandwidths. But traditional TCP is limited in its ability to support high performance transmissions especially in the Grid environment. A transmission protocol for optical network is proposed to invoke high level grid services and support high performance transmissions. The impacts of transmission protocol including signalling delays, throughput control, bandwidth calculation and congestion control are studied.

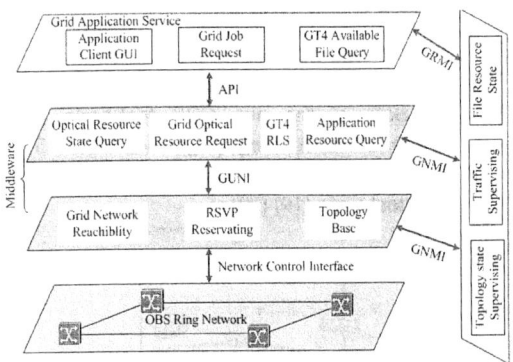

Figure-6 Photonic Grid Technology

**[5] Control for optical network** [22]: The control technologies include routing and signaling protocols, programmability of network, capacity reservation, networking, etc... The evolution of control technologies enable the photonic multicast, CoS, OVPN and potential new optical services to be reality. The control technologies will be integrated with the up-layer networks (e.g. MPLS, GMPLS and T-MPLS). Figure-7 shows the experimental platform on the hierarchical routing in ASON by BUPT, including a scheme deployed to verify hierarchical routing for ASON,

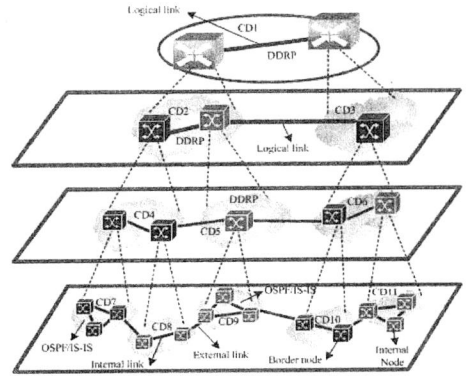

Figure-7 Logical topology of hierarchical network

**[6] Plan for optical network**: Planning and simulation software for optical network is finished, including four layer architecture, classified link resource, guaranteed QoS combined protection and dynamic recovery, Inter-layer mapping (Figure-8 ).

Figure-9 shows the research platform for optical network, mentioned above, finished by BUPT.

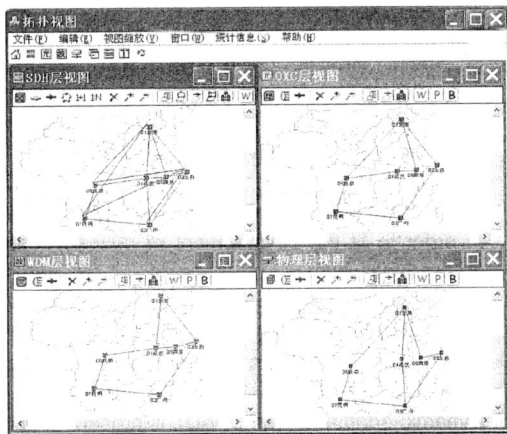

Figure-8 Planning and simulation software for optical network

Figure-9 The research platform for optical network

## 4. Prospect of next generation optical network

There are some hot topics for next generation optical network [23], such as Autonomic Optical Network, Convergence Network, etc.

Although optical network technology has been greatly improved in transmission, switching, management, and

46

so on, intelligence of optical network can not meet the demands of new services. As a result, a new conception of autonomic optical network (AON) is proposed. AON contains three parts or stages: self-aware optical network, self-adaptive optical switching network, and self-organization optical network. In self-aware stage, optical network can cognize physical and statistics characteristics of access optical signal. The characteristics cognized include optical power, signal noise rate (SNR), packet length, packet interim, type of services, etc. Network resources such as topology, node change, link status, route table, can also be cognized by optical network actively. After acquiring the cognized data, optical network process these data in order to get useful control information. This is accomplished by self-learning algorithm. Useful control information can be utilized to optimize optical switching, routing, protection, and so on. The above paragraph describes the whole process of AON. If it comes into true, optical network can realize service self-aware, switching self-adaptive, and network self-organization. Then a self-system is established. This human oriented optical network is the ultimate aim of next generation optical network.

Another hot topic is convergence, such as Optical & Wireless Convergence, Optical & Service Convergence, Optical & Sensing Convergence, and Optical & Resource Convergence.

## 5. Conclusions

Following the rapid increasing of internet, IP service domain the whole networks. For optical network, it is time for service-driven rather than technology-driven. Service-driven network demands new technologies to match IP transmission. In this paper, several technologies such as IP adaptation, PON, OBS, Photonics Grid, Control, and Plan for optical network are introduced and research works about them in China are presented. Finally, a new conception of AON is proposed as next generation optical network.

## Acknowledgement

This research was jointly supported by the National Science Fund for Distinguished Young Scholars ( 60325104), National Natural Science Foundation ( 60572021), and the SRFDP of MOE (20040013001), P. R. China.

## References

[1] www.cnnic.com.cn
[2] http://www.nsfc.gov.cn
[3] http://www.973.gov.cn
[4] http://www.863.org.cn
[5] http://gongguan.jhgl.org
[6] http://www.chinatorch.gov.cn
[7] Yuefeng Ji, Telecommunications Technology, No. 6, 2003, pp.11-14.
[8] Weisheng Hu, et al, COIN2006.Korea
[9] http://www.tnlist.org.cn
[10] http://oclt.sto.bupt.cn, http://oilab.ste.bupt.cn
[11] http://ofcl.pku.edu.cn
[12] http://www.sjtu.edu.cn/base/labs
[13] Chen Xue, et al,Journal of China Universities of Posts and Telecom., vol.12, No.2, 2005, pp.70-75.
[14] Jianli Wang, Proc. SPIE Int. Soc. Opt. Eng. 6022, 60220D (2005).
[15] Mao Qian, CCSA. April, 2006.
[16] C. Qiao et al, J. High Speed Networks, vol. 8, no. 1, Jan. 1999, pp. 69–84.
[17] Zehua Gao, et al, Journal of Optical Communications, vol.26, no.6, Dec. 2005, pp: 255-59.
[18] Jintong Lin, et al, in OFC2006, OWU2.
[19] Hongxiang, Guo, et al., in Proc. OFC2006, OFA6.
[20] Yinghui Qiu, et al, Proceedings of SPIE, APOC'2004, Sep. 2004, vol. 5625, pp. 698-704.
[21] Yuefeng Ji ,COIN2006.Korea.
[22] IYuefeng Ji, et al, ICOCN 2006,China
[23] Yuefeng Ji, et al, ECOC 2006,France.

# Plenary Session
## Special Invited Speakers
**October 25, Wednesday**
**2:00pm – 3:00pm**

**Technology Development of Flat Panel Display (FPD)**
**Prof. Li Dejie, Dept. of Electronic Engineering**
**Tsinghua University, Beijing, China**
李德杰 教授
清华大学电子工程系

**Outlines:**
1) Current State of FPD
2) Paradoxes and Issues in FPD
3) Performance of FPD
4) The Ultimate Display

**Prof. Li Dejie** has been Professor at Tsinghua University, Beijing, China since 1986. Professor Li is also Advisor to several graduate students at Tsinghua University. In addition to academic teaching and supervising graduate students, Professor Li has been heavily involved in optoelectronic research areas, his most recent research focus is on Flat Panel Display, Thin Films, and Solar Cells. Professor Li has published more than 100 papers, he is the author of a book " "Electromagnetic Theory in Microwave and Optoelectronics", Chinese version published by China Electronic Industry, English version by Springer.

# Development of flat panel displays

Li Dejie (Tsinghua University, Beijing 100084)

Recent year production and sale of the flat panel TV (FPTV) sets increase explosively. The global FPTV market exceeds 200 billion us dollars. The liquid crystal TV requirement in 2006 is about 40 millions. The panel production ability is 50 millions. The output of the organic light emitting diodes (OLED) in the second quarter of 2006 is 16 millions, and that of the plasma display panel (PDP) in July of 2006 is about a million. In China in the 1st quarter of 2006 the sale of the LCD TV is about 6 hundred thousands (3.5 times of last 1st quarter).

Despite the output of LCD and PDP increase rapidly, it is too early to talk about which kind of the displays will be dominant in future. From the view point of a consumer, the difference of the displaying performance of various devices is nearly neglectable. The price and other factors will influence the development of FPDs. The issues in development of FPD industry include mainly: the power consumption, the rare metal resources and radiation, and there are some paradoxes about the performances of FPD.

I. Power consumption of FPD

The energy saving of the display is very important for the economic development and social progress. High power devices lead directly huge electricity consumption, and indirectly exhaustion of energy resource and environment pollution. The power will determine living or dieing of these display devices to some extent.

A calculating example:

Brightness is $1000cd/m^2$.

Maximum luminance (for 60 inch panel) is about 2831lm.

Average luminance is 940lm.

The power is 470W for the Efficiency of 2lm/w.

The power consumption is 600 millions kwh for 400 million TV sets (1.5kwh/day for a TV set), that is about 7% of the electricity generation capacity of China, or the electricity generation capacity of 2 Three Gorge hydroelectric power station.

The power consumption limitation has been set up for refrigerator and air-condition. It must be set up for TV before long. The TV with high power will fall into disuse. At present the possible measures may include:

- Limit the quantity of the FPD TVSET with high power.
- Develop FPD TV SET with low power.

The candidates for low power TV sets are FED and OLED. Both of them face some problems of fabrication technology and cost.

II. Radiation in FPD

Radiation includes X-ray radiation and electromagnetic radiation. Only FED has ray radiation. As the anode voltage is less than 10000 volts, the panel glass of 2mm thick can completely absorb the radiation. In FPD the electromagnetic radiation belongs to near-field type. In PDP it is electric radiation. In OLED it is magnetic radiation. In FED it is neglectable because of protection of the anode Al film. In PDP an anti-radiation screen can absorb more than 95% of the electric radiation, but the efficacy will reduce to half of the original value.

By far there isn't transparent magnetic anti-radiation film. The magnetic radiation from OLED with large size must be taken into account. If the working voltage is 5 volt, and the efficacy is 5lm/w, for a 60 inch panel with brightness of $1000cd/m^2$, the peak current is 113 A, and the average current is about 40 A. The magnetic field intensity far exceeds the safety standard in a distance of 3 meters.

III. Rare metal resource

The geological reserve of indium is about 6000 tons in the earth, and it is distributed in Zn, Sn and Cu beds. The exploitable quantity is about 5000 ton, therein 3000 ton exists in China. The output is about 200 ton a year. A majority of it is used in FPD industry. Even if the output does not increase, the resource will be exhausted in 25 years. Some companies have stocked up indium. The measures for this are searching the replacing materials and carrying out research of FPD without indium. The replacing materials include $SnO_2$ (Sb doped) and ZnO (Al or Ca doped). For $SnO_2$ there still isn't an etching method, which limits the use of it. For ZnO the existing problem is the stability.

By far the only FPD without indium is FED.

IV. Contrast about FPD

For TV display the contrast in dark room is more important than that in light room. By far the measured highest contrast of FPD in dark room is 100000:1 from the surface conduction displays. This result is based on a wrong measuring method. For a display the contrast of 200:1 in dark room is good enough for the viewer. The practical contrast for CRT is less than 150:1.

In all exiting display devices, LCD has the highest contrast in light room, and ordinary OLED has the lowest contrast in light room. This contrast of OLED can be improved by an absorbing back electrode at a cost of reducing the luminescent efficiency.

V. So-called ultimate displays

The surface conduction display (SED) is called the ultimate display by its inventor.

The performances include:
- The highest contrast in dark room 100000:1
- The view angle 180 degree
- The power <3w/inch
- The response time <10μs
- Ideal color recovery

Except the contrast in light room, the performances are the best in all flat panel displays. Factors blocking the development of SED with large screen and high brightness include:
- The row current is too large.
- The active process needs too long time and too much electrical power.

For a 36 inch device with a brightness of $350cd/m^2$, the row current is only 200mA that is

50

reasonable. The row drivers are obtainable, and the thickness of the row electrodes is less than 10 μm, which can be processed through screen print. For a 60 inch device with a brightness of 1000cd/m$^2$, the row current will be 2A. The row drivers can't be obtained, and the row electrodes will be 200μm thick that is nearly unrealizable in technology. In the technology of SED there is a so-called active process. The processing current for a pixel is 2mA, and the time is at least 20min.The current is 12A for a row, and the processing time is 20000 minutes for a panel. This is an obstacle in production of SEDs with large size. The possible approach for reducing the row current is adding a film transistor for each emission pixel. The configuration is as following.

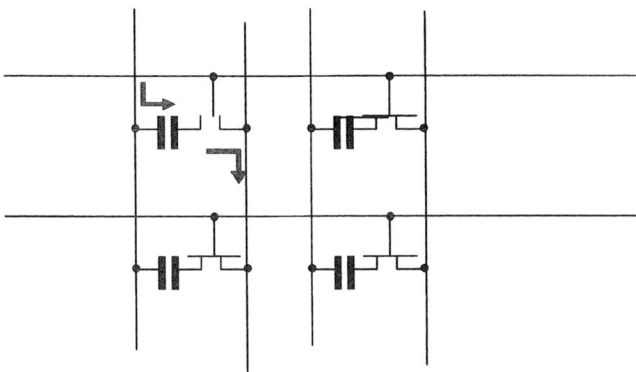

Fig.1 An emitter array with TFT

In this emitter array, there is only the conductance current in the row electrode. The current in the volume electrode is only that of a pixel. The thickness of the row and volume electrodes is less than 1 μm, and can be formed only through film technology.

The required on-off ratio for the TFT is only 1000:1, but the current must reaches 1 mA that is too large for α-Si TFT. The metal oxide TFT (ZnO, InO, SnO, or ITO) can satisfy the requirement for current, but the technology is still not mature.

VI. Summary

The industry of FPD is developing rapidly, but there still exist some problems, such as large power consumption, use of the rare material, ray and electromagnetic radiation etc. It is necessary to develop new kinds of display. The next generation will be energy saving and green.

# Plenary Session
## Special Invited Speakers
**October 25, Wednesday**
**3:30pm – 4:00pm**

**Broadband Access Services Requirements for Optical Transport Network**
**Mr. Zhang Chengliang, Deputy Chief Engineer**
**China Telecom Corporation Limited, Beijing Research Institute**

**Outlines:**
1. DSL broadband access developments in China
2. Changes for metro optical transport network
3. Challenges for long-haul optical transport network
4. Relation between ASON and IP network
5. New generation transport technology

Mr. Chengliang Zhang is Deputy Chief Engineer of Beijing Research Institute, China Telecom Corporation Limited. He is Chairman of Transport Network of China Communication Standards Association. He has been engaged in optical communications since 1992. He was in charge of drafting China national communication standards during 1995-2002 at Research Institute of Telecom & Transmission, Ministry of Information Industry. His current engagement is on ASON (Automatic Switching Optical Network) networking and broadband access technology. He received Bachelor degree from Beijing University of Postal & Telecommunication and Master degree from University of Stathclyde, UK.

# Plenary Session
## Special Invited Speakers
**October 25, Wednesday**
**4:00pm – 4:30pm**

**Dr. Xiongyan Tang, Technical Director**
**CNC System Integration Corporation & CNC Labs**
唐雄燕 副院长, 教授
中国网通研究院
**Optical Access in China: Application Models and Market Prospects**
中国光接入的应用模式与市场前景

1. Development of Broadband Access in China
2. Driving Forces of FTTx
3. Application Models of Optical Access
4. Market Prospects of Optical Access

**Xiongyan Tang** is now the technical director of China Netcom Group (CNC) System Integration Corporation & CNC Labs, an executive director of Beijing Institute of Communications, and a guest professor of Beijing University of Posts and Telecommunications (BUPT). He received his Ph.D degree from BUPT in 1994. He was a post-doctoral fellow at Nanyang Technological University, Singapore from 1994 to 1996, and worked at Technical University of Berlin, Germany as an Alexander von Humboldt research fellow from 1996 to 1997. Since 1998, he served as deputy chief engineer of Beijing Telecom, deputy director of Corporate Development Dept of CNC, deputy director of CNC Labs. His professional fields include broadband communications, optical transmission, access networks and NGN.

# Plenary Session
## Special Invited Speakers
**October 25, Wednesday**
**5:00pm – 5:30pm**

**David Z. Chen**
**PMTS-Technology**
**Verizon**

### Future high bit rate transmission and Ultra Long Haul network applications

This paper outlines our predictions for future optical transmission network applications and potential limitations based upon current technology development. Our emphasis is on the bit rate beyond 40 Gb/s, the technical challenges with fibers, amplifiers, and all electronics and optical components. Also, we are going to take a close look at the IEEE and OIF standard activities to support potential higher bit rate applications.

**David Z. Chen**, B.S. degree in Precision Instruments, Optical Engineering, MS and Ph.D. in Theoretical Physics (Quantum Statistical, Nonlinear, Applied and Computational Optics). He spent five years working on super computers doing advanced numerical simulations in non-linear optics, Higgs Field dynamics and chaos. From 1995 to 2000, he completed the job of building MCI's network control center for voice, data and transport monitoring and provisioning in Dallas Texas. From 2000, Dr. Chen technically led the MCI's advanced optical network technology development. His main focus is on the Ultra Long Haul technology, next generation fibers, 40 Gbps and beyond technology and new modulations. Dr. Chen is a member of OSA, principal member of the technical staff in the Verizon technology organization, NFOEC committee member and EE Adjunct Professor of University of Texas at Dallas

# Plenary Session
# Keynote speech
**October 26, Thursday**
**9:00am – 10:00am**

**Dr. Alan Willner**
**Dept. of Electrical Engineering - Systems**
**Viterbi School of Engineering**
**University of Southern California**
**Los Angeles, California 90089-2565**

**Physical-Layer Challenges in Stable, High-Capacity Optical Communication Networks**

Optical communications has enjoyed explosive growth in terms of technical achievement as well as commercial implementation. This presentation will contain three main sections. Firstly, a broad perspective will be given on some of the future trends in optical communication systems. Secondly, I will describe technical issues related to stable, robust optical networking, including performance monitoring, channel-degrading effects, efficient modulation formats, and switching. Finally, I will discuss adding flexibility and reconfigurability to different aspects of the base optical technologies.

**Alan Willner** received his Ph.D. from Columbia University, has worked at AT&T Bell Labs and Bellcore, and is Professor of Electrical Engineering at USC. He has received the NSF Presidential Faculty Fellows Award from the White House, Packard Foundation Fellowship, NSF National Young Investigator Award, Fulbright Foundation Senior Scholars Award, IEEE Lasers and Electro-Optics (LEOS) Distinguished Traveling Lecturer Award, USC University-Wide Award for Excellence in Teaching, IEEE Fellow, Optical Society of America (OSA) Fellow, and Eddy Paper Award from Pennwell Publications for the Best Contributed Technical Article. Prof. Willner's professional activities have included: President of the IEEE LEOS, Editor-in-Chief of the IEEE/OSA Journal of Lightwave Technology, Editor-in-Chief of the IEEE Journal of Selected Topics in Quantum Electronics, Co-Chair of the OSA Science and Engineering Council, General Co-Chair of the Conference on Lasers and Electro-Optics (CLEO), General Chair of the LEOS Annual Meeting Program, Program Co-Chair of the OSA Annual Meeting, and Steering and Program Committee Member of the Conference on Optical Fiber Communications (OFC). Prof. Willner has 590 publications, including one book.

## KEYNOTE PAPER

**Physical-Layer Challenges in Stable, High-Capacity Optical Communication Networks**

Alan E. Willner
University of Southern California
Dept. of Electrical Engineering
Rm. EEB 538, Los Angeles, California 90089-2565
willner@usc.edu

**Abstract:** *Optical communications has enjoyed explosive growth in terms of technical achievement as well as commercial implementation. This paper will contain three main sections. Firstly, a broad perspective will be given on some of the future trends in optical communication systems. Secondly, technical issues will be described that relate to stable, robust optical networking, including performance monitoring, channel-degrading effects, and switching. Finally, adding flexibility and reconfigurability to different aspects of the base optical technologies will be discussed.*

## Introduction

Optical communications has enjoyed explosive growth in terms of technical achievement as well as commercial implementation. Point-to-point capacity growth has been coupled with efficient all-optical routing, in which an optical data channel can transparently traverse many optical switching nodes and be converted into electrons only when it reaches the final destination. This network must be controlled and managed to within tight tolerances in order to ensure robust operation.

When envisioning the 10-year horizon, there are certain laudable goals that may be pursued, such as higher capacity, efficiency, stability, reconfigurability, flexibility, and security. Specifically, it might be desirable for an efficient and robust future optical network to provide the following functions: (a) optical performance monitoring for enabling self-management of the physical layer of the network, (b) reconfigurability to accommodate a convergence of different types of data traffic over the same network, and (c) "smart" optical signal processing to

complement electronic techniques. Within these three themes, this presentation will describe how integrated optics might be employed towards realizing these functions.

### (a) Optical Performance Monitoring

Today's optical networks are fairly static and are built to operate within well-defined specifications. There are many physical aspects of the network that are not known accurately, are nearly impossible to prevent from varying, and are subject to human error. Moreover, managing an existing optical network or deploying new network nodes requires a fair amount of person-time. At present, we are far from being able to simply plug an optical node into an existing network such that the network itself can allocate resources. In order to enable robust and cost-effective "self-managed" automated operation (see Fig. 1), the network should probably be able to: (i) intelligently monitor the physical state of the network and the quality of propagating data signals (see Fig. 2), (ii) automatically diagnose and repair the network, and (iii) allocate resources, including amplifier gain, signal wavelength, tunable dispersion compensation, electronic equalization, data coding, path determination and channel bandwidth. In order to facilitate widespread adoption, performance monitors should be easily deployable, compact, inexpensive, accurate, stable and reliable.

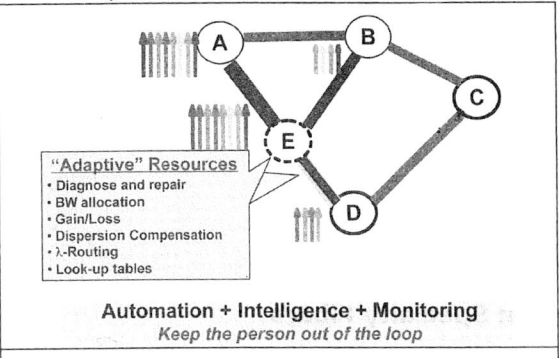

*Figure 1. Self-managed optical network such that physical-layer resources are adaptively allocated.*

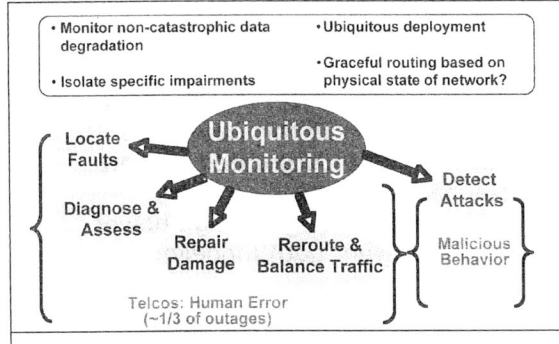

*Figure 2. Various functions that can be enabled by optical performance monitoring.*

## (b) Reconfigurable Systems for Heterogeneous Traffic

The future optical network will probably be used by many users for a variety of applications. Since different applications might have different optimal requirements, it is quite possible that there will be a wide variety of data formats that might wish to be transmitted over a network. It seems fairly inefficient to build a separate optical network to accommodate each application set, but rather have one network to accommodate a wide variety of traffic. One can envision future scenarios for which one network might be required to transmit different modulation formats, data rates, and qualities-of-service. A key challenge is to balance the desire to transport different types of traffic with the need to not over-build the network.

## (c) Smart Optical Signal Processing

The complementary roles of optics and electronics must be evaluated in any future network **task** in order to maximize performance and minimize cost. In general, electronics will be more cost-effective than optics if an individual element must be switched at the data rate. However, there may be several compelling signal processing functions that can take advantage of the inherent high bandwidth of optics, such as wavelength conversion, equalization, synchronization, and correlation (see Fig. 3).

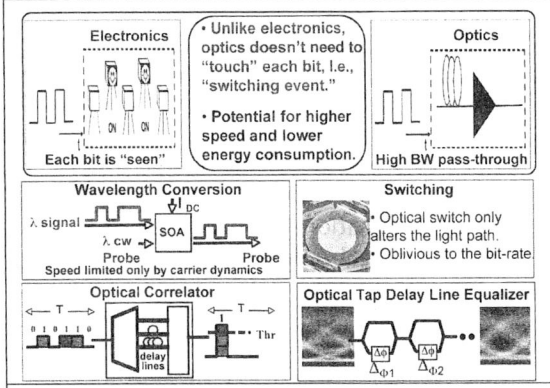

*Figure 4. Different optical signal processing operations that do not require switching an optical element on every data bit time.*

### Further Reading

I. P. Kaminow and T. Li, eds. Optical Fiber Telecommunications IV, Academic Press, 2002.

A.A.M. Saleh and J. Simmons, *IEEE/OSA Journal of Lightwave Technology*, vol. **17**, p. 2431-2448, 1999.

Q. Yu, et. al., *IEEE/OSA Journal of Lightwave Technology*, vol. **20**, pp. 2267-2271, 2002.

A.E. Willner, *OSA Optics and Photonics News*, March, pp. 30-35, 2006.

**Dr. Ming-Jun Li**
**Corning Incorporated**

**Recent Research Activities on Specialty Fibers**

M.-J. Li, G. E. Berkey, X. Chen, J. Coon, S. Gray, J. Koh, S. Li, D. A. Nolan, D.T.
Walton, J. Wang, and L. A. Zenteno
Science and Technology Division, Corning Incorporated, SP-AR-02-2, Corning, NY
14831

Specialty optical fibers have many applications in telecommunication networks, fiber
lasers, instrumentation, sensors, and gyroscopes. Although many different specialty
fiber products are currently available in the market, research and development in
this field are still very active to meet new application demands. In this paper, we
review recent research and development activities at Corning Incorporated on new
specialty fibers such as single polarization fiber, polarization maintaining fiber,
nonlinear fiber and fiber with reduced stimulated Brillouin scattering for high power
laser applications. We discuss fiber design concepts and present both modeling and
experimental results.

Dr. Ming-Jun Li received the B.Sc. degree from the Beijing Institute of Technology,
China, the M.Sc. degree from University of Franche-Comté, France, and the Ph.D.
degree from University of Nice, France.
Dr. Li has been with Corning Incorporated for 15 years and is currently a Senior
Research Associate in Optical Physics and Network Technology Research Group. His
research work is related to new optical fibers for different applications. He holds 22
U.S. patents and has published one book chapter and authored and coauthored over
90 technical papers in journals and conferences. Dr. Li has served as a technical
committee member for OFC, APOC and ITCom, and as an Associate editor for Journal
of Lightwave Technology.

**Michael Kunigonis, Jr.**
**Product Line Manager Access**
**Corning Cable Systems**

### FTTH Network Installation Implementation Innovations

Globally, a perfect storm of bitter competition, new technology, favorable regulation, improved costs and increased service and bandwidth requirements has gripped telecommunication carriers. This has resulted in the largest access network build out since the inception of the telephone and has heralded in the deployment of FTTH. From 2004 to 2006, the home passed cost of FTTH deployments has decreased by 50+%, enabling more than 40 million homes globally to be passed by FTTH. These savings have been increasingly made possible through passive equipment innovations targeted at decreasing installation cost. In this presentation, we will provide an overview of the FTTH market and the forces in play bringing about its deployment. Both single family unit (SFU) and multi dwelling unit (MDU) FTTH network technology, architectures and passive plant will be defined. Significant time will be used to discuss how optical passive plant innovations, such as field connectorization and factory installed termination systems explicitly produce lower installation costs and improve network troubleshooting efficiencies. Many of these innovations are applicable to both SFU and MDU deployments. Challenges discussed include the effective determination of cable pathways, space constraints and termination requirements for the deployment of FTTH to MDU subscribers. Finally, time will be taken to illustrate how carrier deployment challenges are addressed through said innovations.

**Michael Kunigonis** serves as a product line manager supporting Corning Cable Systems' FTTH hardware and equipment product portfolio. In this role, he is responsible for various operation, modeling, product development and marketing activities. Michael currently serves as the FTTH Council Chairman of the International Committee, works in both the United States and various Corning international markets and is active in the promotion and education of fiber-to-the-home.

Joining Corning Incorporated in 2000, Michael worked with Corning Optical Fiber holding several market development and engineering roles responsible for business development, product promotion, various aspects of public policy and end user customer support while holding specific responsibility for Corning's Asia region. Prior, Michael served six years as an engineering officer in the U.S. Army Corps of Engineers. He received his master's degree in Engineer Management from the University of Missouri in Rolla, Missouri, holds a bachelor's degree in Civil Engineering from Rensselaer Polytechnic Institute, Troy, New York and is currently pursuing an MBA at the Kenan-Flagler Business School, University of North Carolina, Chapel Hill.

**Dr. Haiyang Hu**
**Director of Open Lab and Solution**
**Agilent Technologies**

**Characterization of a transmission system on RX, TX, Channel and the Clocking system**

A digital bus is viewed as a communications system even though spans are measured in inches or centimeters. This special speech will address test methods and solutions to evaluate a complete communication system.

**Enabling fast, reliable and cost-effective sensing through highly integrated optical measurement systems.**

Introducing the industry's most reliable, integrated fiber-based distributed temperature sensing (DTS) systems that enable affordable photonic temperature sensing for a wide range of industry applications, such as oil and gas producers, pipeline operators, electrical power distributors and security system operators.

**Hai-Yang Hu** received his bachelor's degree from Zhe-jiang University, China in 1996, and Ph.D degree in 2001 from the Shanghai Institute of Optics & Fine Mechanics, Chinese Academy of Science. He joined Agilent Technologies directly after he got Dr. Degree. During 5 years in Agilent Technologies, he has held a variety of positions at Agilent Technologies, such as application engineer who provide technical support of lightwave & transmission test instruments; technical consultant who deliver consulting and training in Optical and Photonics, Transmission in Telecommunication, High speed digital communication and test, and RF component test. He also has many experiences in test system integration for DWDM, Transceiver and Optical Amplifier test and measurement.

Haiyang Hu is presently working as the director of Open Lab and Solution Center of Agilent Technologies, which is the first in the world to offer complimentary use of Agilent's products and services to our Test & Measurement customers.

# Low-Cost Radio-over-Fiber Network for 2.45GHz In-Building Signal Distribution

L. C. Ong (1), M. L. Yee (2), B. Luo (3) and M. Fujise (4)

1 : Institute for Infocomm Research, Singapore Science Park II, Singapore 117674. ongmichael@i2r.a-star.edu.sg
2 : Institute for Infocomm Research, Singapore Science Park II, Singapore 117674. yeeml@i2r.a-star.edu.sg
3 : Institute for Infocomm Research, Singapore Science Park II, Singapore 117674. luobin@i2r.a-star.edu.sg
4 : Wireless Communications Laboratory, Singapore Science Park II, Singapore 117674. fujise@nict.go.jp

**Abstract** A cost effective in-building signal distribution network is proposed. IEEE 802.11g WLAN system is used to illustrate the range extension a multimode Radio-over-Fiber network can provide. The ROF network allows simplified remote distribution units while the base stations are kept in a central office for ease of maintenance and upgrades.

## Introduction

Wireless communications has been part and parcel in our daily usage for data communications. The increase in wireless networks access communications is expected to continue unhindered in the future due to the convenience wireless provides. The 2.45GHz ISM band is one of the most often used wireless access spectrums. IEEE 802.11g and 802.11n continue to use this spectrum despite it being overcrowded. Wireless local area network (WLAN) will continue to grow and evolve further as an alternative for 100Mbps wired LAN.

In the cellular mobile spectrums, many countries are starting to roll out third generation cellular networks, along side with their second generation networks. The roll-out of 3G cellular networks will see more mobile communications taking place at 2.1GHz, as well as in the 900MHz and 1.8GHz regions. In addition, WiMax IEEE 802.16e expected frequencies to be used include 2.5GHz, 3.5GHz, 5.8GHz as well as 2.3GHz which is already in used in South Korea's variant of the WiMax standard [1]. With the rapid development and deployment of wireless networks, there is an increasing demand for an efficient and cost effective means to distribute RF signals in the 1GHz to 5GHz range of spectrum. Especially in increasingly congested spectrums, spectral utilization becomes increasingly important to avoid radio interference from adjacent or nearby cells. An effective method for improving frequency efficiency is to use microcellular concepts to enable higher frequency reuse. However, microcellular deployment requires a large amount of base stations (BS) to be deployed. This may not be practical in a situation where the cost of the BS is expensive. In addition, when frequent upgrades are required or when new wireless services appear, the effort required to maintain these BS can be overwhelming if there is a large number of BS deployed consistent to pico-cellular coverage requirements. In this paper, we will show that Radio-over-Fiber (ROF) distribution network is an attractive option to distribute wireless signals [2-5].

## Outline of the System

In ROF networks for in-building signal distribution network systems, the number of distribution points is directly proportional to the area of coverage desired and the size of the cell size to be used. In a congested spectrum scenario, small picocells are desirable. Smaller picocells mean lower power required for the distribution points. Since it is not cost effective for a ROF network to only support one type of wireless access service, additional wireless services can be interfaced onto the same ROF network utilizing the common building fiber infrastructure. For this paper, we will use IEEE

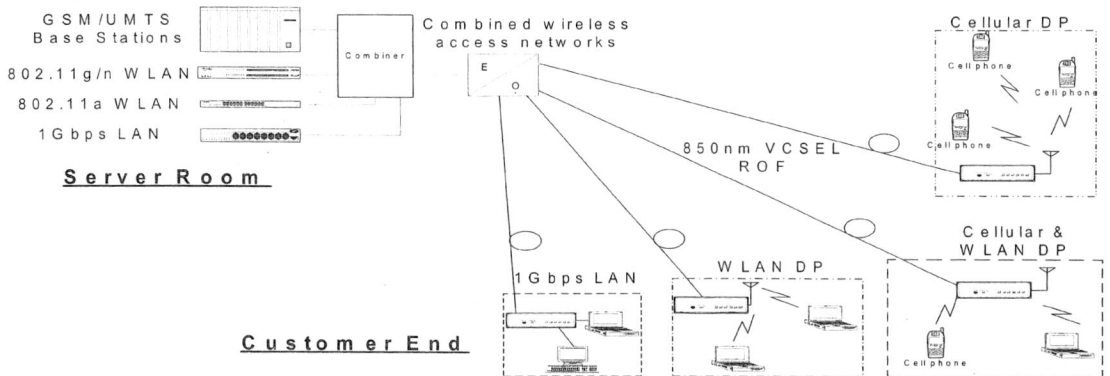

Figure 1: ROF distribution network for multiple wired and wireless access networks.

802.11g as an example of a wireless access network to illustrate the coverage enhancement a ROF system can provide in indoor environment in the 2.45GHz spectrum. This analysis can also apply to other type of wireless access networks such as GSM900, 1800, UMTS, WLAN, DECT and etc. Figure 1 shows an example of a multiple access network. Such a network may one day replace copper in the horizontal distribution of a building telecoms infrastructure.

We setup a test-bed system using three commercially available 802.11g compliant WLAN routers. Each of the WLAN routers are connected to their own ROF transceivers on the local site via SMA connectors. The remote ROF transceiver is linked back to the local site via multimode fiber (MMF) and a measurement laptop is wirelessly associated with the WLAN router via the ROF network. Figure 2 shows the schematic of this WLAN distribution system.

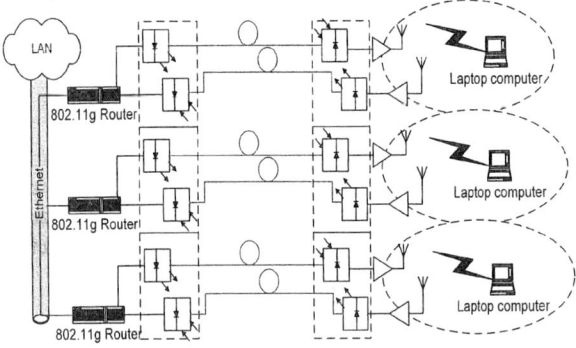

Figure 2: Test-bed for 2.45GHz in-building distribution system.

The RF signal from the router is used to directly modulate the laser diode via intensity modulation. As cost will be important for large deployment networks, we have chosen to use the cost effective Vertical Cavity Surface Emitting Lasers (VCSEL) to be the laser diode [6]. The 10Gbps VCSEL laser is biased at 6mA as it has been experimentally found to be the optimal operating point for analog links. The optical signal is passed into the optical infrastructure. The length of the fiber used in this test-bed is 300 meters. At the remote end, the optical signal is converted back into its electrical representation via a suitable photoreceiver, amplified and transmitted over the air via the antenna. The uplink path is the reverse of the downlink with added considerations to the lower power levels received from the mobile users. Corning's Infinicor SX+ fiber with a modal bandwidth of 2000MHz.km was used.

The signal quality at the remote end of the fiber link should meet WLAN specifications [7] so as to sustain the communications link. We evaluate the ROF transmission performance over MMF for 54Mbps Orthogonal Frequency Division Multiplexing (OFDM) used in IEEE 802.11g.

**Error Vector Magnitude Measurements**

The S-parameters of the ROF link was measured using an Agilent E8364B network analyzer. The S21 of a 1m and 300m ROF channel is shown in Figure 3. The effective bandwidth for the ROF network at 300m was evaluated to be to as high as 7GHz. This shows that the ROF network can be used as an in-building distribution network for multiple high frequency wireless services like WiMax and even 5GHz WLAN access networks. In this IEEE 802.11g ROF experimental system, error vector magnitude (EVM) is used as a measure of the transmission quality for digitally modulated signals, showing amplitude and phase errors.

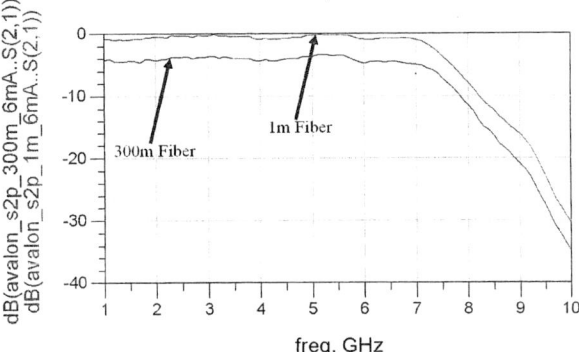

Figure 3: S21 comparison of a 1m and 300m ROF channel.

The most stringent relative constellation RMS error for 802.11g is -25dB or 5.6% at 54Mbps as stated in the standards [7]. Lesser bit rates will have less stringent EVM limits. We used the Agilent E4438 vector signal generator to generate test signal and an Agilent 89600 vector signal analyzer to analyze the EVM of the signal after passing through the ROF channel. For comparison sake, we have measured the effects of the ROF channel on 2.45GHz 802.11g and 5GHz 802.11a signals and how the transmission quality is affected. Figure 4 shows the EVM results of 802.11a and 11g with different lengths of fiber compared with the 5.6% EVM limit shown.

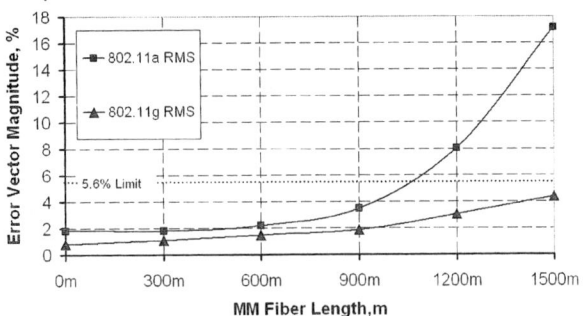

Figure 4: 802.11a and 11g EVM (rms) versus fiber length.

These measurements show that the ROF is able to support 802.11g's 64-QAM OFDM wireless signal for 300m of fiber. Even at 600m of fiber length, the measured EVM did not exceed 5.6% and is only slightly worse off compared with fiber length of 300m.

**Coverage Results for in-building WLAN**

Our office is a realistic scenario of a typical office environment of people and daily work activities. The ROF WLAN distribution network was setup to verify the coverage differences with and without ROF enhancement. All WLAN routers are co-located together in a server room deployed in the bottom left-hand corner of the floor plan in Figure 5. The whole office premise measures about 53m in length and 13m in width. The center portion of the premise consists of cubicle partition of about 1.5m high, with the larger rooms using plaster walls. The mobile station is a laptop with a WLAN surveying software to record the signal strength of the surveyed area. Figure 5a shows the coverage plot of the WLAN coverage with a commercially available off-the-shelf 802.11g wireless router with normal antenna placement. The router was placed on desk level and its coverage was measured. The coverage offered was poor and required more routers to cover the whole premise. The area around the WLAN base station has good signal strength, but at increasing distance, the signal strength drops to less than -90dBm. Therefore a single access point offers poor coverage within this office area. In multiple access network scenario, having to deploy numerous other base stations above the false ceiling or at a distance away may not be practical. Figure 5b shows the coverage plot using optimally positioned antenna. It becomes apparent that the coverage is enhanced. However, the signal quality at the extreme areas of the premise is poor and is not capable of a stable association with the WLAN router. The corridor area of the premise offers slightly better coverage due to better line-of-sight propagation path along the corridor. Figure 5c shows the coverage plot with ROF enhancement. The remote distribution points are connected to the server room via 300m of MMF. Each distribution point is served by its own dedicated WLAN router. Survey measurement shows that WLAN signal was better distributed as expected and coverage has significantly improved. From the indoor scenario, it was shown that ROF is an attractive means for in-building distribution alternative for wireless access networks. Cell area can be kept small to enhance frequency reuse where spectrum is limited. In addition, the ROF link must be capable of simultaneously transmit several wireless access network signals without interference.

On average, each remote distribution point had a cell radius of 3-4 meters radius supporting 54Mbps. However, the effective cell radius is dependant on specific access networks and requirements. In a multiple access network, WLAN is amongst one of the many networks to be distributed via ROF. The ROF network can carry WLAN at 2.4GHz and 5GHz, GSM at 900MHz and 1800MHz, UMTS at 2.1GHz and even WiMax if required.

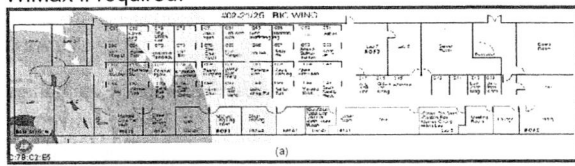

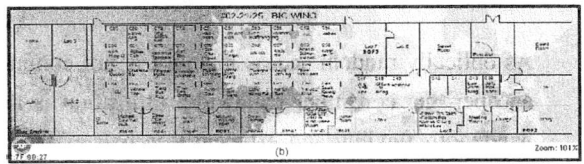

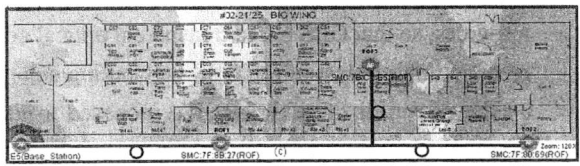

Figure 5: WLAN Coverage plot of (a) WLAN router, (b) WLAN router with optimal antenna, (c) WLAN distributed via ROF. (Darker areas denote better signal strength)

**Conclusions**

We have successfully shown a 2.45GHz in-building distribution network in practical real-world environments. The performance of this ROF network surpasses the performance of current alternatives in terms of performance, cost and weight.

With the increase in wireless traffic and the demand for higher operating spectrums, future generation mobile systems will likely to have smaller and smaller cell sizes. This in turn will allow simplified remote distribution units with the sensitive electronics being kept in a climate controlled environment. These co-located base stations will also simplify radio resources, cell planning and maintenance. In view of these advantages, we propose ROF distribution network for next generation distribution networks.

**References**

1 Hwang S, et al "RoF Technologies for In-building Wireless Systems", APMC 2006, pp. 38-41.
2 Hartmann P, et al "Broadband Multimode Fiber (MMF) Based IEEE802.11a/b/g WLAN Distribution System", MWP 2004, pp. 173-176.
3 Tang P K, et al "PER and EVM Measurements of a Radio-over-Fiber Network for Cellular and WLAN System Applications", *J Lightwave Tech,* V.22, 2004, pp. 2370-2376.
4 Sim C K, et al "Performance Evaluation for Wireless LAN, Ethernet and UWB Coexistence on Hybrid Radio-over-Fiber Picocells", OFC 2005, CD format.
5 Sasaki M, "Trends in Digital Terrestrial Broadcasting", *Broadcasting Technology*, 2003, pp. 7-12.
6 Larsson A, et al "Broadband Direct Modulation of VCSELs and Applications in Fiber Optic RF Links", MWP 2004, pp. 251-254.
7 IEEE Std. 802.11g, "Further Higher-Speed Physical Layer Extension in the 2.4GHz Band," 2003.

# OBS based transparent optical networking policy (Invited)

Xu Anshi, Li Zhengbin

National Laboratory on Local Fiber-Optic Communication Networks & Advanced Optical Communication Systems
Peking University, Beijing, 100871, China
E-mail: lizhengbin@pku.edu.cn

## Abstract

As optically random access memory is not commercially available, and with the fiber being cheaper and cheaper, the core fiber bandwidth utilization is not the key factor of the network, how to use the huge bandwidth flexibly and efficiently is one of the hot topics of optical networks. Optical burst switching (OBS) is thought to be the best way to adapt the bursty traffic, and can be statistically using the optical resources, so that OBS based transparent optical network attracting more and more attentions. However, the precondition of most of the OBS networking solution is based on wavelength conversion, which is limited by the nonlinear of materials. We proposed a novel networking policy by using dual fiber links server model along with traffic spacing mechanism (TSM) by address differentiation, the transmitting IP directly in optical domain for traffic data between physically adjacent nodes, with no necessary burst assembling (NABAN). The theoretical results show that this method can improve network performance and be practically used in fields.

**Key words:** OBS, Data-length Time-lag Product, Dual fiber-Link-Server, traffic spacing mechanism (TSM), no assembling between adjacent nodes (NABAN)

## 1. INTRODUCTION

As optically buffers are not commercially available [1], on the other hand the fiber bandwidth utilization in core network is not the key factor, how to use the huge bandwidth flexibly and efficiently is one of the hot topics of optical networks. Optical burst switching (OBS) [1, 2] can adapt the bursty traffic statistically using the optical resources, OBS based transparent optical network attracts more attentions. However, there are a lot of challenges such as switching technology, assembly algorithm, contenting solution, quality of services, signaling and protocol, which have been discussed extensively in literatures.

We discuss OBS based transparent network. Firstly, we focus on burst contention resolution under the physical nonlinear coefficient limitation of material for wavelength conversion, and the fact of non-buffer network of OBS, dual fiber links server model is introduced as contention avoidance resolution. The mainly contention resolutions in literatures are wavelength conversions, fiber delay lines, deflecting routing, etc. which are vulnerable to network load and may suffer server loss in case of heavy traffic. Then we give a OBS networking by directly transmitting IP packets in optical domain between adjacent nodes (NABAN) in physical topological networks by traffic spacing mechanism (TSM), while the data to other nodes are sent by optical burst switching. The scheme is just based on that the data between neighbourhood nodes need not be optically switched, while data far between remote nodes must be optically switched at intermediate nodes. The IP packets can be transmitted between any spacing of two optical burst. With a generalized optical burst assembling algorithm, which considers the data length and the time duration simultaneity, we give the theoretical results of the proposed networking policy by modelling.

## 2 TRANSPARENT NETWORKING BASED ON OBS

### 2.1 Dual fiber links server model as burst avoidance policy

Now, most of OBS protocols associated with routing and link resource allocation are based on wavelength converters (WC) [2]. The performance of the network is sensitive to the network dimensions or the length of selected routing paths. Without WC devices along the path, the blocking probability of optical burst for the wavelength discontinuity would be too high to make the OBS into practical use, even through the fiber delay lines are used to reduce the burst dropping probabilities, and the physical limitation of WC prevent it not commercial available now. Technically, the aim of adoption of WC is to prevent the wavelength discontinuity, we can use dual fiber links which have the same wavelength at the two fibers acting as partial wavelength conversion.

Dual fiber link (DFL) architecture is show as Fig. 1, where the link connect node i and core node i+1 are two fibers (each fiber is assigned to some wavelengths), i.e. fiber 0 and fiber 1. The parallel fibers are equal before the role. They are not master and slave relations. Both fibers are bi-directional transmission. In the figure, the wavelength $\lambda_0$ is supposed tobe dedicated to control channel. Wavelength $\lambda_1 \ldots \lambda_m$ is for data. If one direction is used as master, the other is for slave. For example, in fiber 0, the transfer direction from left to right is called master use, and the backward transfer direction is slave use.

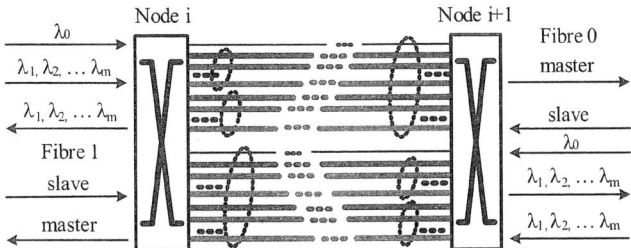

Fig. 1 DFL architecture Dual links server model for OBS networks

The burst control packet is processed at intermediated nodes to configure the switch fabric. If resources are available, the booking was successfully, and the BCP is regenerated and transmitted to the next node. There are two wavelength channels at each wavelength, rather than one, which can partially reduce the wavelength continuity.

## 2.2 Traffic spacing mechanism (TSM) with no assembling between adjacent nodes (NABAN) for transparent networking based on OBS

As we know OBS network is a kind of non buffer network, the burst data blocking probability relates not only the link channels, but also the node numbers along the transmitting path, the longer the path the probability higher. When a node decides to forward burst data to a destination node more than two hops longer, the path is consist of several links. If the links are occupied by burst data from adjacent nodes, the candidate burst data will be blocked or dropped. Along the path, the links of physically adjacent nodes are selected to forward data just the two nodes, when these links are being chose for a long path, they are occupied till the burst data have been transmitted.

As OBS is transparent during the data been transmitted along the path except the source and destination nodes EO/E, there is no necessary to assemble burst data between adjacent nodes, for which there is no switching processing of data. The data of adjacent nodes can be forwarded by IP packets directly as shown following figure, where there are six nodes and five links, if any of the links are chose as path to forward data of adjacent nodes, as say link BC, CD, the blocking probability of other long path will be very high. The data of adjacent nodes can be forwarded directly in optical domain, and the burst data from other nodes can pre-empt the channel, which leads to the IP packets are transmitted among burst data. This kind of transparent networking policy is a traffic spacing mechanism with no assembling between adjacent nodes.

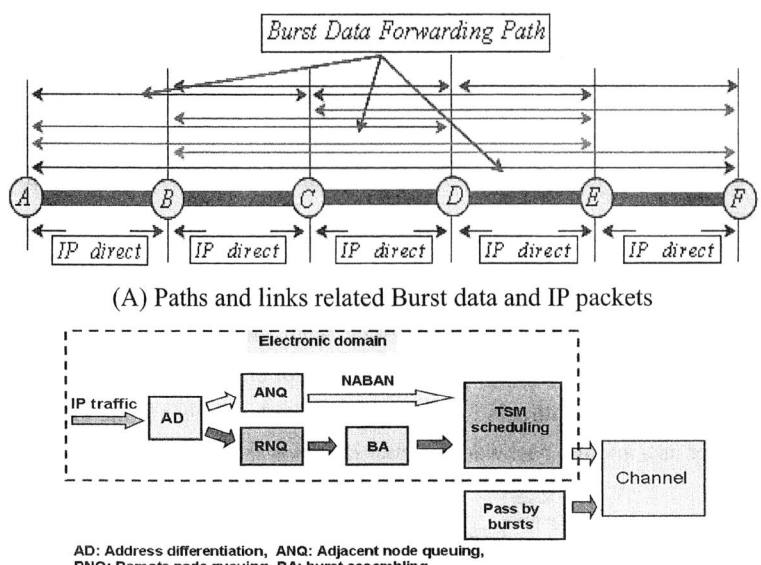

(A) Paths and links related Burst data and IP packets

AD: Address differentiation, ANQ: Adjacent node queuing,
RNQ: Remote node queuing, BA: burst assembling

(B) The procedure of traffic differentiating and IP spaced in OBS

Fig. 2 A novel transparent optical networking without forwarding burst data between adjacent nodes

The function of edge nodes in TSM with NABAN policy may include several blocks. IP packet processing block consists of address differentiate and authentication shown in figure 2. When traffics services loaded by IP arrive, they are distinguished according to their destination nodes. If the destination addresses of the packets are not physically adjacent nodes, they are sent to burst assembling procedure, which will be sent to the destination nodes not adjacent the source node. The IP packets with address adjacent the source node will not be sent to being converged, instead, they are transmitted directly to the next neighbour node by spacing and scheduled into bursts in optical domain instead of Ethernet format. So that the queuing block may be divided into two parts, one is for IP packet directly transmitting to the neighbour destination nodes in optical domain, the other is for assembling optical burst, during which the data length and timer production assembling algorithm is used to adapt the variable and self similar traffics. Then all the data, IP packets or group IP packets i.e. optical burst will be scheduled to suitable wavelength channel for being forwarding. The main functions of core nodes may include OBS signalling processing, resource scheduling and so on based on dual fiber links server model.

## 3. PERFORMANCE EVALUATION

Now, we give the evaluation result of the performance of proposed networking model or policies. The mixed policy is demonstrated by using dual fiber link server model in to avoid wavelength converters, and adapt TSM with NABAN of direct forwarding IP packets in optical domain between adjacent nodes in physically topology networks while transmitting optical burst data among the other nodes. To evaluate the performance of the policy, we compare it with the policy without double link servers and no IP packet directly to next neighbor nodes. A six nodes loop topology of two fiber as core links server model with 4 wavelength for each fiber is selected to demonstrate the blocking or dropping probability, the end to end delay, the throughput and link utilization.

It is assumed that the requests for wavelengths as Poisson process with rations $\lambda$, and service time $b$. The link is dual fibers with total $2w$ wavelengths. When a burst arrives, there are *two* wavelength channels to go ahead. It is clear that a burst is dropped only if both channels are busy. The channel busy probability equals to the link utilization given by $\mu_C = \rho_C \times (1 - P_{DF-1})$, where $\rho_C = \lambda \times b/2$ is the total offered load, and $P_{DF-1}$,

# High Performance Burst-Mode Upstream Transmission for Next Generation PONs

X.Z. Qiu, Y.C. Yi, P. Ossieur, S. Verschuere, D. Verhulst, B. De Mulder, W. Chen,
J. Bauwelinck, T. De Ridder, B. Baekelandt, C. Melange, J. Vandewege

Ghent University, INTEC/IMEC, Sint-Pietersnieuwstraat 41, 9000 Gent, Belgium, xingzhi@intec.ugent.be

**Abstract** *This paper presents the effective optical gain achieved with FEC in 1.25 Gbit/s GPON uplink yielding -36 dBm burst-mode receiver sensitivity and 25 dB dynamic range. A low cost embedded fiber monitoring system measures fiber attenuation and locates deteriorations without interference with TDM(A) network operation or power budget penalty. Finally, a 10 Gbit/s WDM/TDMA long reach PON system is introduced.*

## Introduction

After the successful development of a burst-mode PMD (Physical Media Dependent) chip set [1]-[5], a 1.25 Gbit/s GPON (Gigabit Passive Optical Network) uplink has been integrated and validated at the UGent INTEC_design Lab. This was the first public demonstration of ITU-T GPON-compliant operation at 1.25 Gbit/s and has shown very promising performance [6] [7]. To further minimize the cost per subscriber, a FP (Fabry-Perot) laser is preferred for the ONT (Optical Network Termination). As the FP multimode spectrum causes MPN (Mode Partition Noise), the ITU-T Recommendation G.984.2 proposes to use FEC (Forward Error Correction) to reduce the associated penalty. The effective optical gain achieved with FEC allows for a longer physical reach or a higher split ratio. However, the effective optical gain G in burst mode operation was unknown and is not specified in G.984.2. The first part of this paper presents how we studied MPN in the GPON upstream channel and quantified the improvements after implementing FEC in burst-mode 1.25 Gbit/s transmission.

By sharing the OLT (Optical Line Termination) and the fiber plant between many subscribers, PONs are quite cost-effective in offering large scale broadband connectivity. Once the PON is in use however, an operator does not want to shut it down to locate a fiber plant irregularity. With increasing numbers of subscribers and services, and more fiber in the last drop, there is a growing need for low cost fiber plant monitoring. Many operators monitor the data traffic and BER (Bit Error Ratio) in optical fiber networks, but physical layer monitoring is not so widespread because of its hitherto high implementation and operation costs. Classic OTDR (Optical Time Domain Reflectometry) techniques are invasive, require the network to be shut down, and/or rely on expensive equipment that cannot be embedded in low cost ONTs. The second part of this paper reports an innovative FIBERMON[TM] system designed for low-cost physical layer monitoring in operational fiber networks. FIBERMON[TM] operates in the background, and can provide round-the-clock verification of the

integrity of the fiber plant, and a quick and precise identification and localization of link deterioration without interference with ongoing services [8].

Finally, the EU-funded IST project PIEMAN (Photonic Integrated Extended Metro and Access Network) [9] is introduced. The project performs ambitious physical layer research into a future broadband optical access and metro system with capacity and reach well beyond what is achievable today.

## GPON uplink performance with FEC

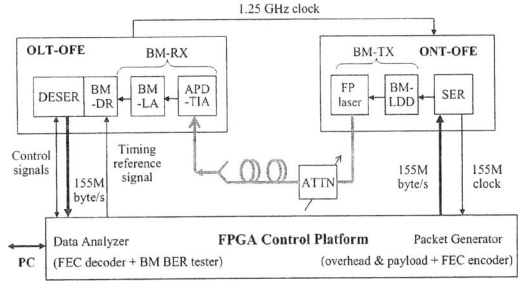

*Fig.1 GPON 1.25 Gbit/s uplink measurement setup*

Figure 1 depicts the complete GPON 1.25 Gbit/s upstream burst-mode PMD chip set designed by INTEC_design, and the first GPON uplink demo setup including FEC. The minimum setup contains one BM-TX (Burst-Mode Transmitter) at the ONT-OFE (Optical Front End), and a BM-RX (BM Receiver) followed by a BM-DR (Data Recovery) chip at the OLT. The BM-TX contains a cheap FP laser diode and a BM-LDD (Burst-Mode Laser Diode Driver) [5]. The ONT SER (Serializer) adds a programmable delay with bit accuracy to align the launched bursts with the ongoing traffic at the OLT. The 155 Mbyte/s parallel data outputs from an FPGA-based packet generator are serialized to generate a 1.25 Gbit/s burst-mode data stream after a FEC encoder with Reed-Solomon code RS (255, 239). The BM-RX contains a high sensitivity burst-mode APD-TIA (Transimpedance Amplifier) and a wide dynamic range BM-LA (Limiting amplifier) designed for instantaneous packet amplitude recovery [4]. The

the bursts drop rate by one of the two fiber, as $P_{DF\_1} = [\rho_C(1-e^{-\rho_C})]/[1+\rho_C(1-e^{-\rho_C})]$     (1)

As channels busy is independent each other, then we can get dual fibers are busy

$$P_{DF} = [\rho_C(1-e^{-\rho_C})]^2 / [1+\rho_C(1-e^{-\rho_C})]^2 \qquad (2)$$

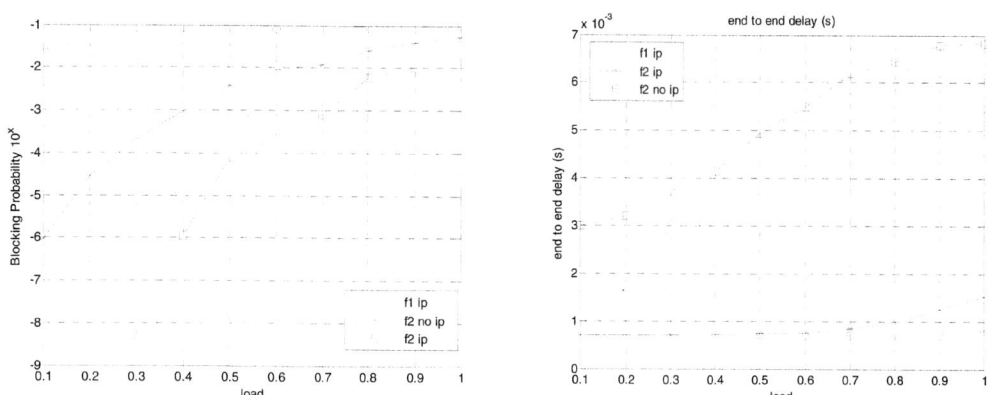

Fig. 3 Comparing of the blocking probability and the end to end delay of the integrated transparent policy with no TSM in six node loop network architecture

The source is admixture of http, ftp, IPTV, email and P2P. Figure 3 gives the performance of a six nodes loop structure, where the TSM with NABAN has the lowest data blocking probability. Accordingly, the end to end of TSM with NABAN is almost the same as directly transmitting IP traffic between adjacent nodes. There are lots of projects left to be investigated, such as the theoretical analysis of TSM with NABAN, the insights of dual fiber server model in non buffer network and no wavelength converters, the cost assessment, etc.

## 4. CONCLOSION

In conclusion, a novel transparent optical networking policy is analysed, without FDL, wavelength conversions and deflection routing, the novel networking policy TSM with NABAN is investigated, and the performance of integrated policy is demonstrated in a six node loop networks, which has many good features, and will be discussed deeply in the future works.

## ACKNOWLEDGEMENT

This work is supported by the National Natural Scientific Foundation of China (No. 60372025), 863 Program (No.2005AA122310), and Fund Project of Chinese Ministry of Education (No. 2002000102).

## References

1. A. Huang, L. Xie, Z. Li, and A. Xu: Time-Space Label Switching Protocol (TSL-SP) –A new paradigm of network resource assignment, Photonic Network Communications, vol. 6, no. 3, (September 2003), pp. 169-178.

2. Huang A, Xie L, Li Z, Lu D, Ho P, Optical Self-Similar Cluster Switching (OSCS) - A Novel Optical Switching Scheme by Detecting Self-Similar Traffic, Photonics Network Communications, Vol. 10, No. 3. (November 2005), pp. 297-308.

3. Yuan Chi, Huang Junbin, Li Zhengbin, Xu Anshi, A Novel Burst Assembly Algorithm for OBS networks—Based On Data-length Time-lag Product, 2005 Asia-Pacific Conference on Communications, Perth, Western Australia, 3 – 5   October 2005, p319~323

BM-DR IC contains the high-speed blocks of the line termination and includes the DESER (Deserializer) function. It performs the upstream retiming i.e. the CPA (Clock Phase Alignment) and the burst alignment via delimiter detection. The FPGA-based data analyzer measures the BM-BER from recovered incoming packets after a RS (255, 239) decoder.

FP lasers (wavelength around 1310 nm) were used for experiments to investigate the effective optical gain with FEC. The effective optical gain G of the system employing FEC is defined as the difference of optical power at the receiver input, with and without FEC, for a BER = $1 \times 10^{-10}$. No notable MPN penalty is observed even with 20 km G652 fiber, when the central wavelengths of the FP lasers used are within the zero dispersion window of the fiber (1302~1322 nm). However, ITU-T G.984.2 specifies the ONT transmitter operating wavelength in the range of 1260-1360 nm, so in worst case the MPN penalty cannot be negligible. To emulate such a condition by lack of a suitable 1360nm FP laser, 3.1 km of G653 DSF (Dispersion Shifted Fibre) with dispersion of -20 ps/nm.km at 1307 nm was inserted into the uplink. This is equivalent to the dispersion of a 1360 nm FP laser over 12 km G652 fibre with dispersion of 5 ps/nm.km. Fig. 2 plots the measured BM-BER (at APD gain=6) of the GPON uplink with 3.1 km DSF fiber where B1 and B2 denote OLT-OFE Board #1 and Board #2 respectively. This illustrates the penalty caused by the MPN, and also the error correction capability of the RS (255, 239) code.

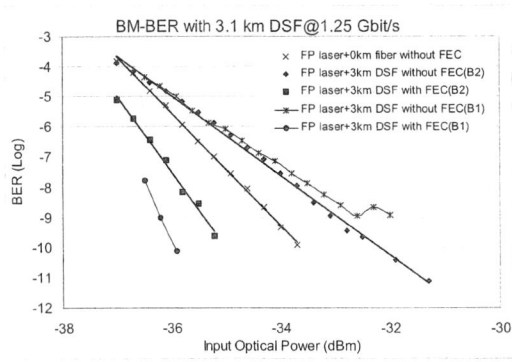

*Fig.2 Measured BM-BER with & without FEC*

Table 1 Different effective optical gain G

| Fibre (km) | FEC gain | RX sensitivity (dBm) |
|---|---|---|
| 20km (G652) | no FEC | -33.6 (DFB) |
| 20km (G652) | G=2.6 dB | -36.2 (DFB+FEC) |
| 20km (G652) | no FEC | -33.0 (FP) |
| 20km (G652) | G=2.7 dB | -35.7 (FP+FEC) |
| 3.1 (G653 DCF) | no FEC | BER floor > $10^{-9}$ (B1) |
| 3.1 (G653 DCF) | G >> 3 dB | -36 (OLT #B1) |
| 3.1 (G653 DCF) | no FEC | -32.1 (OLT #B2) |
| 3.1 (G653 DCF) | G=2.9 dB | -35 (OLT #B2) |

Table 1 summarizes the measured effective optical gain G on the GPON uplink. A gain of 2.6~2.9 dB was found which depends on two factors: the slope of the BM-BER curve and the distribution of the errors (burst or discrete). Note that the slope of the BM-BER curve is very sensitive to the performance of the upstream PMD components such as the BM-LA and BM-DR. Any small difference due to settings, tolerances and DC offset will have influence on the characteristic of the error distribution. To conclude, experiments prove that the FEC with RS (255, 239) code can reduce the penalty due to MPN when using a cheap FP laser in the upstream direction. However the BM-RX overload specification is not altered by FEC gain. This guarantees both the high sensitivity and wide dynamic range of the BM-RX, which allows to extend the PON reach or to nearly double the number of subscribers in a given PON network.

**A low cost fiber monitoring system**
An embedded, non-intrusive fiber monitoring system called FIBERMON[TM] [8] has been developed for low-cost physical layer monitoring in operational fiber networks. Two major benefits that cannot be met by classic OTDR, nor by any alternative "embedded" OTDR system that was published, are:
1). The cost of the FIBERMON[TM] system is low. This is especially critical for PON ONTs. The analog front-end electronics of FIBERMON[TM] can be embedded inside every subscriber ONT as shown in Fig. 3, and there is no need for additional optical components such as an optic coupler or an OTDR PIN photodiode.
2). The presence of the FIBERMON[TM] system does not have any negative impact on the upstream optical power budget and network/system performance. The FIBERMON[TM] system measures and detects optical reflections but it does not interfere with the data traffic carried by the optical fiber plant, nor with the MAC (Media Access Control) functionality.

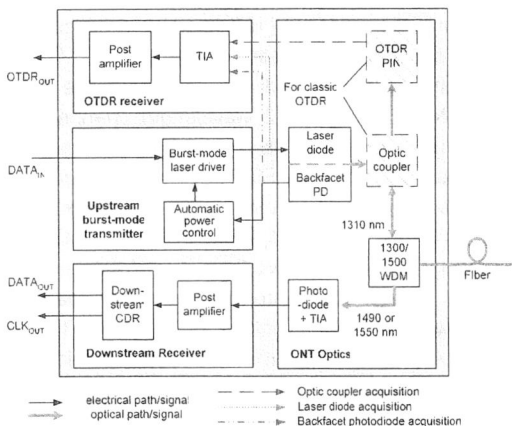

*Fig.3 OTDR receiver integrated into an ONT without additional optical components*

The FIBERMON[TM] system can make full use of

resources that are already available in an operational fiber network, such as the data laser diode(s), and the data light bursts themselves. An optical echo signal caused by an optical burst or pulse can be extracted electronically from both a laser diode or a laser back facet photodiode, without any need to physically interrupt the optical link or to introduce extra optical components. For this purpose, dedicated OTDR RX front-end electronics (a slow speed TIA and a post amplifier) can be integrated into the burst-mode laser driver chip, feeding a small off-chip DSP (Digital Signal Processing) unit. The OTDR measurements are performed at the exact wavelength and modal distribution of the data pulses, and so can be very accurate.

The complete OTDR measurements can be performed passively, without the need to inject specific OTDR signals into the network. As is proven by recent INTEC_design research, the optical echo's caused by ongoing burst-mode data transmissions contain suitable information to perform OTDR measurements. Innovative FIBERMON[TM] signal processing allows one to extract complete OTDR curves by combining optical reflection measurements taken in different time windows and originating from different data bursts. In the background, and completely transparent for any MAC function, FIBERMON[TM] composes valuable information such as the fiber attenuation as a function of distance, the location of abrupt changes in the optical attenuation or reflection as caused by connectors, breaks, fiber joints, stress points etc., and the strength and nature of such changes. The only requirement on the system is the occasional occurrence of a "dark" time window after data transmission, during which optical reflections can be observed.

### Next generation 10 Gbit/s WDM/TDMA PON
The PIEMAN project proposes a new photonic communication system [10] integrating access and metro into a single network with a reach of 100 km between customers and the major service node. PIEMAN is aiming at 32 wavelength channels and symmetric 10 Gbit/s data rates in both downstream and upstream direction. EDFAs (Erbium-Doped Fiber Amplifiers) will be employed at the local exchange and as a preamplifier in front of the PON OLT as shown in Fig. 4, to achieve the required optical power budget. Up to 512 customers will share each 10 Gbit/s wavelength. Currently a 10 Gbit/s BM-TIA and a BM-LA with high sensitivity and wide dynamic range are under development (by IMEC / INTEC_design). A major challenge is achieving upstream burst mode operation of a high-split, amplified PON at 10 Gbit/s with 100km reach.

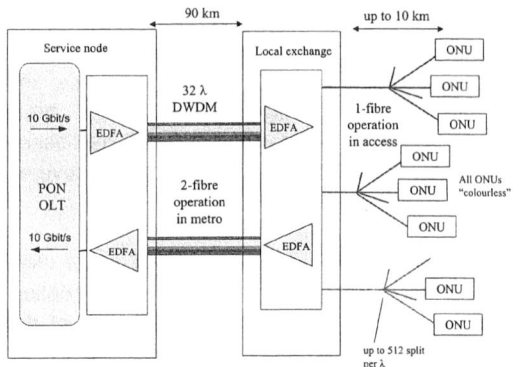

*Fig.4 The PIEMAN WDM/TDMA system architecture*

### Conclusions
Several advanced techniques are presented in this paper to improve the performance of next generation high speed high split PONs. We demonstrated for the first time that an APD based BM-RX can simultaneously achieve -36 dBm sensitivity and 25 dB dynamic range by employing FEC with RS (255, 239) code when a cheap FP laser is employed. Secondly, the low cost, embedded FIBERMON[TM] system is introduced, designed for non-intrusive fiber plant monitoring, especially in PON drop sections after the split. Finally, an ultra high capacity hybrid WDM/TDMA network is introduced with 100km reach, based on 32 DWDM wavelengths. Each wavelength is shared by up to 512 customers and carries 10 Gbit/s, yielding an aggregate 320 Gbit/s data rate and connecting up to 16384 subscribers.

### Acknowledgment
The authors would like to thank Paolo Solina of Telecom Italia Lab for the Reed Solomon decoder IP and Heinz Krimmel of Alcatel R&I Stuttgart for supporting the FEC experiment. We also would like to acknowledge the European IST (Information Society Technologies) committee for supporting the PIEMAN research activities and the PIEMAN partners (BT, Alcatel, Siemens, CIP and UCC) for their cooperation.

### References
1. X.Z. Qiu et al, JLT, Vol. 22 (2004) pp. 2498-2508.
2. P. Ossieur et al, EL, Vol. 40 (2004), pp. 447-448.
3. J. Bauwelinck et al, EL, Vol. 40 (2004), pp. 501-502.
4. P. Ossieur et al, JSSC, Vol. 40 (2005) pp.1180-1189.
5. J. Bauwelinck et al, JSSC, Vol. 40 (2005) pp 1322-1330.
6. X.Z. Qiu et al, ECOC 2004, pp. 398-399.
7. D. Verhulst et al, OFC 2005, Anaheim, OFI2.
8. J. Vandewege et al, "Embedded, non-intrusive Fiber Monitoring System", Patent pending.
9. http://www.ist-pieman.org/
10. R. P. Davey et al, JLT. Vol. 24, 29-(2006).

# Restoration Mechanism and Maximum Dual Failure Restorability of Span-restorable Optical Networks

Ling Zhou

Swiss Federal Laboratories for Materials Testing and Research (EMPA), CH-8600 Duebendorf, Switzerland

ling.zhou@empa.ch

**Abstract** *A survivable network is commonly designed to be* 100% *restorable to any single span failure. The restoration mechanism of networks with span restoration is not as straightforward as that of networks with dedicated path protection. The restoration of span-restorable networks can be conducted in real time or preplanned and cross-connected immediately when any failures occur. The disadvantages and advantages of these two mechanisms are discussed. A linear programming formulation to maximize average dual-failure restorability of a network is developed. A case study analyzes how much average dual-failure restorability can be improved for span-restorable networks.*

## Introduction

Networks designed with fully restorability to single span failures, expressed by $R_1 = 1$, can be found in a down state when node failures and multiple failures occur. According to the concept of 搯ost likely paths to failure?[1], the simplest combinations of elemental failures that lead to an outage state in any system with redundancy will dominate the overall availability. Dual-span failures can then be the next dominant failure modes in a network with $R_1 = 1$. Therefore, it is important to analyze the influence of dual failures on span-restorable networks. Dual failure restorability ($R_2$) is usually a measure to check the impact.

In the paper, the concept of $R_2$ is explained first. Then we analyze the disadvantages of three restoration models for span restoration, which has been illustrated in [2, 3]. A simple and reliable restoration approach is proposed. After that we introduce an integer linear programming (ILP) formulation to achieve maximum $R_2$ under a given investment. In the case study, concrete experimental data manifest the flaw of the old restoration mechanism. Conclusions are given at last.

## Dual-span failure restorability

According to [3], the individual 揹ual-span failure restorability? $R_2(i, j)$ of a pair of failed spans $i$ and $j$ is defined as the ratio of the restored capacity to the total working capacity on spans $i$ and $j$ when both spans fail.

$$R_2(i,j) = 1 - (N_i + N_j)/(w_i + w_j) \qquad (1)$$

where $w_i$ and $w_j$ are the working capacities of spans $i$ and $j$, respectively, and $w_i + w_j$ is not zero; $N_i$ and $N_j$ are the individual non-restorable capacities of spans $i$ and $j$, respectively. Once we have the $R_2(i, j)$ values of each individual failure combination $(i, j)$, the average dual-failure restorability $R_2$ of a network can be defined as a weighted form to reflect the overall demand-impact [3]:

$$R_2 = \sum_{i \neq j}(w_i + w_j) R_2(i,j)/ \sum_{i \neq j}(w_i + w_j) \qquad (2)$$

## Restoration mechanism of span restoration

Span restoration is a typical scheme in which restoration paths can be found in real-time or preplanned. The restoration path-set may comprise all distinct routes bridging the end-nodes of a particular failed span instead of one replacement route only for all failed traffic. The limitation of hop-limits or physical length limits may be applied to the distinct restoration routes in the implementation.

For a network with dedicated path protection or 1+1 automatic protection switching (1+1 APS), there is a dedicated protection path to provide full protection for each working path. The restoration mechanism of span-restorable networks is not as straightforward as that of 1+1 APS-protected networks. For span restoration, three technical models with various levels of adaptability for the restoration process were developed in [2, 3] and were referred to as: Static, partly-adaptive and fully-adaptive behavior.

These models are all based on k-shortest path (KSP) routing behavior for the basic single-failure response model [4, 5]. By KSP algorithm, each restoration path for one failed span is found in real time according to the rule of taking the shortest feasible path first with its maximum non-zero flow. The next shortest restoration path is then checked on its feasibility with the remaining spare capacity and so on, until all the failed traffic can be restored. The mechanism of KSP routing is something of an unregulated behavior. There is usually a constraint length for a service path when the quality of signals is considered. In practice, only a set of backup paths within a limited length can be chosen as the qualified restoration paths.

Intuitionally, we can conjecture that KSP algorithm can not guarantee to find all the necessary restoration paths within the qualified backup routes. Considering the disadvantage of KSP algorithm, we suggest that the restoration paths of each span should be taken directly from any optimized result. All the ILP formulations about span restoration can generate the complete feasible restoration flows of single- or dual-span failures depending on the optimization objective. These restoration routes should be computed and planned in advance. When a failure occurs, the restoration paths can be cross-connected in real-time according to the pre-determined plan. Even if the number of backup routes is restricted, the optimized results can certainly guarantee the restorability delivered by the ILP formulation.

## ILP formulation to Maximize $R_2$

Dual-span failures can occur simultaneously or within a certain interval, i.e. non-simultaneously. The latter situation can be depicted as follows: after the first span failure occurs, the second span failure appears before the repair to the first failure is finished. In [2, 3], a proposed ILP formulation aiming to reach dual failure maximum restorability (DFMR) is only suitable for the simultaneous dual-failure situation and doesn挟 distinguish the failing sequence of two failed spans. In [6], a formulation with the same objective, called single-failure pre-plan (SFPP), was proposed to address the both situations. SFPP imitates the behavior of static pre-plan model. The following formulation, called optF1F2, imitates the behavior of fully-adaptive model. Both SFPP and optF1F2 formulations indicate the failing sequence explicitly.

**optF1F2:**     Minimize $\displaystyle\sum_{(i,j)\in S^2|i\neq j} N(i,j)$     (3)

Subject to:

• Ensure full single failure restorability:

$$\sum_{p\in P_i} f_i^{\,p} = w_i \,,\qquad\qquad i\in S \quad (4)$$

• Un-restored working capacity:

$$N(i,j) = w_i + w_j - \Bigg[\sum_{p\in P_i}(f_i^{\,p}+f1_{i,j}^{\,p})(1-\delta_{i,j}^{\,p}) +$$

$$\sum_{p\in P_j} f2_{j,i}^{\,p}(1-\delta_{j,i}^{\,p})\Bigg],\qquad \forall(i,j)\in S^2|i\neq j \quad (5)$$

• Dual failure restoration flow maximums:

$$\sum_{p\in P_j} f2_{j,i}^{\,p}(1-\delta_{j,i}^{\,p}) \le w_j \,, \qquad\qquad (6)$$

$$\sum_{p\in P_i}(f_i^{\,p}+f1_{i,j}^{\,p})(1-\delta_{i,j}^{\,p}) \le w_i \,, \qquad (7)$$

$$\forall(i,j)\in S^2|i\neq j$$

• Basic spare capacity necessary to ensure $R_1$=1:

$$\sum_{p\in P_i}\delta_{i,k}^{\,p}\cdot f_i^{\,p} \le s_k \,, \qquad \forall(i,k)\in S^2|i\neq k \quad (8)$$

• Additional spare capacity necessary to enhance $R_2$:

$$\sum_{p\in P_i}\Big[(f_i^{\,p}+f1_{i,j}^{\,p})\cdot\delta_{i,k}^{\,p}\cdot(1-\delta_{i,j}^{\,p})\Big]+$$

$$\sum_{p\in P_j} f2_{j,i}^{\,p}\cdot\delta_{j,k}^{\,p}\cdot(1-\delta_{j,i}^{\,p}) \le s_k$$

$$\forall(i,j,k)\in S^3|i\neq j, i\neq k, j\neq k \quad (9)$$

• Budget restriction:

$$\sum_{k\in S} c_k\cdot s_k \le B \,, \qquad\qquad (10)$$

where

$i$, $j$, $k$ are the specific spans. $i$ denotes the first failed span and $j$ the second failed span. $k$ denotes a span which is neither span $i$ nor $j$;

$S$ is the set of spans in the network;

$P_i$ is the set of eligible routes for the restoration of span $i$;

$w_i$ is the number of working capacity units on span $i$ (input integer parameter);

$c_k$ is the cost of each unit of capacity on span $k$ and can be the length of the span (input integer parameter);

$\delta_{i,k}^{\,p}$ is 1 if the $p^{th}$ restoration route of span $i$ crosses span $k$, and 0 otherwise (input binary parameter);

$s_k$ is the number of spare capacity units placed on span $k$ (non-negative integer variable);

$N(i,j)$ is the number of non-restored working units under dual failure of spans $(i, j)$. The expression $(i, j)$ reflects the sequence of the two failed spans, i.e. span $i$ is the first failed span and $j$ the second (non-negative integer variable);

$f_i^{\,p}$ is the restoration flow assigned to the $p^{th}$ backup route of span $i$ as a single isolated-span failure scenario (non-negative integer variable);

$f1_{i,j}^{\,p}$ is still the restoration flow of the first failed span $i$, which is found after the second failure appears (non-negative integer variable);

$f2_{j,i}^{\,p}$ is the restoration flow of the second failed span $j$, which is found with the remaining spare capacity after the restoration of the first failure (non-negative integer variable);

$B$ is a budget limit for total spare capacity cost (input integer parameter).

The objective of optF1F2 is to minimize the total non-restored capacity. According to Eq. (1) and (2), its objective is in fact to maximize the network average dual failure restorability $R_2$. If the two span failures occur non-simultaneously, the network operator can restore the first single failure according to the preplanned single-failure restoration assignment, i.e. the result of $f_i^{\,p}$. The spare capacity on the backup routes of span $i$ is seized for its restoration. When the second failure span $j$ occurs and hits some backup routes of span $i$, i.e. $\delta_{i,j}^{\,p}=1$, the initially seized spare capacity on these broken backup routes can be released now. The network operator then follows the computed flow $f1_{i,j}^{\,p}$ to restore the first failure as much as possible within the remaining spare capacity. After this operation, he can restore the second span failure according to $f2_{j,i}^{\,p}$. Certainly, there must be two pre-determined plans depending on the result of optF1F2. One is for the restoration of single span failures or the restoration of the first failure in a dual-span failure scenario; the other deals with dual span failures.

A direct and economical application to enhance $R_2$ can be implemented by setting the budget restriction with the minimum cost for the network. Basic spare capacity allocation (basic SCA) formulation in [7] can find the minimum necessary spare capacity for a span-restorable network with full restorability to single span failures.

## Case study

Dual-span failure restorability analysis is conducted on three test networks, respectively. Their basic characteristics are listed in Table 1.

Table 1: Test networks characteristics

| Test nets | Nodes | Spans | Average nodal degree | Ave length per span (km) |
|---|---|---|---|---|
| US net | 19 | 28 | 2.95 | 930.9 |
| n9s15 | 9 | 15 | 3.33 | 183.7 |
| n11s20 | 11 | 20 | 3.64 | 154.7 |

Table 2: Computations with all possible backups

| Test nets | Average | | $R_2$ | Improved $R_2$ (%) optF1F2 |
|---|---|---|---|---|
| | basic SCA | KSP fully-adp | optF1F2 | |
| US net | 0.630 | 0.655 | 0.659 | 4.6 |
| n9s15 | 0.559 | 0.668 | 0.684 | 22.3 |
| n11s20 | 0.640 | 0.689 | 0.696 | 8.6 |

Table 3: Computations with limited backups

| Test nets | Num of backups | Average | $R_2$ | $R_1(i) < 1$ spans |
|---|---|---|---|---|
| | | optF1F2 | fully-adp | |
| US net | 10 | 0.770 | 0.751 | {6,10,12,28} |
| | 10_all | 0.796 | 0.770 | none |
| n9s15 | 5 | 0.687 | 0.661 | {5,8,14} |
| | 5_all | 0.721 | 0.703 | none |
| n11s20 | 5 | 0.772 | 0.747 | {13,16,18} |
| | 5_all | 0.793 | 0.774 | none |

Tabel 2 shows the network average $R_2$ results of optF1F2 optimization and fully-adaptive model based on KSP routing, when all possible backup routes are allowed. Table 3 shows the similar results with some restrictions on the maximum number of backup routes for each span. The input parameter $B$ of optF1F2 is provided by basic SCA. Because the other two KSP-based models, i.e. static and partly-adaptive models, obtain lower $R_2$ than fully-adaptive model, we only list the result for the fully-adaptive model considering the limited paper size. In the last column 撓mproved $R_2$ (%) optF1F2?of Table 2, the $R_2$ results of optF1F2 are compared with those of basic SCA. By exactly applying the optimization results from optF1F2, we can improve $R_2$ as much as 22.3% in the studied test networks. We also observe that KSP-based models can work properly when all possible backup routes are exploited. The column 撓umber of backups? of Table 3 shows the restrictions on the backup routes.

? 0? or ? ? means that at most ten or five backup routes for each span are allowed in the computations. ? 0_all? or ? _all? means that all possible backup routes are allowed but the spare capacity allocaton on each span is kept the same as the result from basic SCA with ten or five backup routes. The flaw of KSP-based models appears when the number of backup routes is restricted. Three to four spans cannot even reach full restorability to single failures i.e. $R_1 < 1$. For the rows with ? 0_all? and ? _all? KSP-based models don拱have such problem because all possible routes are provided in the computations. In Tables 2 and 3, optF1F2 always reaches a higher $R_2$ than the KSP-based fully-adaptive model and guarantees fully restorability to single-span failures.

## Conclusions

The restoration mechanism based on KSP algorithm is problematic and unreliable when not all possible backup routes are allowed. The new introduced formulation optF1F2 aiming to maximize dual-failure restorability can provide an optimal result, guarantee to fully restore single failures and guide network operators to make restoration explicitly for both simultaneous and non-simultaneous dual failures in span-restorable networks.

## References

1 M. Willebeek-LeMair and P.Shahabuddin. Approximating dependability measures of computer networks: An FDDI case study. IEEE/ACM Transactions on Networking, 5(2):311? 27, April 1997.

2 M. Clouqueur. Availability of Service in Mesh-Restorable Transport Networks. PhD thesis, University of Alberta, Canada, 2003.

3 W. D. Grover. Mesh-Based Survivable Networks, Options and Strategies for Optical, MPLS, SONET, and ATM Networking. New Jersey: Prentice Hall PTR, 2004.

4 D. A. Dunn, W. D. Grover, and M. H. MacGregor. A comparison of k-shortest paths and maximum flow methods for network facility restoration. IEEE Journal on Selected Areas in Communications, 12(1):88? 9, January 1994.

5 M. H. MacGregor and W. D. Grover. Optimized k-shortest paths algorithm for facility restoration. Software-Practice & Experience, 24(9):823? 34, September 1994.

6 Optimization of Path Availability of Span-restorable Optical Networks, L. Zhou and M. Held, SPIE Photonics Europe, April 2006, Strasburg, France.

7 M. Herzberg and S. Bye. An optimal spare-capacity assignment model for survivable networks with hop limits. In Proc. IEEE GLOBECOM? 4, pages 1601☐ 1607, San Francisco, CA, USA, November 1994.

# Performance Limitations of an Optical Heterodyne CPFSK System Impaired by Polarization Mode Dispersion in a Single Mode Fiber

M. S. Islam (1),    S. P. Majumder (2)

(1): Institute of Information and Communication Technology,
Bangladesh University of Engineering and Technology, Dhaka-1000, Bangladesh
(2): Department of Electrical and Electronic Engineering
Bangladesh University of Engineering and Technology, Dhaka-1000, Bangladesh
E-mail: mdsaifulislam@iict.buet.ac.bd,   spmajumder@eee.buet.ac.bd
Tel: 880-2-966 5602, Fax: 880-2-861-3026

**Abstract** *The degradation of bit error rate performance of an optical heterodyne CPFSK system caused by signal phase distortion due to polarization mode dispersion (PMD) is evaluated analytically in a single mode fiber. The results at a bit rate of 10 Gb/s show that the effect of PMD is found to be significant at higher modulation index.*

### Introduction

Optical continuous-phase FSK (CPFSK) may be an attractive modulation format for future multi-channel optical system. However, at high bit rate operation of such systems in conventional single mode fiber, the major limitation is the group velocity dispersion (GVD), unless a dispersion compensation scheme is used [1]. Further, PMD presents a unique challenge for high-speed optical system because the induced pulse spreading is a frequency-dependent statistical parameter that varies randomly over time. Consequently, an incident linearly polarized wave may be subjected to random phase fluctuation which causes spectral broadening of the transmitted signal that leads to bit pattern corruption and higher bandwidth requirement, and eventually contributes to BER deterioration, performance variation or system fading even at moderate bit rates [2]-[9].

The effect of PMD has been the subject of considerable research interests during the past few years and the developed ideas are briefly described in a series of publication [3]-[9]. The simulation results on the effects of PMD on an amplified IM-DD system is reported in Ref. [3] as a function of the instantaneous DGD. The effect of PMD on the conditional bit error rate performance of heterodyne FSK system is presented [6]. Recently, the performance of optical DPSK system with direct detection receiver is reported in presence of PMD [8], [9] as a function of mean DGD.

In this paper, we provide an analytical approach to evaluate the BER performance limitations of an optical heterodyne CPFSK with delay-demodulation system impaired by PMD as function of instantaneous DGD. The method is based on the linear approximation of the output phase of a linearly filtered angle-modulated signal such as the CPFSK signal. The probability density function (pdf) of the random

phase fluctuation due to PMD and group velocity dispersion (GVD) at the output of the receiver is determined analytically. The conditiona BER performance and power penalty suffered by the system due to PMD and GVD   ( for BER of $10^{-9}$ ) are evaluated at a bit rate of 10 Gb/s using the moments of the random phase fluctuation at the receiver output in the presence of receiver noise.

### Receiver model

The block diagram of a CPFSK delay demodulation receiver is shown in Fig.1. The received optical signal is combined with the optical signal output from the local oscillator and the two signals are detected by photo-detector. The output of the photodetector which is an intermediate frequency (IF) signal is amplified by the receiver pre-amplifier and then filtered by a gaussian filter with center frequency set to the IF. The bandwidth of the IF is kept twice the bit rate for optimum demodulation. The output of the filter is then demodulated using a delay line discriminator. The output of the discriminator is then filtered by a low-pass filter (LPF) and fed to a sampler followed by a comparator.

### Theoretical analysis

The electric field input to the fiber is given by,

$$E_1(t) = \sqrt{2P_T} \, \exp[\, j2\pi f_c t + j\phi_s(t)][c_1.e_1] \qquad (1)$$

$$E_2(t) = \sqrt{2P_T} \, \exp[\, j2\pi f_c t + j\phi_s(t)][c_2.e_2] \qquad (2)$$

where, $f_c$ is the carrier frequency, $P_T$ is the transmitted optical power and the CPFSK modulating phase $\phi_s(t)$ is given by

$$\phi_s(t) = 2\pi \, \Delta f \int_{-\infty}^{t} I(t)dt \quad + \phi_n(t) \qquad (3)$$

$$I(t) = \sum_{k=-\infty}^{\infty} a_k \, p(t - kT) \qquad (4)$$

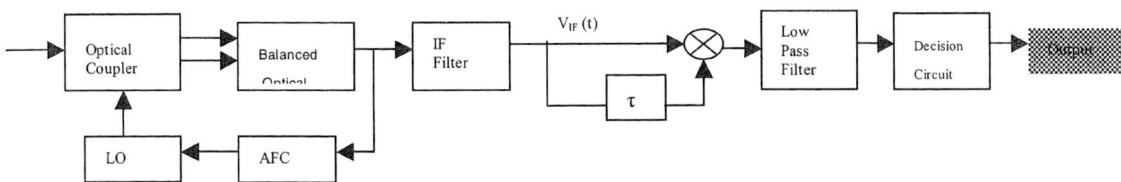

Fig.1: Block diagram of a heterodyne CPFSK delay demodulation receiver

with $a_k = \pm 1$ is the $k$th bit of random NRZ bit pattern of the $k$th information, $\Delta f$, the peak frequency deviation, $p(t)$, the elementary pulse shape of duration T seconds and $\phi_n(t)$, the phase noise of the transmitting laser. Here, $c_1, c_2$ represent unit vectors and $e_1, e_2$ represent the two principal states of polarization respectively.

The electric field output of the fiber is then given by

$$E_{01}(t) = \sqrt{2P_T} \exp[j2\pi f_c t + j\phi_s(t)][c_1.e_1] \otimes h_1(t) \quad (5)$$

$$E_{02}(t) = \sqrt{2P_T} \exp[j2\pi f_c t + j\phi_s(t)][c_2.e_2] \otimes h_2(t) \quad (6)$$

where $h_1(t)$ and $h_2(t)$ are the inverse Fourier transform of the fiber low pass transfer function of a non-dispersion shifted lossless fiber $H_1(f)$ and $H_2(f)$ respectively, which include the effect of PMD and group velocity dispersion. The expression for $H_1(f)$ and $H_2(f)$ are given by,

$$H_1(f) = \sqrt{\alpha} \exp[j2\pi f(-\frac{\Delta\tau}{2}) - j\gamma(\pi fT)^2] \quad (7)$$

$$H_2(f) = \sqrt{1-\alpha} \exp[j2\pi f(\frac{\Delta\tau}{2}) - j\gamma(\pi fT)^2] \quad (8)$$

Where $\alpha$ is the PMD power splitting ratio, $\Delta\tau$ represents the DGD between the two PSPs and now, assuming linear phase approximation, the output electric fields for two polarization states can be given as,

$$E_{01}(t) = \sqrt{2P_s} \exp[j2\pi f_c t + j\phi_{01}(t)][c_1.e_1] \quad (9)$$

where

$$\phi_{01}(t) = 2\pi \Delta f \int_{-\infty}^{t} \sum_k a_k \, \mathrm{Re}[\, p(t-kT) \otimes h_1(t)] dt \quad (10)$$

$$\phi_{02}(t) = 2\pi \Delta f \int_{-\infty}^{t} \sum_k a_k \, \mathrm{Re}[\, p(t-kT) \otimes h_2(t)] dt \quad (11)$$

The output of the IF filter can be expressed as,

$$v_0(t) = 2R_d\sqrt{P_s P_{Lo}} \, a(t)\cos[2\pi f_{IF} t + \phi_0'(t)] + n(t) \quad (12)$$

Following the delay demodulation (delay time τ) and low pass filtering (wide enough to pass the signal undistorted), the low pass filter (LPF) output is given by

$$v_{out}(t) = A(t)\cos[\Delta\phi_0(t,\tau)] \quad (13)$$

$A(t)$ is the amplitude and the differential output phase of $v_{out}(t)$ of the LPF can be expressed as,

$$\Delta\phi_0(t,\tau) = 2\pi f_{IF}\tau + \Delta\phi_0'(t,\tau) + \Delta\theta_n(t,\tau) \quad (14)$$

Say, a 'mark' is transmitted, thus $a_0 = 1$

$$\Delta\phi_0'(t,\tau) = 2\pi\,\tau\,\Delta f\, q(t) + 2\pi\,\tau\,\Delta f\sum_{k\neq 0} a_k q(t-kT) \quad (15)$$

$$\Delta\phi_0'(t,\tau) = \overline{\Delta\phi_0} + \xi \quad (15a)$$

where

$$q(t) = \frac{1}{\tau}\int_{t-\tau}^{t} g_0(t)dt \quad ; g_0(t) = \frac{1}{2}[g_1(t)+g_2(t)] \quad (16)$$

and

$$g_1(t) = \mathrm{Re}[\, p(t) \otimes h_1(t) \otimes h_{IF}(t)] \quad (16a)$$

$$g_2(t) = \mathrm{Re}[\, p(t) \otimes h_2(t) \otimes h_{IF}(t)] \quad (16b)$$

In delay demodulation receiver the data decisions are based on the polarity of $v_{out}(t)$. Assuming a 'mark' is transmitted (say $a_0 = 1$) and under ideal CPFSK demodulation condition,

$$2\pi f_{IF}\,\tau = (2n+1)\frac{\pi}{2}; \quad n \text{ is an integer and}$$

$2\pi\,\Delta f\tau = \frac{\pi}{2}$ for NRZ data.

Now, applying the above conditions, the phase of $v_{out}(t)$ at the sampling instant $t_0$ can be written as

$$\Delta\phi_0(t_0,\tau) = 2\pi f_{IF}\,\tau + \frac{\pi}{2} - \frac{\pi}{2} + \frac{\pi}{2}q(t_0) + \frac{\pi}{2}\sum_{k\neq 0} q(t_0-kT) + \Delta\theta_n(t_0,\tau) \quad (17)$$

The first two terms in (17) provide the expected phase change during the demodulation interval τ corresponding to the ideal situation, the next three terms, represent the undesired contribution to the phase due to PMD and GVD and the last term, $\Delta\theta_n(t_0,\tau)$ represents the phase distortion due to receiver noise.

We define the IF SNR $\rho = V_0^2 / 2\sigma_n^2$. For a given IF SNR, the conditional BER probability $P(e|\xi)$ conditioned on a given value of $\xi$ and instantaneous DGD $\Delta\tau$, can be obtained following Ref. [10].

So for a given value of differential group delay (DGD), $\Delta\tau$ the conditional BER can be expressed as,

$$BER(\Delta\tau) = \int_{-\infty}^{\infty} P(e|\xi)\, P_\xi(\xi)\, d\xi \quad (18)$$

where $P_\xi(\xi)$ is the pdf of $\xi$. The above integration can be carried out by Gauss-quadrature rule.

**Results and discussion**

Following the analytical approach, the BER performance results for heterodyne CPFSK with delay-demodulation are evaluated at a bit rate of 10 Gb/s with several values of the instantaneous DGD and fiber lengths. The plots of conditional BER versus received optical power Ps are shown in Fig.2 and Fig. 3 for modulation index 0.5 and 1.0 respectively.

Results show that the BER is highly degraded when the DGD is higher for a given fiber length and system suffers a significant amount of power penalty due to the effect of PMD. It is observed that the penalty at BER =$10^{-9}$ is increasing with increasing values of the instantaneous DGD.

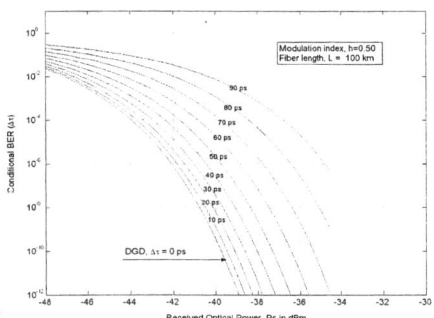

Fig. 2: Plots of conditional BER versus received power, $p_s$ for heterodyne CPFSK receiver impaired by PMD (modulation index, h = 0.50)

The amount of penalty is found to be approximately 0.70 dB, 1.25 dB, 2.5 dB and 4.50 dB for corresponding instantaneous DGD of 20 ps, 30 ps, 50 ps and 70 ps respectively when the modulation index is 0.50. It is further observed from Fig. 3 that the penalty is significant at higher modulation index such as h=1 compared to h=0.5, e.g. the penalty is approximately 5.80 dB when instantaneous DGD is 70 ps for h=1.0

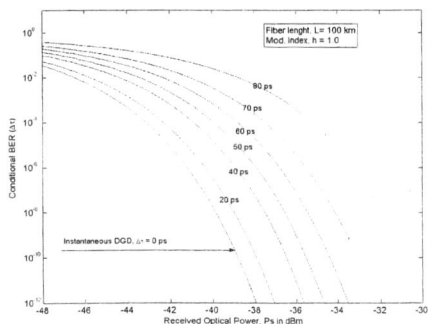

Fig. 3: Plots of conditional BER versus received power, $p_s$ for heterodyne CPFSK receiver impaired by PMD ( modulation index, h = 1.0)

The plots of penalty due to other modulation schemes as reported in [7], where penalty is evaluated against

mean DGD, are also shown in Fig. 4 for comparison. It is noticed that penalty suffered by heterodyne detection CPFSK with modulation index 0.50 and 1.0 higher up to 40 ps than that of NRZ-OOK and DPSK systems. But at higher instantaneous DGD ( i.e., above 40 ps), penalty suffered by heterodyne CPFSK is less than that of OOK and DPSK systems.

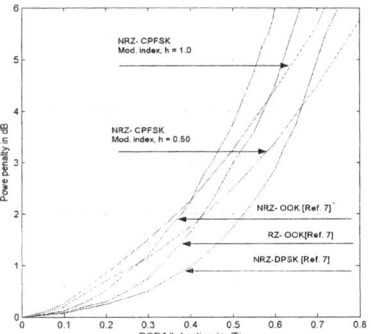

Fig.4: PMD-induced power penalty as a function of DGD/bit duration for NRZ- and RZ-OOK, NRZ-DPSK and heterodyne NRZ-CPFSK system.

**Conclusions**

An analytical approach is presented to evaluate the impact of polarization mode dispersion on the BER performance of heterodyne CPFSK system. The results show that the penalty due to PMD suffered by the CPFSK system is significantly higher at higher modulation index and higher value of the differential group delay between the two polarization modes.

**References**

1  A. F. Elrefaie et al, IEEE Photon. Technol. Lett., vol. 3/1 ( 1991), 71-73
2. C. D. Poole et al, IEEE photon. Technol. Lett, vol, 3(1), 1991, 68-70
3.  C. D. Angelis et al, Journal of optical communications, vol.16/5 (1995), 173-178
4.  C. De Angelis et al, IEEE J. Lightw. Technol, vol. 10 /5 (1995), 552-555.
5.  F. Bruyere et at, Optical Fiber Technology, vol. 2/2 (1996,) , 269-280
6.  S. P. Majumder et al, Conference proceedings, Telecommunication global conference 2000, GLOBECOM'00, vol.2 (2000), 1233-1236.
7.  Jin Wang, et al, J. Lightw. Technol., vol.22/2 (2004), 362- 371.
8.  Chongjin Xie et al, 31$^{st}$ European conference on optical communication, 2005, ECOC, vol.3 (2005) 345-346.
9.  H. Yoon et al, IEEE Photon. Technol. Lett. Vol. 17/12 (2005), 2577-2579.
10. S. P. Majumder et al, IEEE Photon. Technol. Lett, vol. 7/10 (1995), 1207-1209

# Ultrafast Bit and Byte addressing of All-Optical Memory based on Microring Resonators for Next-Generation Optical Networks

Yingyan Huang and Seng-Tiong Ho

Northwestern University, Dept of ECE, Evanston, IL, 60208. sth@northwestern.edu

**Abstract** *We present a novel ultrafast all-optical memory scheme that has the potential to meet the demanding requirement for 100Gb/s all-optical data storage. The all-optical memory integrates an all-optical ring resonator flip-flop with an innovative GMCI (gain manipulation of coupler interference) based all-optical switch for ultrafast READ and WRITE. The potential device performance advantages include very low operating power for the memory element (<50μW per bit), low read-write power for the GMCI based all-optical switch (<0.5-5mW), ultrafast operation (10-100Gbit/s), very compact size (<10μm x10μm per bit), addressable at bit or byte level, and monolithically integrated on InP chip.*

## Introduction

The advancement of packet-switched optical communication requires all-optical memory that can store and retrieve data at ultra-high rate (>100Gb/s). Most current research implementations use an optical storage loop to store the entire data stream. [1] The optical storage loops are typically large in size (km-long) with finite storage time of up to only a few seconds. Such storage loops are also typically not bit addressable and too preliminary for practical application. It would be desirable to develop ultrafast all-optical memory with long or perpetual (static) storage time that is addressable at bit or byte level. Beside applications to all-optical packet switching, such memory will be a key enabling element for ultrafast all-optical signal processors or computers. Below, we shall refer to such memory device as Ultrafast Bit/Byte-Addressable All-Optical Static (UBAOS) memory. In this paper, we present a novel ultrafast all-optical memory scheme that has the potential to meet the demanding requirement for 100Gb/s all-optical data storage. The all-optical memory integrates an all-optical ring resonator flip-flop with an innovative GMCI (gain manipulation of coupler interference) based all-optical switch for ultrafast READ and WRITE.

## Working Principle of the All Optical Memory Unit

The proposed all optical memory units consist of three main parts: the ultra-fast write /reset, the static storage and the ultra-fast read, as shown in Fig. 1. The main functions of the memory unit include:

SET: The input signal is at wavelength $\lambda_L$ close to 1550nm and consists of many bits. The ultra-fast set part works as an AND gate with ultra-high speed of 10-100Gb/s. When the writing clock and the signal both equals to "1", a pulse will be transferred to the storage part; otherwise, no pulse will be generated. By adjusting the writing clock, we can select a certain bit in a multi-bit pulse train and store it in a particular memory unit.

STORAGE: The storage part consists of a Mach-Zehnder Interferometer (MZI) with one arm coupled to a microdisk, which is partially filled with saturable absorber. In brief, the two arms of the MZI have a phase difference of π so the two beams will interfere destructively at the output without the presence of the ring [2]. For input power supply beam within certain intensity range, two scenarios are possible. State "0": the absorber is at ground state initially so the small portion of light coupled into the ring will get absorbed away. State "1": the absorber is at excited state initially so the intensity inside the cavity will build up quickly. The on-resonance CW beam will then experience a phase shift of π. As a result, the two beams of the MZI will add constructively at the end and produce a constant output, resulting in a quasi-CW beam entering the port "input to reading".

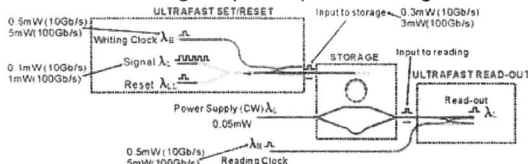

Fig. 1 Components of the all-optical memory unit

RESET: The memory will retain its status until the RESET pulse at wavelength $\lambda_{LL} > \lambda_L$ comes in. The RESET pulse is at longer wavelength so it will deplete the active medium inside the ring and brings it back to the ground state. The memory will then be reset to initial "0" state.

READING: The reading part follows the same principle as the set part. Basically, when the reading clock and the "input to reading" are both "1", the output end will have a pulse "1" coming out with the same pulse width as that of the reading pulse. Otherwise the output will be zero.

## Storage: Principle and Simulation

The storage part utilizes the bi-stability of a microdisk cavity with saturable absorber. First we discuss the case where a waveguide is coupled to a lossy disk, as shown in Fig.2a.

Assuming the input wavelength is on resonance with the disk, if the electrical field coupling efficiency between the disk and the waveguide is $-j\sqrt{T}$, and the absorption coefficient of the absorber part is $\alpha$ per micron, the input/output relation will be:

$$E_{out} = E_{in}[\sqrt{1-T} - Te^{-\alpha L} - \sqrt{1-T}Te^{-2\alpha L} - (\sqrt{1-T})^2 Te^{-3\alpha L}......]$$

$$= E_{in}(\sqrt{1-T} - \frac{Te^{-\alpha L}}{1-\sqrt{1-T}e^{-\alpha L}})$$

where L is the total length of the absorption region per round trip.

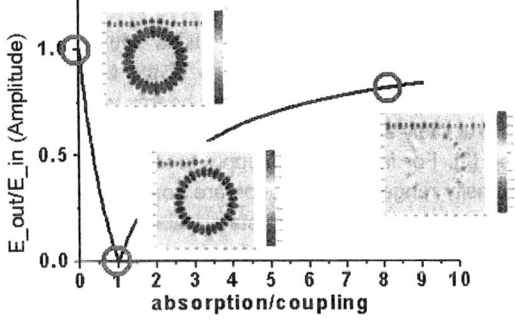

(a) Waveguide coupled to lossy ring

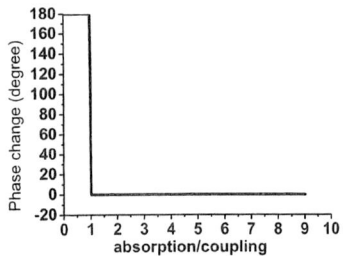

(b) Amplitude change at different $\alpha$

(c) Phase change at different $\alpha$

Fig. 2 Input/output relations for a waveguide coupled with lossy ring

As an example, we plotted out the input/output amplitude and phase change as a function of the ratio between absorption loss and coupling $(1-e^{-2\alpha L})/T$ for T=0.05. The result is shown in Fig. 2. There are three scenarios of interest as shown in thumb nail pictures in Fig.2b: 1st scenario: when absorption loss is zero, the output amplitude will be the same as the input amplitude, but there will be a $\pi$ phase shift. 2nd scenario: when absorption loss equals coupling loss, the output amplitude will be zero (critically damped). 3rd scenario: when absorption loss is much larger than coupling loss, the output amplitude will be close to the input amplitude. The phase change is zero.

Since the absorptive medium inside the ring is saturable, we will be able to switch between the 1st and 3rd scenario for input power supply intensity within certain range. The existence of bistability requires $C = a_0 L/(4T) > 4$ [3], where T is the coupling between the waveguide and the cavity. When T is smaller (cavity Q higher), it is easier to get bistability but the cavity bandwidth will be narrower.

For certain structure (T, $\alpha_0$), the input intensity $I_{in}$ needs to satisfy the following two equations:

$$\frac{I_{ring}}{I_{sat}} = (\frac{\alpha_0}{\alpha} - 1); \quad \frac{I_{ring}}{I_{sat}} = kT/(1-e^{-\alpha L}\sqrt{1-T})^2$$

where (assuming $I_{in} = k \cdot I_{sat}$) $I_{ring}$ is the intensity inside the cavity. Bistability occurs only when there are two cross points between those two curves. Fig. 3 illustrated a set of typical curves.

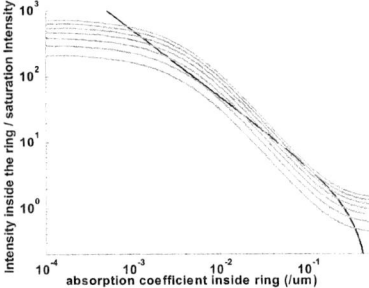

Fig. 3 Bistability curves for 5um diameter micro disk with T=0.1, $\alpha_0 = 0.5/\mu m$, and refractive index 3.

The complete structure of the storage part is shown in Fig. 4. Without the "input to storage" pulse, the "input to reading" will be zero. With the "input to storage" pulse, the ring will induce a $\pi$ phase shift and cause the two arms of the MZI to add coherently, and the "input to reading " will be "1". Fig. 5 shows the status of the memory when it stores "1" and "0" respectively.

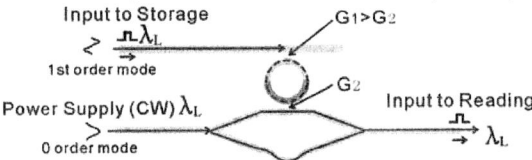

Fig. 4 Storage part of all optical memory unit

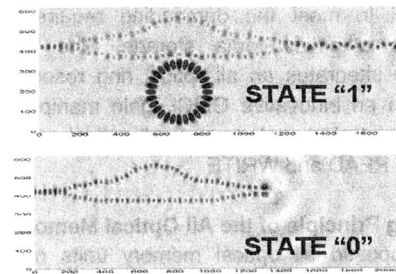

Fig. 5 FDTD simulation of "1" and "0" states

**Ultra-fast Write/Reset and Read: Principle and Simulation**

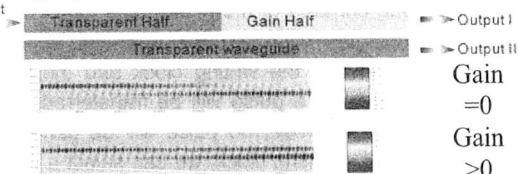

Fig.6 Directional coupler with one arm transparent and one arm half transparent

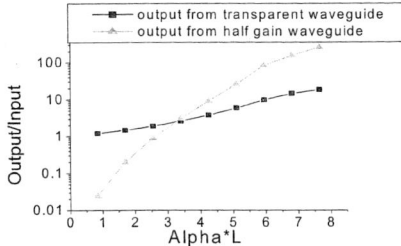

Fig. 7 Power output from two waveguides as a function of αL

The basic structure of a GMCI based device is shown in Fig. 6, the directional coupler has one transparent arm and the other arm half transparent, half with gain medium. The gap of the coupler is chosen so that the total length of the structure will be one full coupling length for $\lambda = 1.55 \mu m$. As a result, when the gain coefficient in top half waveguide is zero, power output from output II would be the same as the power input, while power output from output I would be zero. (Fig.6 middle). Now we increase the gain coefficient $\alpha$ of the top waveguide and plot the output power from those two waveguides as a function of αL, where L is the coupler's length. The simulation is done by FDTD method so the complex gain and coupling effect can be accurately modeled. Note that in this illustrative simulation, we keep the intensity of the input much lower than medium saturation intensity, so the gain coefficient is the one given by initial inversion levels.

When $\alpha=0$ so αL=0, the power output from the transparent waveguide is 1 and the output from active waveguide is zero. When αL increases as the gain goes higher, the output from the active waveguide (output I) will increase faster than output II. After αL value reaches 3, the power output from active waveguide is actually higher than the original input power. (Fig.6 bottom) The physical reason behind this is the same as gain-guided semiconductor lasers. When certain sections of the photonic circuit have gain, the light will have a tendency to be confined in that section. As shown in the simulation result, when the gain goes higher, the confinement will be stronger.

The structure of the GMCI based ultra-fast writing part is shown in Fig.8 for which the right half of lower arm is active and the upper arm is transparent. To function as the ultra-fast write part, an optical multi-bit pulse train at wavelength $\lambda_L$ is sent into the left side of the lower waveguide. This beam then couples to upper waveguide and exit the right side of the upper waveguide. This input beam will be referred to as the "signal" beam. When a writing clock pulse at $\lambda_H < \lambda_L$ is sent into the left side of the upper waveguide, it will partially couple to the saturable absorber region and excite the active medium. The result is that the signal beam now sees gain at the last half of the lower waveguide, which causes part of its energy to remain at the lower waveguide and exit the lower waveguide on the right. The net effect is a pulse output when both signal and writing clock is equal to "1".

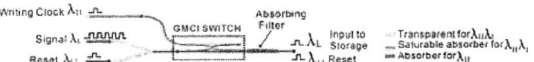

Fig.8 GMCI based ultra-fast write/reset

The structure of the ultra fast reading part follows the same principle as the write part. (Fig. 9) The "input to reading" beam at $\lambda_L$ will exit from the right side of lower waveguide when reading clock is zero. When a writing clock pulse at $\lambda_H$ is sent into the left side of the lower waveguide, it will partially couple to the saturable absorber region and excite the active medium. The result is that the "input to reading" signal beam now sees gain at the last half of the upper waveguide, which causes part of its energy to remain at the upper waveguide and exit the upper waveguide on the right. The net effect is a pulse output with duration the same as that of the reading clock pulse. The output from the right side of the upper waveguide will also contain reading clock pulse at $\lambda_H$, which will be filtered away by an active section with bandgap higher then $\lambda_L$ and lower then $\lambda_H$ (the absorbing filter in Fig.9).

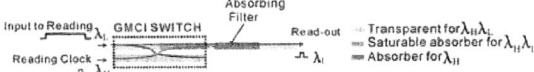

Fig.9 GAMCI based ultra-fast reading part

**Conclusions**

In this paper, we illustrate a nanophotonic ultrafast all-optical memory device that has the potential to meet the following properties: (1) Capable of operating at ultrafast (>10-100Gb/s) data rate. (2) A perpetual (static) all-optical memory monolithically integrated on chips. (3) Having low memory operating power per bit (~50μW/bit) (4) Bit or byte addressable all-optically at 10-100Gb/s with low peak switching power (0.5-5mW or pulse energy of 50fJ). (5) Capable of composing output as synchronized pulse train at 10-100Gb/s. (6) Potentially scalable to addressing >100-1,000 Bits (or >10-100 Bytes) on a cm-size chip.

**Acknowledgement**

The work is supported by NSF under Award No. ECS-0501589, by NSF MRSEC program under grant DMR-0076097, and by the NASA Institute for Nanoelectronics and Computing under Award No. NCC 2-1363.

**Reference**

1 Georgios I. Papadimitriou, etc, "Optical Switching: Switch Fabrics, Techniques, and Architectures", Journal of Lightwave Technology, Vol.21, No.2, pp384-403, Feb. 2003.
2. Yingyan Huang, Guoyang Xu, and Seng-Tiong Ho, " An Ultra Compact Optical Mode Order Converter", to be published in Photonics Technology Letter.
3. Hyatt M. Gibbs, "Optical Bistability: Controlling Light with Light", Academic Press, Inc. 1985.

# Implementation of Token-based Optical Burst Switching Ring Network Node and Testbed Using Fixed Transmitter and Tuneable Receiver

Hongxiang Wang (1), Gan Wen (2), Yuefeng Ji (3)

1 : Beijing University of Posts & Telecommunications, P. O. Box 128, No. 10 Xitucheng Road, Haidian District, Beijing, China, wanghx@bupt.edu.cn

2 : wengan1982@tom.com, 3 : jyf@bupt.edu.cn

**Abstract** *Token protocol is introduced to resolve the burst collision in OBS network. Token-based OBS ring network node using fixed transmitter tunable receiver is implemented, and a ring network testbed with three nodes is constructed. Experimental results show burst blocking is eliminated in the token-based OBS network.*

## Introduction

With the explosive growth of IP data traffic in Internet, wavelength routed optical network based on circuit switching is becoming inappropriate for the emerging optical Internet. Optical packet switching maybe an alternative solution to meet this demand, but it is not mature enough to be implemented at this moment. Optical burst switching (OBS), combining the advantages of optical circuit switching and optical packet switching, is proposed as a promising data service oriented technology to provide an all-optical layer for future optical Internet [1].

In OBS network, several packets with the same attributes from the upper layer are grouped into an data burst. Each data burst has a control packet. Data burst and control packet are transmitted on separate wavelength. Control packet is transmitted slightly ahead of a time, called offset time, to reserve resource for data burst. By this way, data burst can go through the network all-optically without E/O/E conversion. Comparing with circuit switching based optical network, OBS technology can offer higher bandwidth efficiency and higher throughput. So far, much work [2, 3] has been done to promote this encouraging technology. However, some challenges are still underdone, in which collision and contention of data burst is vital to be resolved. Because there is no end-to-end bandwidth reservation, thus no guarantee of bandwidth, and the limitation of optical devices, burst collision occurs when two or more data bursts attempt to reserve the same wavelength on the same link at the same time. Efficient contention resolution method is essential to the application of OBS technology. Up to now, many studies have been reported on burst collision and contention resolution in OBS network, such as deflection routing [4], fiber delay line, wavelength converter pool, smart wavelength assignment [5], and so on. Considering that the ring topology is widely used in optical network, the token protocol designed for ring network can be introduced in OBS network for burst collision. In this paper, a token-based OBS node with fixed transmitter and tunable receiver for ring network is implemented

successfully to avoid the data burst collision at the receiver side, and an OBS ring network (OBSRN) testbed with three such nodes is built.

## OBSRN Node Structure

In OBS mesh network, there are two kinds of nodes: core node and edge node. For OBS ring network, each node acts as both core and edge node. All OBSRN nodes are connected by WDM links. Figure 1 shows the structure of OBSRN node, which mainly consists of an optical processing unit and an electric processing unit. The electric processing unit is used for Gigabit Ethernet (GE) access, burst assembling/disassembling, scheduling, and token controlling, as well as control packet producing, parsing, and transmitting. The optical processing unit is used for the wavelength adding and dropping of data burst and control packet, which is controlled by token protocol.

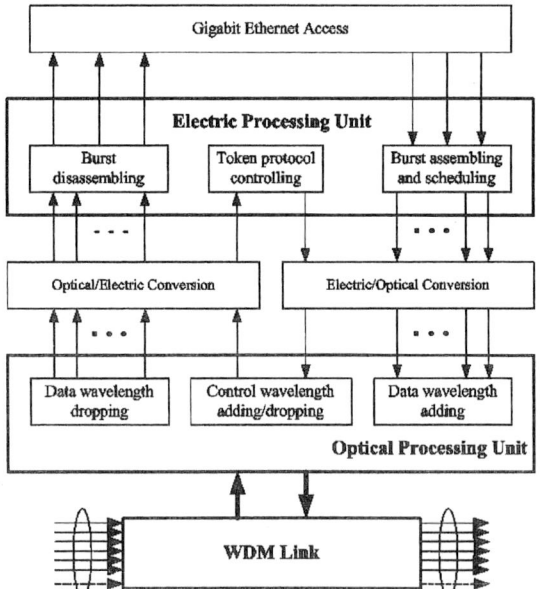

*Figure 1: Node structure of OBS ring network*

The design of an effective optical processing unit is

essential to OBSRN node. Figure 2 shows the design of optical processing unit, which adopts fixed transmitter and tuneable receiver. By using this optical processing design, wavelengths adding and dropping of both control packet and data burst can be realized efficiently. In OBSRN node, control wavelength in 1310 nm is separated from data burst wavelengths and dropped. WDM optical signal of data burst is divided into two parts: one enters a tuneable optical filter after an Erbium doped fiber amplifier (EDFA) with power equalization; the other is input to an passive optical device name optical add/dropping multiplexer (OADM), which adds one wavelength while blocking the same wavelength at the same time. Finally, data burst wavelengths are multiplexed with the control wavelength, and are transmitted into the ring fiber WDM links.

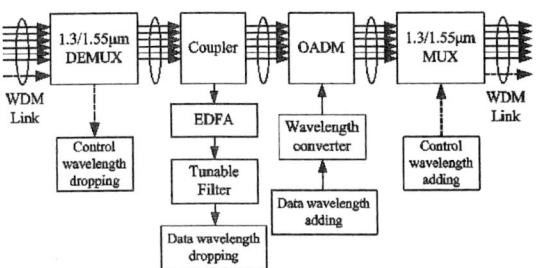

Figure 2: Optical processing Unit in OBSRN node

Figure 3 shows the functional diagram of electronic processing unit in OBSRN node. When data stream arrive at local node, the control wavelength is intercepted and transformed to electronic signal and the control information is parsed subsequently. The bursts whose destination is local node are dropt. According to control information, the wavelengths which transmit these bursts are found, then the receive module is tuned to these wavelengths and drops the bursts under the controlling of control module. The packets are disassembled from burst in the dropt wavelength, and switched to their destination Gigabit Ethernets.

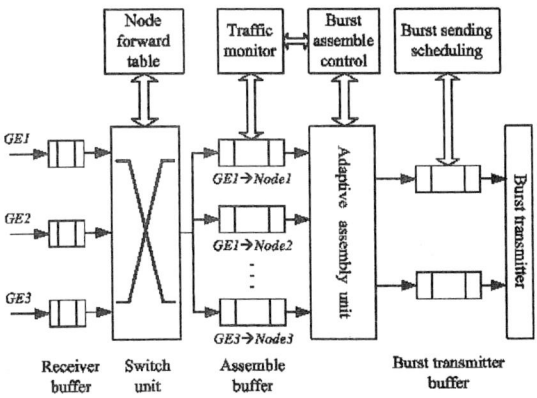

Figure 3: Electronic processing Unit in OBSRN node

In transmit part, packets from three gigabit Ethernets are switched to three queues according to their destination address. Packets in one queue are transmitted among these three Ethernets and the other two queues connect to assembly module. In assembly module, packets are assembled into data burst and control packet is created based on the data burst. Control packet is transmitted to control module, and inserted into the control packet queue from upstream node. Then the whole control packet queue is transmitted in control wavelength. For every node, there are two home wavelengths used to transmit data burst to other two nodes on the OBS ring respectively. In order to transmit the local data burst, the two home wavelengths are cut down firstly and then the local data burst are transmitted in the two home wavelengths respectively according to their destination node. Along the whole OBS ring, data burst is transmitted in all optical filed while control packet is converted to electronic signal at every node it passes through. Each OBSRN node implemented has three Gigabit Ethernet ports and two WDM ports, which makes it more flexible for burst transmitting.

To solve the burst contention problem, token scheme is used in the network, in which the utilization of each receiver is uniquely identified by a token. One source node can send the burst to other nodes if and only if it captures the token of destination node. The controlling flow for burst transmitting is shown in Figure 4.

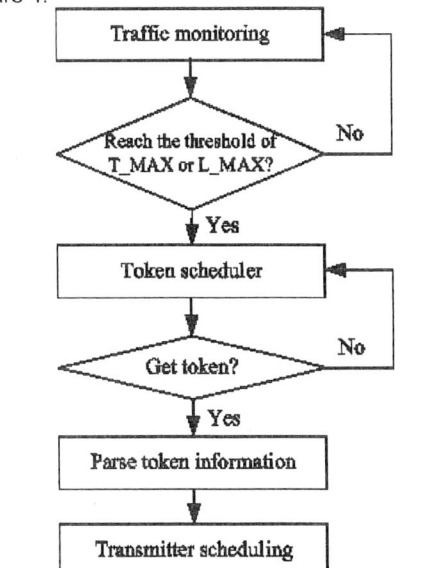

Figure 4: Controlling schedule flow for burst transmitting

### OBS Ring Network Testbed

The unidirectional OBSRN testbed constructed contains three nodes designed above, interconnected by fiber link, as shown in Figure 5.

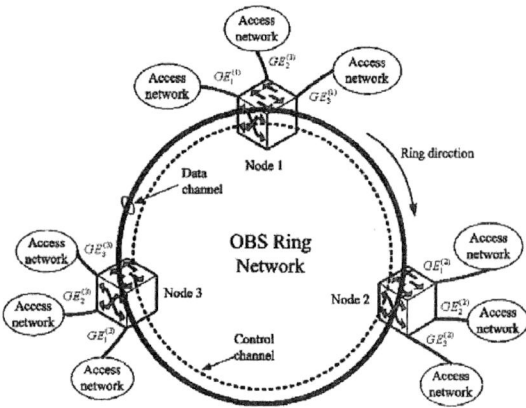

Figure 5: Testbed of OBS ring network

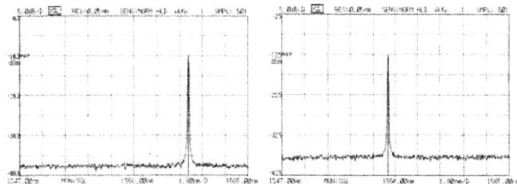

Figure 6: Wavelength dropping at an OBSRN node

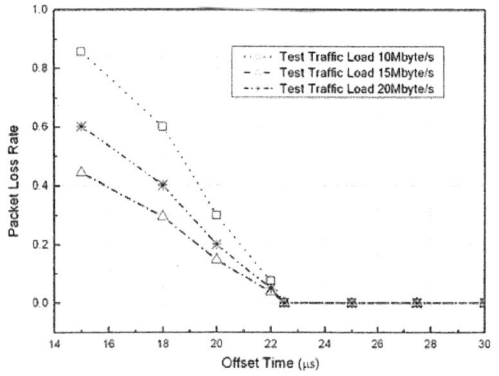

Figure 7: Packet loss rate with different offset time and different traffic load (frame size = 790 bytes, in node 2)

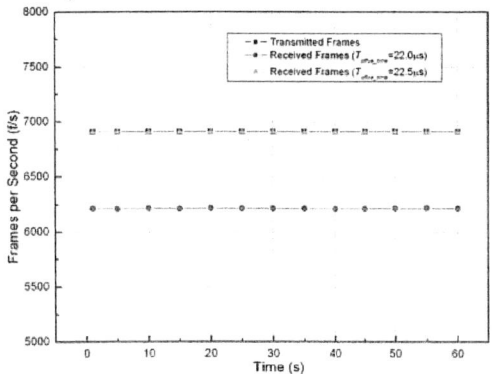

Figure 8: Network behavior after the system runs into the stable status (frame size=1,518bytes, test traffic load=10Mbytes/s, in node 2)

Figure 6 shows the example of wavelength dropping in OBSRN node 2. By using the testbed constructed,

optical burst loss rates with different network configuration are measured. Experimental measuring results are shown in Figure 7 and Figure 8. Figure 7 shows that when the offset time is smaller than 22.5µs, packet loss phenomenon can be clearly observed. Moreover, packet loss rate decreases with test traffic load increasing. That is because the unsuitable offset time will only cause the same number of incorrectly received packets under different traffic load. Thus, a heavier traffic load will lead to lower packet loss rate. Figure 8 shows the packet loss phenomenon when the offset time is selected as 22µs and when frame size is set as 1,518 bytes in Node 2. It is clear that approximately 20% frames are incorrectly received due to the excessively small offset time. It is a sharp contrast that the packet loss rate is only about 10% when the frame size is 790bytes. Figure 8 shows the network behavior under the stable running status for Node 2. It is clear that when the traffic load of the analyzer is set to 10Mbytes/s, the input traffic is equal to its output traffic which indicates that there isn't any packet loss in Node 2.

## Conclusions

Token-based OBS ring network node by using fixed transmitter and tunable receiver is implemented, and a ring network testbed with three nodes is constructed. The performance of the three-node OBS ring network testbed is measured. Measurement results show that optical burst loss can be eliminated with offset time larger than 22.5 microseconds in the token-based OBS ring network.

## Acknowledgment

This work is partially supported by the National Science Fund for Distinguished Young Scholars (No. 60325104), National Natural Science Foundation of China (No. 60572021), the SRFDP of MOE (No. 20040013001), P. R. China, and Fujitsu Research and Development Center Co., Ltd. Thanks for the great help.

## References

1 C. Qiao et al J. of High Speed Networks, 8 (1999), 69-84

2 Yijun Xiong et al IEEE J. Select. Areas Commun., 18 (2000), 1838-1851

3 C. Qiao et al IEEE Commun. Mag., 1 (2000), 104-114

4 Ching-Fang Hsu et al J. of High Speed Networks, 14 (2005), 341-362

5 M. Düser et al ECOC 2000, paper Tu 4.1.4, 2000, pp. 23-24.

# G-TEP: A GMPLS Testing and Emulation Platform

Weiqiang Sun, Xuejuan Xie, Mingzhi Zhao, Yaohui Jin, Weisheng Hu, Wei Guo, Da Feng, Zhengyu Wang
State Key Lab of Advanced Optical Communication System and Networks
Shanghai Jiao Tong University, 800 Dongchuan Road, Shanghai 200240, P.R.China
E-mail: sunwq@sjtu.edu.cn

**Abstract:** *We report G-TEP which includes a GMPLS tester of highly dynamic GMPLS networks, and an emulated GMPLS network with up to 200 nodes targeted at future nation-wide deployment of ASON in China.*

## Introduction

The introduction of an intelligent distributed control plane into traditional statically provisioned transport networks greatly simplifies network operation and management. Among different control plane solutions, generalized multi-protocol label switching (GMPLS) has become the primary choice. Although it has been widely accepted that ASON can provide dynamically provisioned connections (or switched connections: SC), it is still unknown either to the industry or the academia on how dynamic a GMPLS network can actually be. As more and more applications over ASON are emerging, it is worthwhile to investigate the performance of GMPLS networks under high traffic load. Meanwhile currently deployed GMPLS networks are small in scale, and the problems of deploying GMPLS networks in national or international scale are still to be explored. In this paper, we introduce G-TEP, a GMPLS testing and emulation platform, to test the dynamic utmost of GMPLS networks and to discover scalability issues of large scale deployment.

## G-TEP: A GMPLS Testing and Emulation Platform

The platform is made up of two blocks: ASBAT as a GMPLS tester and GAP as a large scale GMPLS emulation platform.

### ASBAT – A Simple Burst ASON Tester

ASBAT, or A Simple Burst ASON Tester, is a standard compatible tool to explorer the dynamic capabilities of GMPLS networks. As shown in Fig.1, ASBAT local and remote entities are placed at the ingress and egress sides of the GMPLS network under test, and communicate with GMPLS network through OIF User Network Interface (UNI). ASBAT has a full implementation of GMPLS signaling (RSVP-TE) which acts as UNI-C entity. It has demonstrated inter-operability with ASON devices from major vendors in China, including Huawei, Fiberhome and ZTE [1].

To perform highly dynamic testing, a traffic generator is configured with each ASBAT local entity to generate setup/release requests according to input parameters from Graphical User Interface (GUI). All the resources (such as UNI port availability) are managed by resource manager. Traffic generator randomly select source and destination from resource manager, and forward a connection request consisting of a {srcport, destport, holdtime} tuple to connection manager. Connection manager will then setup a connection

entry for the request and initiate a UNI signaling process to setup the connection.

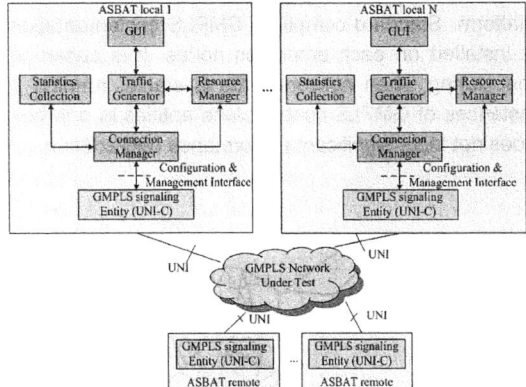

*Fig.1 ASBAT – A GMPLS Network Tester*

Upon completion of a signaling process, an indication will be sent back to connection manager. In case a connection is successfully setup, connection manager will in turn modify the connection status and start a timer for this connection. It will also notify the resource manager that the corresponding resource has been allocated and can no longer be used by other connections. In case any setup failure occurs, connection manager will remove the connection entry, release the reserved resource and wait for the next connection request. The established connection will stay active until its timer expires. The connection manager will initiate a connection release signaling process to delete the connection. The connection setup/release delay is collected by statistics collection box and displayed on GUI.

To emulate real application scenario, multiple ASBATs local can be placed anywhere in the network and generate request concurrently. In case of centralized testing, i.e., network nodes are geographically located in the same lab, ASBAT local and remote entities can be placed in one box. It is measured that the overhead of ASBAT running on a Pentium 4 2.8GHz CPU and 1 Gigabytes memory is usually less than 5 ms and has a mean around 2 ms, which may cause less than 1% overhead in a typical 2-hop GMPLS network.

### GAP – GMPLS emulation Platform

As predicted by China Telecom (CT), as more as 400 ASON nodes are needed to cover major regional cities in China. However, currently deployed GMPLS networks are small in scale [2]. To possibly reveal the scalability issues of large scale GMPLS deployments, we report a GMPLS emulation platform named GAP. As shown in Fig.3, GAP is made up of 25 computers acting as emulation nodes and 1 computer as configuration and management station. Each of the computers has a dual core Pentium 4 processor running at 2.8GHz and 2 Gigabytes memory. The nodes inside a rack are connected through fast Ethernet and inter-rack connection is GBE on fiber links to avoid bottleneck. Fig.2 shows an emulated network with 69 backbone nodes covering major cities in China. Fig.3 shows a photo of the emulation platform. Standard compliant GMPLS implementation is installed on each emulation nodes. It is observed that during when network state is stable, running 8 instances of GMPLS control plane entities in one box does not incur significant performance degradation.

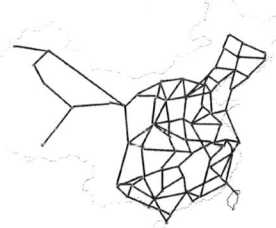

Fig.2 An emulated optical backbone network covering 69 cities in China

**Experiment result and conclusions**

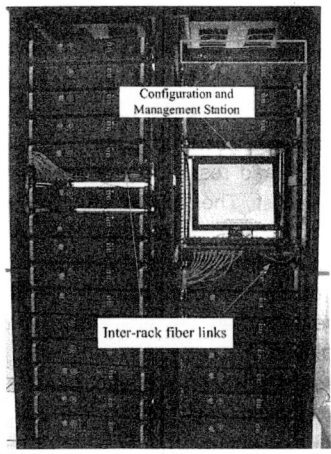

Fig.3 Photo of the emulation platform made up of 25 computers

Experiment was carried out on G-TEP to evaluate the dynamic provisioning capability of GMPLS networks. In the experiment, ASBAT send out setup/release requests periodically according to input traffic load parameters. The setup/release delay is recorded as well as blocking probability. During the whole process of experiment, no call blocking is observed. The connection setup delay at different load (i.e., different arrival interval and holdtime combination) is shown in Fig.4 (a) and (b). It can be seen that higher load leads to longer setup delay and higher delay variance. In Fig.5 the setup delay is plotted against traffic load. Fig.6 shows that setup delay increase almost linearly

when the number of hops increases.

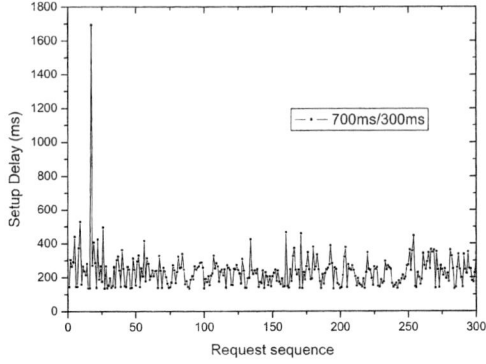

(a) load =700ms/300ms

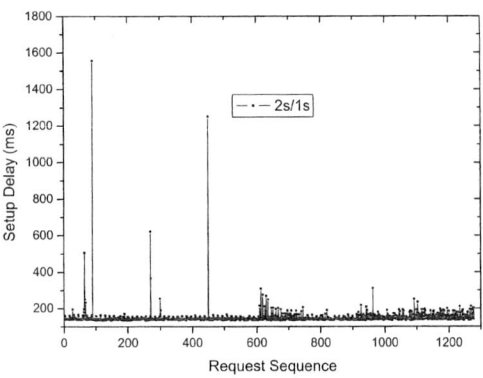

(b) load = 2s/1s
Fig.4 Setup Delay vs. time

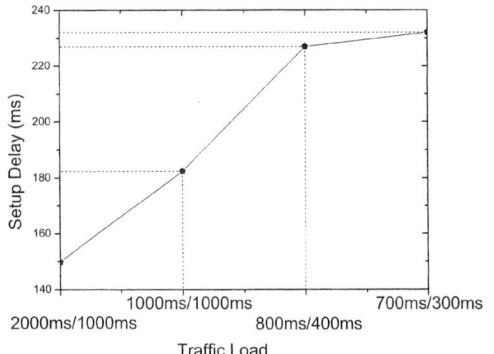

Fig.5 Setup Delay vs. traffic load (2 Hops)

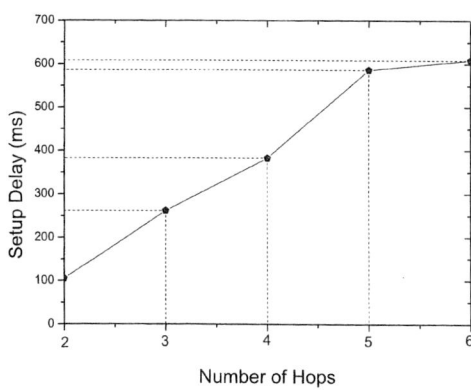

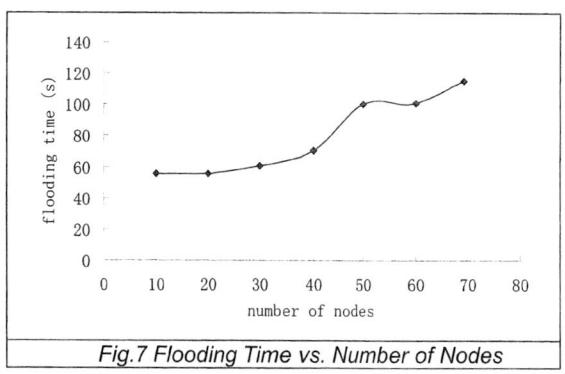

Fig.7 Flooding Time vs. Number of Nodes

*Fig.6 Setup delay vs. number of hops (Load = 3s/1s)*
Network synchronization delay is also measured on G-TEP. It is observed that link state synchronization of the single-layered networks takes considerable time to complete and also generates considerable control plane overhead. Fig.7 shows that the variation of the OSPF flooding time when the nodes in the network changes. The flooding time increases with the nodes in the network almost linearly. Future work includes multi-domain routing and signaling, large scale protection and restoration.

This work is supported by China 863 program and NSFC.

**Reference**
[1] X. Wei et al., Demonstration of GMPLS-Controlled Dynamic Point-to-Multipoint Trees in Optical Networks, ECOC'05
[2] S. Okamoto, et al, "Nationwide GMPLS Field Trial Using Different Types (MPLS/TDM/Lambda) of Switching Capable Equipment from Multiple Vendors," OFC05 PDP-40, March 2005
[3] W. Guo et al., Optical Grid Experiment on 3TNET in China, FON held at OFC'06

# Phase noise due to ASK amplitude un-equality and Kerr nonlinearity in optical ASK-DPSK, ASK-DQPSK systems

Yang Aiying, Xiao Yao, Sun Yunan

Beijing Institute of Technology

Room 236, Qiushi Building, Beijing Institute of Technology, Beijing, 100081, P. R. China

Email: yangaiying@bit.edu.cn

**Abstract** *Phase noise due to ASK amplitude un-equality and Kerr nonlinearity is first investigated in this paper. Then, for the phase modulation path, the eye-diagram closure at the receiver end is observed with the impact of the phase noise. Finally, by analyzing the eye-opening penalty varying with extinction ratio for ASK-path and DPSK/DQPSK path signal at the balance receiver end, it reveals that the optimal extinction ratio varies with the input power. The conclusion is quite different from that the optimal extinction ratio is constant without consideration of the phase noise.*

## Introduction

Quaternary combined amplitude and differential phase-shift keying (ASK-DPSK) [1, 2] and 8-ary combined amplitude and differential quadrature phase-shift keying (ASK-DQPSK) [3], taking advantages of both ASK and differential phase-shift keying are drawing quite some attention. Compared to quaternary-differential phase-shift keying (DQPSK) and 8-level differential phase-shift keying (8-DPSK) respectively, ASK-DPSK and 8-ary ASK-DQPSK require less components and thus simpler to implement at the receiver end [4-5]. A great deal work has been reported on this subject [1-7], but phase noise introduced by ASK amplitude un-equality and Kerr nonlinearity is seldom researched to the author knowledge. In this paper, we calculate the phase noise due to ASK amplitude un-equality and Kerr nonlinearity. Then, by numerical simulation, *the eye-diagram closure for the phase modulation path at the receiver end is observed with the impact of the phase noise*. Finally, by analyzing the eye-opening penalty varying with extinction ratio (ER) for ASK-path and DPSK/DQPSK path signal at the balance receiver end, it reveals that the optimal extinction ratio (ER) varies with the initial power; while in [4] the optimal ER is constant without consideration of the phase noise.

## ASK-DPSK and ASK-DQPSK transmitted signal

As in [4], the ASK-DPSK transmitter in Fig. 1(a) consists of a Mach-Zehnder modulator (MZM) and a phase modulator (PM) in series. The electrical drive signal $a(t)$ generated from the bit sequence $a_k$ modulates the amplitude of the optical signal from the continuous-wave laser, resulting in two different amplitudes $a$ and $b$. The electrical drive signal $b(t)$ generated from the differentially encoded bit sequence $b_k$ modulates the phase of the optical signal, such that two phase angles 0 and $\pi$ exist. The amplitude ratio (AR) $b/a$ of the signal points is related to the extinction ratio (ER) of the optical signal by $10\log(b^2/a^2)$.

The ASK-DQPSK transmitter in Fig. 1(b) consists of two MZM and one PM in series. Here, the electrical drive signal $a(t)$ generated from the bit sequence $a_k$ modulates the amplitude of the optical signal from the continuous-wave laser such that there are two amplitudes $a$ and $b$. The second MZM and the PM build up a DQPSK transmitter. The electrical drive signals generated from the differentially encoded bit sequences $b_k$ and $c_k$ modulate the phase of the optical signal such that there are four phase angles 0, $\pi/2, \pi, 3\pi/2$.

The time-domain raised cosine (RC) impulse shapes both in the ASK-DPSK and ASK-DQPSK transmitter and for all references have impulse responses

$$h(t) = \begin{cases} 1 & ,|t| \leq \frac{T}{2}(1-\alpha) \\ \cos^2\left[\frac{\pi}{4}\frac{2|t|-T(1-\alpha)}{\alpha T}\right] & ,\frac{T}{2}(1-\alpha) < |t| < \frac{T}{2}(1+\alpha) \\ 0 & ,|t| \geq \frac{T}{2}(1+\alpha) \end{cases}$$

(1)

Where $T = 1/R_s$ is the symbol duration, $\alpha$ is the roll-off factor.

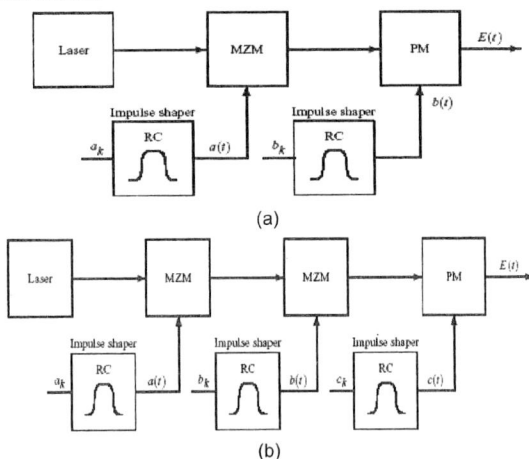

Fig.1 (a) ASK-DPSK transmitter;   (b) ASK-DQPSK transmitter.

The electrical field of the optical signal at the output of the transmitter can be written as

$$E(t) = \sqrt{P}A(t)\exp\{j[\omega_0 t + \phi(t)]\}$$

(2)

86

Where $\omega_0$ is the optical carrier frequency, $\sqrt{P}$ is the peak power of the CW laser, $A(t)$ and $\phi(t)$ stand for the modulated amplitude and phase respectively. At the time instants $t = nT, n = 0,\pm1,\pm2\cdots$ ,

$$A(nT) = A_n \in \{a,b\}$$ with $$b > a > 0$$ ,

$$\phi(nT) = \phi_n \in \{0,\pi\}$$ and

$$\phi(nT) = \phi_n \in \{0,\pi/2,\pi,3\pi/2\}$$ for DPSK and

DQPSK respectively with the symbol duration $T$. The amplitude ratio (AR) $b/a$ of the signal points is related to the extinction ratio (ER) of the optical signal by $10\log(b^2/a^2)$.

## ASK-DPSK and ASK-DQPSK received signal

In this paper, the structure for ASK-DPSK and ASK-DQPSK receivers is shown in Fig.2 [4]. In Fig. 2(a), the ASK-DPSK receiver has an ASK path with a photodiode for direct detection and a DPSK path with a delay & add filter (DAF) and a balanced detector with two photodiodes. Binary sampling & decision devices estimate the bit sequences in both paths. Similarly, the ASK-DQPSK receiver in Fig. 2(b) has an ASK path and a DQPSK path. The ASK path again uses a single photodiode and a binary sampling & decision device, whereas the DQPSK path needs two DAF, two balanced detectors and two binary sampling & decision devices. The optical receiver filters are 2nd order Gauss band-pass filters and the electrical receiver filters are 3rd order Bessel low-pass filters.

For the phase path, the resulting current at the sampling instants is

$$I(nT) = \frac{1}{2}kI_0 A_n A_{n-1}\cos(\phi_n - \phi_{n-1}) \qquad (3)$$

Where $k$ is the proportionality factor, $R$ is the responsivity of the photodiode, and $I_0 = RP$. For the estimation $\hat{b}_{P,n}$, it is just necessary to decide whether the electrical signal in (3) is positive or negative.

It is investigated that, the greater the amplitude ratios b/a, the greater the electrical eye opening in the ASK path, but the smaller the electrical eye opening in the DPSK path or DQPSK path, respectively, and vice versa [1]. In [4], $b/a$ is optimized in order to achieve minimum BEP in the presence of optical noise.

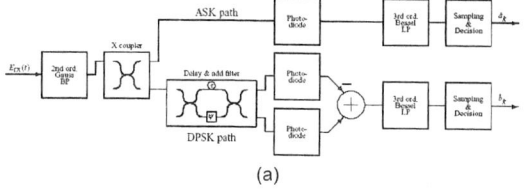

(a)

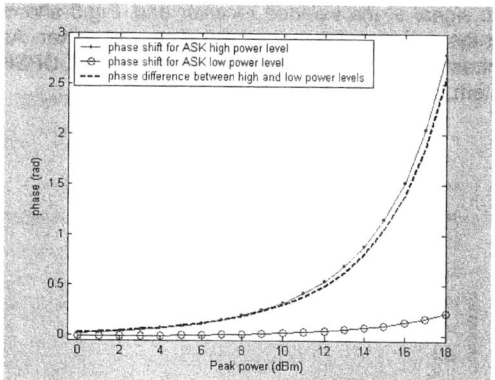

(b)

Fig.2 (a) ASK-DPSK receiver (b)ASK-DQPSK receiver

## Phase noise caused by Kerr nonlinearity in ASK-DPSK, ASK-DQPSK systems

In a single channel optical communication system, the phase noise is induced by ASK amplitude un-equality and self-phase modulation. To obtain the phase noise, the nonlinear Schrödinger equation for single channel is numerically solved using split-step Fourier method, and the chromatic dispersion of SMF is compensated by DCF. In the simulation, $b/a$ is set to be 3. The pulse shape is Gaussian. To take NRZ impulse shaping for example, the phase noise due to ASK-amplitude un-equality and SPM is shown in Fig.3. It is not hard to understand that, the higher the input power, and the higher the phase noise. In real optical communication case, the emergence of 1 and 0 bits is pseudo, so the phase difference between ASK modulation 1 and 0 bits is a kind of noise added to the phase modulation path.

Fig.3 Phase noise caused by Kerr nonlinearity and amplitude un-equality for ASK-DPSK, ASK-DQPSK signal.

It can be derived from equation (3), for ASK/DPSK, the phase noise will lead to the eye closure for DPSK path signal, and the eye-opening penalty will increase therefore. As shown in Fig.4, the received DPSK eye closure is observed when the peak power of input signal is 10dBm. It is reasonable to predict that, the bit decision error of 1 and 0 will occur at the phase demodulation path when the phase noise is over $\pi/2$. For ASK/DQPSK, the bits of two orthogonal paths will mixed up when the phase noise is over $\pi/4$.

87

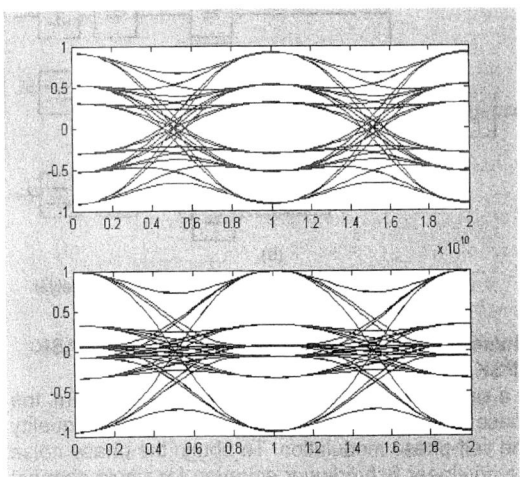

Fig. 4 Eye closure for DPSK path signal at the receiver end. Top: transmitted DPSK-path eye-diagram; Bottom: received DPSK-path eye-diagram

## Optimum AR for ASK-DPSK, ASK-DQPSK systems

Phase noise added to the phase path will cause eye-closure of DPSK and DQPSK signals. To have the minimum BEP for ASK-DPSK, ASK-DQPSK systems with NRZ impulse shaping, the optimum AR is researched. The parameters are chosen as the same in [4]. By analyzing the *eye-opening penalty varying with extinction ratio for ASK-path and DPSK/DQPSK path signal at the balance receiver end,* Fig.5 shows that the optimum AR varies with the input power. As contrast, the optimum AR in [4] is 2.5 for ASK-DPSK system, and 2.1 for ASK-DQPSK system.

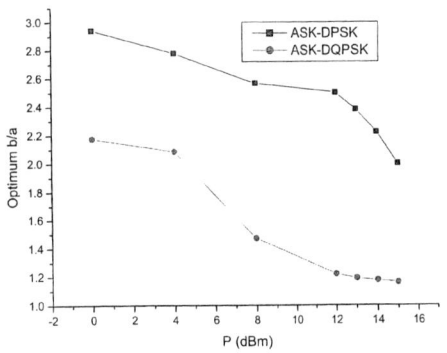

Fig.5 Optimum AR for ASK-DPSK and ASK-DQPSK system

## Conclusions

In ASK-DPSK and ASK-DQPSK optical communication systems, ASK amplitude un-equality and Kerr nonlinearity in optical fiber introduces phase noise added to the differential phase shift keying path. For the phase modulation path, the eye-diagram closure at the receiver end is observed with the impact of the phase noise. By analyzing the eye-opening penalty varying with extinction ratio for ASK-path and DPSK/DQPSK path signal at the receiver end, it reveals that the optimal extinction ratio varies with the input power. It is quite different from the reported conclusion that the optimal extinction ratio is constant without consideration of the phase noise.

## Acknowledgement

The project was supported by Open Fund of Key Laboratory of Optical Communication and Lightwave Technologies, Beijing University of Posts and Telecommunications, Ministry of Education, P.R.China

## References

1 M. Ohm et al, IEEE Photon. Technol. Lett., 15, Jan. 2003, 159–161
2 X. Liu et al, Proc. ECOC'03, Th 2.6.5
3 S. Hayase et al, Proc. ECOC'03, 2003, Th 2.6.4
4 M. Ohm et al, SPIE APOC'04, 2004, 5625-5638
5 M. Ohm et al, ITG-Fachtagung Photonische Netze, Leipzig, Germany, May 2005, 211-217
6 Moshe Nazarathy et al, JOURNAL OF LIGHTWAVE TECHNOLOGY, VOL. 24, NO. 5, MAY 2006, 2248-2260
7 Torger Tokle et al, Proc. ECOC'05, Vol. 6, Th 4.1.6

**Dr. Jurgen Michel**
**Massachusetts Institute of Technology, USA**

### Building Blocks for Electronic- Photonic Integrated Chips

The complete integration of photonic devices into a CMOS process flow will enable low cost photonic functionality within electronic circuits. To reach this goal, technologies and design tools necessary to fabricate an electronic-photonic integrated circuit have to be developed. Some of the crucial building blocks are high performance photodetectors and fiber-to-chip couplers. We will present the current status of waveguide integrated Ge p-i-n photodetectors, fabricated on a SOI platform. We will furthermore present fiber-to-waveguide couplers with less than 0.6dB coupling loss. Challenges and opportunities for CMOS integrated photonic devices will be discussed.

**Jurgen Michel** is a Principal Research Scientist in the Microphotonics Center at the Massachusetts Institute of Technology. He manages and conducts research projects in silicon-based photonic materials and devices and agglomeration issues in silicon processing. His main focus is currently on on-chip WDM devices, coupling and packaging issues in silicon photonics, and self-assembly of nano dots.

Prior to joining MIT in 1990 he was Postdoctoral Member of Technical Staff at AT&T Bell Laboratories, studying defect reactions and defect properties in semiconductor materials. He was educated in Germany and earned his diploma in Physics at the University of Cologne and his doctorate in Applied Physics at the University of Paderborn. He has authored or co-authored more than 120 scientific papers.

# CAD for Photonic Devices and Circuit

Chenglin Xu
School of Information Science and Engineering
Shandong University
Jinan, Shandong, 250100, P.R. China

## Abstract

As the development of photonics industry, there is a growing demand for accurate and efficient CAD tools for designing photonics devices and circuits. Compared with the EDA tools in R/F microwave and microelectronics, however, photonics CAD tools is still in its early stage. Following the successful paths of R/F microwave and microelectronic industries, we have developed a photonic circuit simulator. It breaks the traditional ideologies of simulation tools in the market. First of all, it is no longer method orientated and it is problem orientated. Different methods will be used for different problems. Secondly, it is in a hierarchical structure, which divided the design problem in four different layers and each layer focus on specific problems. It allows designer to decouple a big complex problem in to a number of small simpler problems. Thirdly, a lot of design knowledge is built inside the pre-defined libraries so that the new designers do not have to start from the scratch. Therefore, the demand on user's knowledge is much lower and the learning curve is much shorter.

## Introduction

In the earlier days, photonic designs mainly relied on designers' experience and based on simple back-on-envelope calculation. As the advance of photonic technologies, especially the emerging technology, such as photonic integration, however, the design problem becomes much more complicated than ever before. As a result, simple and approximate calculation cannot meet the design requirement and much more complicated and accurate design tools are needed. In the past two decades, significant efforts have been devoted to develop powerful design tools for photonics devices and circuits and a lot of simulation algorithms been invented. Among them, the most important ones are Fourier beam propagation method(BPM)[1], finite-difference vector BPM[2], coupled mode theory (CMT)[3], and finite-difference time-domain (FDTD) method[4], as well as method of line[5]. Nowadays, those methods are widely used in the design and simulation of various photonic devices and circuits and they are built in quite a few commercial design and simulation software packages, such as APSS[6], BeamProp[7], and OptiBPM[8]. However, those methods have their own advantages and disadvantages and hence can solve particular problem only. For instance, the FDTD is a powerful method and it can trace the light propagation in all direction in time-domain, whereas it is very time consuming and requires a lot of computer memory. The BPM, on the other hand, is a very efficient method, however, the backward reflection cannot be taken into account due to its unidirectional propagation. In similar fashion, most commercial software package are method oriented and can solve particular problem only. Photonics designers, however, are no longer satisfied with the conventional trial-and-error design approach, which is very costly and inefficient. They want enjoy the luxury like those designers in R/F microwave and microelectronics industries, where all the design work are done on computers through computer modeling and simulation, which provides a complete total solution. Over the past decade, we keep our research focus on developing a powerful photonic design tool to provide a complete total solution to photonic designer.

## Hierarchical model

To isolate and simplify the problems, we adopt four-level hierarchical model, Material, Waveguide, Device, and Circuit as shown in Fig. 1 by analyzing the unique characteristics of photonic devices and circuit.

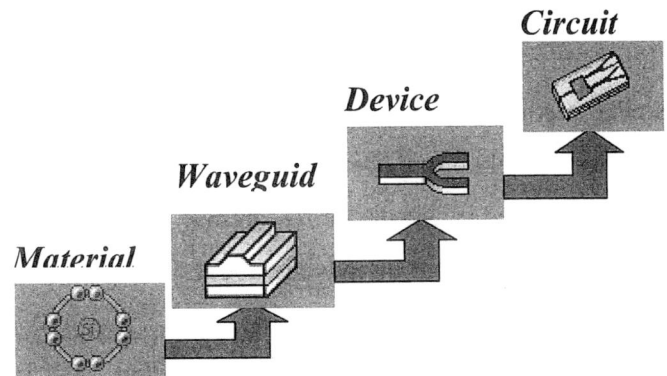

Fig. 1 Hierarchical structure of the circuit simulator

In the hierarchical structure, different problems are isolated and addressed at different levels, which are linked together seamlessly and only useful information is passed from one level to another. In the Material level, material properties, such as optical properties, thermal optical properties, electro-optical properties, material gain, etc. are addressed. In the Waveguide level, all properties related to wave guidance are investigated. In addition to its fundamental properties, which are the modal index and modal profile, as well as its near and far-field pattern, other associated waveguide properties are also investigated. Those include waveguide dispersion, polarization dependence, bending loss, and coupling loss with other waveguides. Built on the Waveguide is the Device module. The "devices" here we are talking about may be different from the devices in the real world. They are fundamental building blocks which perform basic functionalities in a photonic circuit. They can be as complicated as an arrayed waveguide grating (AWG) device, or as simple as a straight waveguide. In the Device level, the propagation behaviors of lightwave are investigated by different methods. Those properties, such as transmission and reflection, are all included in a scattering matrix of a device at different wavelength. In addition, the field pattern, which colorfully visualizes the light propagation inside a device, is also available whenever it is needed. The devices can be integrated together to form a circuit in the Circuit level, where the overall performance of the whole circuit can be evaluated.

## Embedded methods

In the academic world, researchers emphasize more on simulation methods. In the industrial world, however, engineers are more interested in solving their problems regardless of whatever method used. In the earlier days and even nowadays, most simulation and design tools, including those commercial products are method orientated since they are initially developed by academic researchers. The real serious users face a high entrance barrier due to significant knowledge about the simulation method itself is required in order to use the design tools properly.

Having taken the users' concern into consideration, we proposed a new methodology for simulation and design of photonics devices and circuits. In the new approach, we emphasize more on the problem itself rather than underline algorithm used to solve those problem. The simulation methods are embedded and they can be picked automatically based on application itself and user's interest. For instance, a waveguide junction as shown below in Fig. 2 can be simulated by the efficient BPM if the backward reflection is not user's concern. If the user is interested in the reflection, the more rigorous FDTD will be used at the expense of longer computation time and more computer memory.

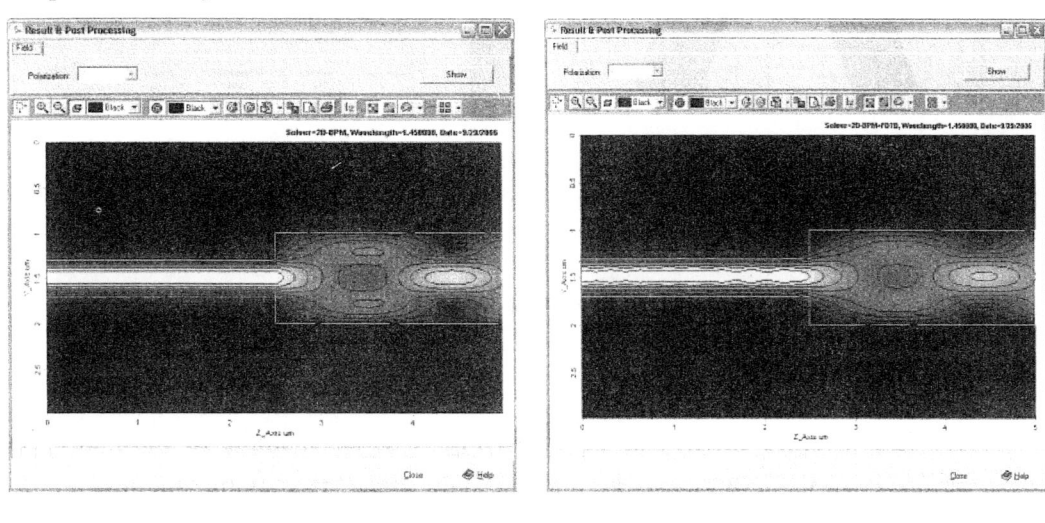

(a) without reflection by BPM          (b) with reflection by FDTD

Fig. 2 Simulated field patterns for a waveguide junction

## Built-in knowledge

Due to complexity of a problem, designing a photonic device or circuit may require all sorts of knowledge, which maybe a challenge for a new designer. Therefore, some design libraries with built-in knowledge will make a new designer's life easier to start and make an experienced designer's life simpler. We have proposed the design library idea following IC and R/F counterparts. In the cell library, we build in all available knowledge, including geometric configuration and underline simulation algorithm, and users may not need to know all the details and start from scratch. In addition, we classify all different design parameters into different categories and expose users those only under users' control. For instance, an AWG as shown in Fig. 3 (a) is a very complex structure. To design such a device, extensive knowledge is required. The layout alone requires significant effort. With the built-in knowledge, the design

93

parameters are itemized and the user only need to check the list from the table shown in Fig. 3 (b).

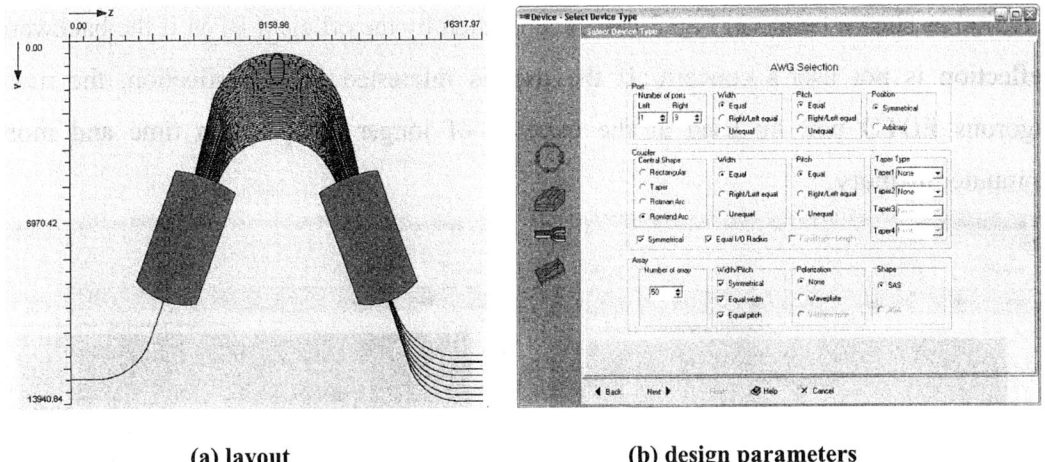

<table>
<tr><td>(a) layout</td><td>(b) design parameters</td></tr>
</table>

**Fig. 3 Built-in knowledge for an AWG device**

# Design examples

We take a lattice filter[9] as shown in Fig. 4 as an example to show how the design resource can be used repeatedly without extra effort. The whole circuit is a quite large structure considering only FDTD can be used for such structure. Significant computer money and simulation time is required for simulating such structure as a whole, especially for 3D simulation which may not be feasible in regular PCs. By examining the circuit, we identify that there is only one "device", the ring coupler, and the whole circuit can be formed by cascading a number of it together.

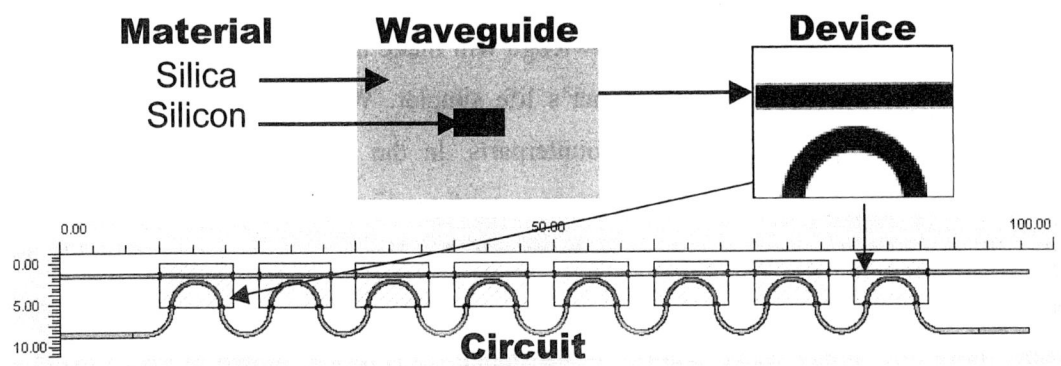

**Fig. 4 Four levels decomposition of the lattice filter**

Simulation of the ring coupler, which is only a few micrometers by a few micrometers, is a trivial task. It can be done easily even for 3D simulation on a regular

PC. Once the device results are available, a circuit can be constructed and simulated very easily. Shown in Fig. 5(a) is the simulation result and the filtering phenomenon is clearly visible. For the sake of comparison, the reported experimental result is also shown in Shown in Fig. 5(b). It is observed that the simulated result is in very good agreement with the experimental result.

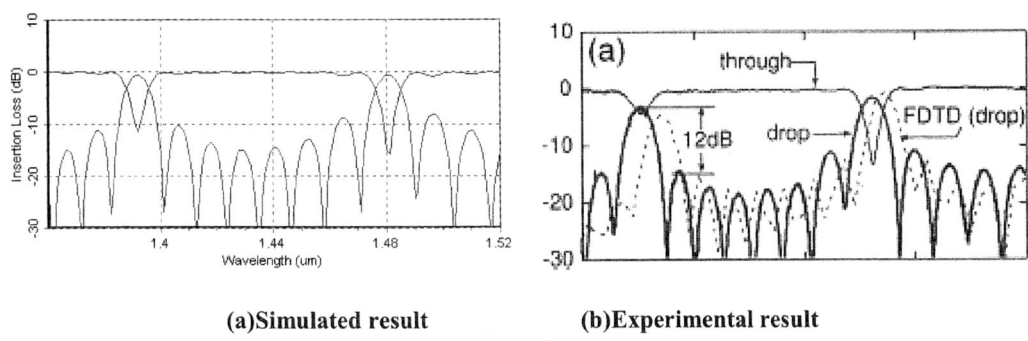

**(a)Simulated result          (b)Experimental result**

**Fig. 5 spectral response of the lattice filter**

# Acknowledgement

The author would like to thank Prof. Wei-Ping Huang at McMaster University for his guidance and collaboration during the past 17 years, and former and current Apollo's employees for their team-work with the author.

# Reference:

[1]. M. D. Feit and J. A. Fleck, Jr., "Light propagation in graded-index optical fibers," Appl. Opt., vol. 17, no. 24, pp. 3990-3998, 1978.

[2]. W. P. Huang, C. L. Xu, S. T. Chu, and S. K. Chaudhuri, "a finite difference vector beam propagation method: analysis and assessment," IEEE J. Lightwave Technol., vol. 10, no. 3, pp. 295-305, 1992.

[3]. W. -P. Huang, "Coupled-Mode Theory for Optical Waveguides: An Overview." Invited paper in J. Opt. Soc. Am. A., vol. 11, no. 3, pp. 963-983, 1994.

[4]. S. T. Chu and S. K. Chaudhuri, "a finite-difference time-domain method for the design and analysis of guided-wave optical structures," IEEE . Lightwave Technol., vol. 5, no. 12, pp. 2033-2038, 1989.

[5]. R. Pregla, "The method of lines for the unified analysis of microstrip and dielectric wave-guides," Electromagnetics, vol. 15, no. 5, pp. 441-456, 1995.

[6]. APSS, Apollo photonics solution suite, Apollo Inc., www.apollophotonics.com

[7]. BeamProp, Rsoft Design Group, www.rsoftdesign.com

[8]. OptiBPM, Optiwave Inc., www.optiwave.com          Koji Yamada, *at el,* "Silicon-wire-based ultrasmall lattice filters with wide free spectra ranges," *Optics Lett.,* vol. 28, No. 18, pp. 1663-1664, Sept. 2003.

# Recent development of ion-exchanged glass waveguide technology[1]

Hao Yinlei, Wang Minghua, Li Xihua, Yang Jianyi, Jiang Xiaoqing, Zhou Qiang

(Department of Information Science & Electronics Engineering,
Zhejiang University, Hangzhou, 310027, China)

**Abstract:** A brief review is given on the physical mechanism and the state-of-the-art of glass-based integrated optic devices by ion-exchange technology, after which an comprehensive introduction is made on our recent research work conducted on glass-based waveguide devices manufacturing, including ion-exchange technology investigation, device configuration optimization, and novel devices development, Emphasis being put on development of glass based integrated optical splitter for optical communication networks application. In the last part, a prospect is given on glass based hybrid optical waveguide, which is regarded as a promising solution to low-cost, high-performance, multifunctional integrated optic devices, for it combines the advantages of glass waveguide and functionalities of another material.

**Key words:** ion exchange, glass, waveguide.

## 1. Introduction

Glass waveguide made by ion-exchange has long been regarded as one of most promising candidate for building integrated optics devices, due to its compatibility with optical fibers, cost-effectiveness, low propagation loss, convenience of high concentration rear-earth ions doping, ease of integration into system, and environmental stability as well. Therefore, ion-exchange technology for glass waveguide fabrication have been investigated intensively since 1970s, when Izawa and Nakagome[1] demonstrated the first ion-exchanged waveguide of $Tl^+$ ions in silicate glass containing oxides of sodium and potassium.

In the last thirty years, decades of researching teams over the world show intensive interest to researching and developing glass based ion-exchange technology. Thanks to collaboration of scientists and industrialists, performance of glass based waveguide devices have been improved continuously. Up to recent years, several kinds of glass integrated optics devices made by ion-exchange have been brought out from the laboratory into the realm of practical application, integrated optical splitters for optical communication network is a case in point. At the same time, novel integrated optic devices based on ion-exchanged glass waveguide have been being devised, manufactured, and employed in various fields, optical sensing, for instance.

Up to present year, ion-exchange has grown into one of important platform

---

This work was supported by the by the Key Fund of Natural Science Foundation of China, under Grant 60436020; by Natural Science foundation of Zhejiang Province, under Grant Y105088, and the Natural Science Foundation of China, under Grant 60477018.

technology, consistent with plasma enhanced chemical vapor deposition (PECVD), for integrated optic waveguide devices.

## 2 Physical mechanism of ion exchanged waveguide

The most distinguished structural characteristic of glass is the lack of long-range regularity. An important feature attribute to the uniqueness of the glass structure is the flexibility of its chemical composition. Many properties of glass such as molar volume, thermal expansion coefficient, refractive index, etc. can be approximately described using the law of additivity of constituents, which reflect the intrinsic nature of the respective constituents in proportion to their contents.

In the glass network, there are many stable sites for network modifier ions. Alkali ions that are easily thermally activated can move from one stable site to another within glass. Such a movement of alkali ions within a glass structure enables us to replace alkali ions near the surface of a glass by other doping ions of the same valence, i.e. $Na^+$ by $Ag^+$, $Na^+$ by $K^+$, $Na^+$ by $Tl^+$, etc.

$$A^+ NO_3^- + B^+ glass^- \rightarrow A^+ glass^- + B^+ NO_3^- .$$ (1)

Where $A^+$ and $B^+$ denote doping ion (ion from ion source) and indigenous ion (ion originally in the glass), respectively.

This replacement of indigenous ion by doping ion partially modifies the composition of the glass, and hence its properties. Refractive index change due to this replacement can be estimated by HSD model[2].

$$\Delta N = \frac{\chi}{V_0} ( \Delta R - \frac{R_0 \Delta V}{V_0} ) .$$ (2)

Where $\chi$ is the fraction of indigenous ion replaced by the doping ion, $V_0$ is the volume per mole of oxygen atoms, $R_0$ is the refraction per mole of oxygen atoms, and $\Delta R$ and $\Delta V$ are the changes in these quantities as a result of ion replacement. The index change represent by equ. (2) is caused by two factors: the first term arises as a result of the difference in the ionic polarizability of the exchanging ions part, and the second term represents the contribution due to the change in the molar volume of the glass caused by the difference in ionic radii of the two ions.

Based on the glass refractive index modification mechanism stated above, channel waveguide can be manufactured, by ion exchange on glass substrate, given waveguide pattern are formed on the masking film as ion-exchange barrier. There are two kinds of ion-exchange configuration, molten salt ion-exchange[1] and metallic film ion-exchange[3], dependent on the doping ion source, the former process being used more often. For molten ion-exchange process, there are several processing configuration, as listed in [4], Of these, a sequence consisting of thermal exchange from a salt melt, with or without the presence of an applied electric field, followed by field-assisted burial and thermal annealing, has been shown to produce waveguides. Three representative ion-exchange processing configuration are given in Fig.1, including (a) thermal exchange from molten salt, (b) electric field-assisted thermal exchange from molten salt, (c) electric field-assisted burial.

Concentration profile of doping ion is critical for designing of ion-exchanged glass waveguide devices. Utilizing the second Fick Law, and invoking the continuity equation for ionic flux, the time evolution of doping ion concentration can be derived as[4,5].

$$\frac{\partial C_A}{\partial t} = \frac{D_A}{1-(1-M)C_A}\left[\nabla^2 C_A + \frac{(1-M)(\nabla C_A)^2}{1-(1-M)C_A} - \frac{q\vec{E}_{ext}\cdot\vec{\nabla}C_A}{kT}\right]. \tag{3}$$

Where $D_A$ and $D_B$ are the self diffusion coefficient of A$^+$ and B$^+$, respectively, and $M=D_A/D_B$ is their ratio. $C_A$ is the concentration of $A^+$ ions, normalized with respect to the saturated concentration, the saturated concentration depends on the composition of the glass substrate and the melt, as well as ion-exchange temperature. Its except value is generally unknown, but this problem is overcome by setting $C_A=1$ at the surface of the substrate and using a normalized value for $\Delta n_0$ that is determined experimentally. $E_{ext}$ is the applied electric field. $T$, $k$ and $q$ are the absolute temperature, Boltzmann's constant, and the electron charge, respectively.

The first term on the right hand side of (1) is the contribution to $\partial C_A / \partial t$ arising from the concentration gradient of the $A^+$ ions. The second term arises from the internal electric field due to the local distribution of dissimilar ions. When there is an externally applied field, as in the field assisted burial step, the third term representing ion drift must be included as well.

Solving (3) subject to the boundary conditions produces the $A^+$ concentration profile.

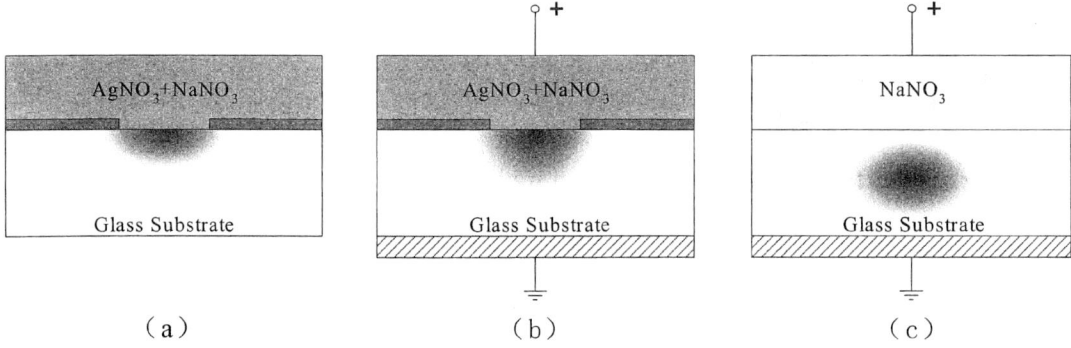

(a)          (b)          (c)

Fig. 1. Glass waveguide manufacturing configurations: (a) thermal exchange from molten salt, (b) electric field-assisted thermal exchange from molten salt, (c) electric field-assisted burial.

## 3 Development of ion-exchange glass waveguide

Over the last three decades, extensive investigation has been made on theoretical and technological problems related to ion-exchange glass waveguide manufacturing.

Theoretical investigation related to ion exchange falls into several aspects, which include: (1) Ion-exchange between glass and ion source (molten salts or metallic film)[1,3]; (2) Dependence of glass optical parameters on its composition[2]; (3) Ion diffusion mechanism in glass and its dependence on glass composition and microstructure; (4) Light guiding characteristics of graded index waveguide formed by ion diffusion; (5) Influence of ion exchange and annealing on waveguide characteristics, including mode field profile, coupling loss, and propagation loss.

In respect of technological investigation, experimental research on the following issues has been carried out. (1) Ion-exchange properties of various kinds of glass, silicate glass, phosphate glass (usually doped with rare earth ions), borate glass; (2) Various ion-exchange process, including molten-salt ion-exchange, metallic film ion-exchange, ionic masked ion-exchange, field assisted ion-exchange and annealing; (3) Various exchanging ion part: $Ag^+/Na^+$、 $Li^+/Na^+$、 $Cs^+/Na^+$、 $Rb^+/Na^+$、 $K^+/Na^+$、 $Tl^+/Na^+$、 $Cu^+/Na^+$; (4) Various kinds of waveguide, including: planer waveguide and channel waveguide, multi-mode waveguide and mono-mode waveguide, passive waveguide and active waveguide, and waveguide with grating[6], etc; (5) Glass based composite waveguide, which including: passive/active glass waveguide[7], glass/polymer waveguide, glass/oxide waveguide, glass/semiconductor waveguide.

Based on the theoretical and technological investigations stated above, more than ten kinds of waveguide configurations have been manufactured, which includes: directional coupler, Mach-Zehnder interferometer, Michelson interferometer, ring resonator, symmetric and asymmetric Y-branch, multi-mode interferometer, grating, and so on. Employing the functionality of these configurations, more than a decade of integrated optical devices has been manufactured on glass substrate, including: optical splitter, filter, multiplexer/demultiplexer (Arrayed waveguide grating, Bragg grating, etc.), polarization splitter, thermo-optical modulator, variable optical attenuator, laser and amplifier, as well as sensors such as gyroscope, displacement sensor, chemical/biological sensor[8]. In Tab. 1 configuration and application of glass are listed.

Tab. 1. Configuration and application of glass waveguide.

| Configurations | Applications |
| --- | --- |
| Direction Coupler | MUX/DMUX, Polarization Splitter |
| Grating | Filter, Coupler, Interconnect |
| Mach Zehnder Interferometer | Sensor, Modulator, VOA |
| Michelson Interferometer | Sensor |
| Ring Resonator | Filter, Sensor |
| Cascaded Y-branches | Power Splitter |
| N×N Coupler | MUX, Splitter |
| Asymmetric Y-branch | MUX/DMUX |
| Asymmetric Mach Zehnder | MUX/DMUX |
| Taper transition | Sensor |

In the rapidly overspreading optical communication network, especially the deployment of FTTH program, low-cost and high-performance optical devices, optical splitter for instance, is in great demand. This potential demand, together with the unique cost-effectiveness of ion-exchange waveguide technology, attracted continuously attention from researchers, and at the same time, inspired great passion in industrialists. Up to now, optical splitter has been industrialized and commercialized in Teemphotonics[9] and Color-chip[10], respectively, and employed in optical communication network.

In china, investigations have been being continuously conducted since late 1970's by several teams at research institute and college. Up to present years, industrialization of glass based ion-exchanged integrated optical devices is still under way.

# 4 Research at Zhejiang University

Since the year of 2002, glass waveguide technologies have been being conducted at institute of microelectronics & optoelectronics, at Zhejiang University[11-15]. And a project for industrialization of glass based optical splitter has been started up in 2004, in collaboration with Zhejiang Nanfang Communication Group. In these years, investigation on glass based waveguide manufacturing technology has been magnificently improved in several aspects.

## 4.1 Investigation on ion-exchange technology

High insertion loss is one of bottle-neck problem in glass based waveguide devices industrialization. Burying the waveguide core into underneath the glass substrate surface, by ion exchange with electric field assisted, is key to this problem. Buried waveguide are advantageous over surface waveguide in three aspects: the first one is a dramatically decreased propagation loss, for much weaker scattering at glass surface defects; the second one is a much lower coupling loss, due to a mode field matching that of optical fiber; the last one is a reduced polarization dependent loss, due to its higher symmetry of refractive index profile in vertical and horizontal directions, with respect to the glass substrate surface.

Intensive investigations for making buried optical waveguide in glass substrate have been carried out in our laboratory. Various processes for burying, including field assisted ion-exchange, field assisted annealing, ionic masked ion-exchange, has been attempted, using several kinds of optical glasses (BK7, B270, Pyrex, etc.) as substrate. We find few optical glass brand commercially available is suitable for buried waveguide manufacturing, for not optimized for ion-exchanged devices. Using specially designed glass substrate, by specially designed ion-exchange process, deeply buried waveguide has been manufactured, Fig. 2 shows microstructure of the buried waveguide cross-section under optical microscope, at transmission working mode.

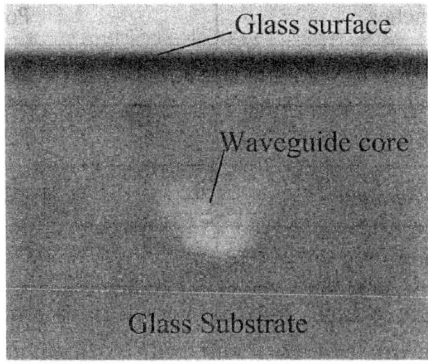

Fig. 2. Cross section of buried waveguide by ion-exchange.

## 4.2 Device configuration optimization

For convenience of pigtailing, output waveguides on optical splitter chip are usually spaced at the same pitch as that of fiber array, typically $250\mu$ m or $127\mu$ m. Ordinary cascaded Y-branch splitter has a configuration as shown in Fig. 3 (left), in which branching waveguide is leveled before being branched further. This requires a large devices length, especially for the case of $1\times16$ and $1\times32$ branching. To make device more compact, and hence optimize device performance, we applied another cascaded branching method shown in Fig. 3 (right), in which only the output waveguides are leveled with respect to the input waveguide, while other branching waveguide are tilted with respect to the input waveguide. By adopting this method, device length can be significantly reduced, while the maximum radius of curvature keeps the same as that in the ordinary method. For example, a device length of 32 mm, 84% the length of ordinary cascaded Y branching splitter, could be realized for a $1\times16$ splitter[11,12].

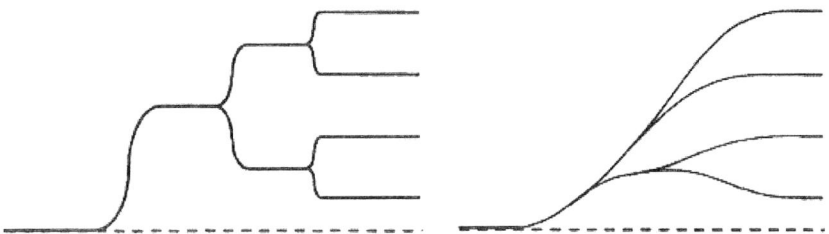

Fig. 3. Geometry of ordinary (left) and new designed (right) cascade splitter.

Another improvement in device configuration is truncated Y-branching structure, as shown in Fig. 4, which avoid dispersion of sharp angle formation in photolithography and mask patterning steps, hence improves uniformity of Y-branching shape.

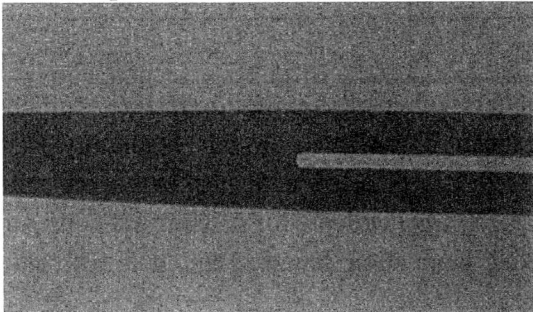

Fig. 4. Truncated Y-branching structure.

## 4.3 Optical devices manufacturing

Using ion–exchange developed in our laboratory, and the optimization method stated above, cascaded Y-branching $1\times4$, $1\times8$, $1\times16$, $1\times32$ splitters have been manufactured[13], parameters being listed in Tab. 2. Key parameters, insertion loss and polarization dependent loss, of these splitters are approaching the corresponding level of products from Teemphotonics and Color-chip. We are now doing further experiment for industrialization of these splitters. Fig. 5 is pigtailed optical splitter manufactured in our laboratory.

Fig. 5. Glass based power splitter by ion-exchange technology.

Tab. 2. Parameters of ion-exchanged glass splitters.

| Parameters | | Splitters | | | |
|---|---|---|---|---|---|
| | | 1×4 | 1×8 | 1×16 | 1×32 |
| Operating Wavelength (nm) | | 1260-1360 & 1520~1570 | | | |
| Insertion Loss (dB) | Maximum | 8 | 12 | 16 | 20 |
| | Optimum | 7.7 | 11 | 14.5 | 17.5 |
| Uniformity (dB) | Maximum | 0.8 | 1.0 | 1.5 | 2.0 |
| | Optimum | 0.2 | 0.7 | 0.6 | 1.6 |
| PDL (dB) | Maximum | 0.1 | 0.1 | 0.1 | 0.15 |
| | Optimum | 0.07 | 0.05 | 0.06 | 0.1 |
| Return Loss (dB) | | >55 | | | |

Besides optical splitters, other glass based optical waveguide devices manufactured by ion-exchange have been investigated in our laboratory[14,15]. A typical one is Multimode interference (MMI) coupler, a device based on self-imaging effect. MMI coupler is attractive for many applications, for its compact configuration, low excess loss, wide band-width, relaxed fabrication requirement, and realization of multi-input and multi-output.

MMI based devices are usually strongly confined, which made the phase error higher for larger order modes in the case of large-port-count, and in turn degrades imaging quality of MMI devices. Our investigation suggest that weakly confined MMI coupler split light with higher uniformity, and is suitable for multi-channel devices. Moreover, optimized performance, splitting smoothness for instance, could be achieved, utilizing graded index waveguides with suitable refractive index profile.

By employing the ion-exchange on glass substrate, 1×8 and 8×8 weakly confined MMI devices have been manufactured. Splitting uniformity as low as 0.27dB is obtained for the access waveguide at the centre of input plane, see Fig. 6. In contrast, the uniformity degrades, in the case of access guide close to edges of input plane. Further study is under way to improve this uniformity.

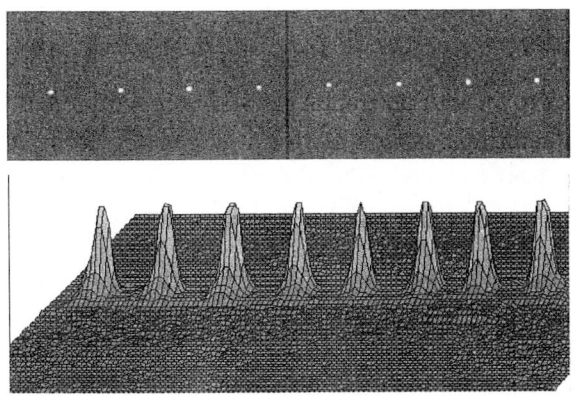

Fig. 6. Near-field pattern of is glass substrate weak-confined Multimode Interference 8×8 coupler, and the three-dimension field intensity.

## 5 Prospective review for glass based ion-exchange technology

Glass ion-exchange process has grown into an important and competitive technology, for integrated optical devices manufacturing, thanks to investigation of hundreds of scientists in last thirty years. Since ion-exchange is promising for low-cost waveguide devices manufacturing, many scientists are continuously showing intensive interest toward application of devices based on this technology, to develop novel devices with high performance, or create new applications for these devices. In recent years, glass based composite waveguides have become one of focus of many research teams, and show potentiality of being developed into an important branch of glass waveguide.

Integrated optical devices are being developed toward the destination of more functionalities, higher integrated density, higher performance, smaller dimensions, and lower cost. These requirements cannot be fulfilled simultaneously, using single material for device manufacturing. Glass is a kind of excellent optical material, but suffered from inherent deficiency of monotonic functionality, as integrated optic material. By hybrid integration, making composite waveguide configuration using two or more kinds of material of different properties, High performance devices might be realized, for taking advantage of properties of different material.

Investigation have been conducted on glass based composite integrated optical devices in the past years, for the aim of developing novel integrated optical devices using this low-cost technology. Several kinds of composite waveguide have been manufactured, including rare earth doped/undoped glass[7,16], polymer/glass[17,18], III - V semiconductor/glass[19], $TiO_2$/glass[20], $SiO_2$/glass[20], ZnS/Glass[21], etc. These composite waveguides are attractive for the combining the advantage of the two kinds of materials. On the one hand, ion-exchanged glass waveguide possesses low-cost, high thermal and mechanical stability, as well as easy coupling to fibers, hence an ideal platform; on the other hand, the other material have desirable birefringence properties, (electro-optic properties, thermo-optical properties, magneto-optical properties, acousto-optical properties, optical non-linearity, etc.), or can be used to make light source, optical detector, optical amplifier, which give the device special functionality. Combination of glass and the other functional material gives more functionality to integrated optical devices, which is desired for many applications.

# 6 Concluding remarks

Ion-exchange has grown into an important and competitive technology for integrated optical devices manufacturing, especially for low-cost devices. At the same time, by developing novel devices based on this technology, more and more high-performance optical devices could be devised and put into application.

# References

[1] T. Izawa, H. Nakagome, Optical waveguides formed by electrically induced migration of ions in glass plates[J], Appl. Phys. Lett., 1972, 21(12):584-586.

[2] S. D. Fantone. Refractive index and spectral model for gradient-index materials [J]. Appl. Opt., 1983, 22(3): 432-440.

[3] A. Belkhir, Detailed study of silver metallic film diffusion in a soda-lime glass substrate for optical waveguide fabrication[J], Appl. Opt., 2002, 41(15):2888-2893.

[4] A. Tervonen, A general model for fabrication processes of channel waveguides by ion exchange[J], J. Appl. Phys., 1990, 67(6):2746-2752.

[5] B. R. West, P. Madasamy, N. Peyghambarian, S. Honkanen, Modeling of ion-exchanged glass waveguide structure[J], J. Non-cryst. Solid., 2004, 347: 18-26.

[6] S. Blaizea, L. Bastarda, C. Cassabnees, et al., Ion-exchanged glass DFB lasers for DWDM, www.teemphotonics.com/documents/ technical_articles/2002_11.pdf.

[7] Pratheepan Madasamy, S. Honkanen, D. F. Geraghty, and N. Peyghambarian, Single-mode tapered waveguide laser in Er-doped glass with multimode-diode pumping[J], Appl. Phys. Lett., 2003, 82(9):1332-1334.

[8] I. M Hoffmann, Integrated optics for communication networks, http://www.hft.e-technik. uni-dortmund.de /en/forschung/int_op/index.html.

[9] http://www.teemphotonics.com/documents/product/splitter/1xNPlanarSplitter.pdf.

[10] http://www.color-chip.com/ColorChipSplitter.pdf.

[11] Shanying Lu, Xihua Li, Xiaoqing Jiang Minghua Wang, Compact design of 1×16 cascaded Y-branch splitter[J], Chinese Optics Letters, Vol.2, No.8, pp.459-461, 2004.

[12] Lu Shanying, Jiang Xiaoqing, Wang Minghua, Compact design of planar 1×N branching splitter, SPIE 5279, 2004, 56-61..

[13] Huilian Ma, Minghua Wang, Xihua Li, Yigang Xu, Nanxin Song, Study and fabrication of 1×8 MMI optical power splitter in glass material[J], Acta Photonica Sinca, 2002, 31(5):580-583.

[14] Yin R, Jiang X Q, Yang J Y and Wang M H Structure with improved self-imaging in its graded-index multimode interference region[J], J. Opt. Soc. Am. B, 2002, 19, 1301-1303.

[15] X. Li, X. H. Li, X. Q. Jiang, et al., Fabrication of an 8×8 Multimode Interference Optical Coupler on a BK7 Glass Substrate by Ion-Exchange[J], Chinese Phys. Lett., 2004, 21(8):1556-1557.

[16] S. D. Conzone, J. S. Hayden, D. S. Funk, A. Roshko, D. L. Veasey, Hybrid glass substrates for waveguide device manufacture[J], Opt. Lett., 2001, 26(8): 509-511.

[17] M. Alshikh Khalil, G. Vitrant, P. Raimond, P. A. Chollet, F. Kajzar, Optical parametric amplification in composite polymer ion exchanged planar waveguide[J], Appl. Phys. Lett., 2000, 77(23):3713-371.

[18] Alain Morand, Celia Sanchez-P`erez, Pierre Benech, Sma¨ıl Tedjini, Dominique Bosc, Integrated Optical Waveguide Polarizer on Glass with a Birefringent Polymer Overlay[J], IEEE J. Photon. Tech. Lett., 1998, 10(11): 1599-1601.

[19] Matthieu Nannini, Etienne Grondin, Arnaud Gorin, Vincent Aimez, and Jean-Emmanuel Broquin, Hybridization of III-V Semiconductor Membranes Onto Ion-Exchanged Waveguides[J], IEEE J. OF Select. Top. Quantum Electron., 2005, 11(2):547-554.

[20] Abliz Yimit, Axel G. Rossberg, Takashi Amemiya, Kiminori Itoh, Thin film composite optical waveguides for sensor applications: a review[J], Talanta 2005, 65:1102–1109.

[21] I. S. Mauchline, G. Stewart, Glass integrated-optic channel-dropping filter[J], Opt. Lett.,1993, 18(7):500-502.

# Guided Modes in a Slab Waveguide with an Anisotropic Dispersive Plasmonic Core

Guoan Zheng

Department of Electronic and Optical Engineering, Zhejiang University, HangZhou, China, 310027

E-mail: guoanzheng@msn.com

**Abstract** *Light transmission along a dielectric waveguide with an anisotropic dispersive plasmonic core is investigated and the subwavelength guidance characteristic is given. It is found that the different orientations of the optical axis of the anisotropic plasmonic core will lead to different dispersion relation curves of the guided modes.*

## Introduction

Light transmissions through subwavelength device have become the focus of extensive study [1]. Such attention is related to the numerous potential nanophotonic applications, e.g. optical data storage, near field optical microscopy and bio-photonics [2]. The fast/slow, forward/backward and TM/TE mode characteristics of a plasma slab have been briefly reported [3, 4]. The dispersion relations of surface plasmons were investigated for single and multiple layered metal films including a metal-dielectric-metal structure [5]. The guided modes and associated attenuation characteristics of metal-insulator-dielectric slab waveguides have also been reported [6, 7]. In the past few years, increasing interests have been devoted to the study of the surface plasmon polariton modes, i.e., the electromagnetic wave confined to the interface between materials with positive and negative permittivity [8-11]. More recently, the guided modes in the slab waveguide with a left-handed material is studied in [12-15]. However, previous work on the fundamental guided dispersion characteristics is still incomplete. In particular, discussions related to the anisotropic of the plasmonic media are insufficient, which can be significant design guidelines for subwavelength guidance.

In this paper, we investigated the dispersion relation of guided modes in a slab waveguide with an anisotropic plasmonic core whose permittivity tensors is partially negative [14]. It is shown that different orientation of the optical axis of the anisotropic plasmonic core will lead to different dispersion curves of the guided modes. In addition, if the orientation of the optical axis is chosen appropriately, the group velocity of some guided modes can approach zero. Our result may have some potential application in compact nanophotonic devices.

## Determination of the guided modes

We first consider the dielectric constant of a uniaxial media as follow (the media we consider here is non-magnetic, e.g. $\mu = 1$)

$$\overline{\overline{\varepsilon}} = \begin{pmatrix} \varepsilon_\perp & 0 & 0 \\ 0 & \varepsilon_\perp & 0 \\ 0 & 0 & \varepsilon_\parallel \end{pmatrix} \qquad (1)$$

The tensor is given in principle coordinate (x-y-z system). The optical axis is along the z direction with dielectric constant $\varepsilon_\parallel$. For a conventional uniaxial media, $\varepsilon_\perp$ and $\varepsilon_\parallel$ are both positive. However, $\varepsilon_\parallel$ can be negative and obey the frequency dependence of the well-known Drude model, or the plasmonic model (note that $\varepsilon_\perp$ is a positive constant) [16]. The idea to achieve $\varepsilon_\parallel < 0$ in optical frequency is realized in Ref. [17, 18]. To just give one example [13, 16], for the composite of 10% of SiC nanospheroids with an aspect ratio of 1/2, aligned with their shorter axis along the x axis and embedded in quartz, for a wavelength of $CO_2$ laser of 12 $\mu m$, we obtain

$$\varepsilon_\parallel = -2.7 + 6 \times 10^{-4} i, \varepsilon_\perp = 1.6 + 1 \times 10^{-5} i.$$

We consider a slab waveguide with a partial negative permittivity core, which is shown in Fig.1. The thickness of the core is $d$ and the cladding layers are air. The angle between the optical axis of the core and the z axis is $\theta$ (note that the optical axis is in plane x-z).

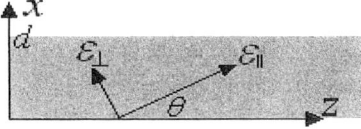

*Fig.1 Slab waveguide of thickness d. The cladding layers are air.*

All the modes in Fig.1 have TM polarization with the magnetic field perpendicular to the wave propagation direction. In order to guide the field in the core, the field in the dielectric must be evanescent in the x direction. We write the field solutions for the TM modes as follow [12]

$$\vec{H}_1 = \vec{y} A_1 e^{\alpha x} e^{ik_z z} \quad x < 0 \qquad (2a)$$

$$\vec{H}_2 = \vec{y}[A e^{ik_{2x}^i x} + B e^{ik_{2x}^r x}] e^{ik_z z} \quad 0 \le x \le d \qquad (2b)$$

$$\vec{H}_3 = \vec{y} A_3 e^{-\alpha x} e^{ik_z z} \quad x > d \qquad (2c)$$

106

The subscript 1,2 and 3 in Eq.(2) stand for lower cladding, core and upper cladding respectively.

Based on Ampere's law, we can obtain the electric displacement D as follows:

$$D_{1x} = \frac{k_z}{\omega} A_1 e^{\alpha x} e^{ik_z z} \ ,$$

$$D_{1z} = -\frac{\alpha}{\omega i} A_1 e^{\alpha x} e^{ik_z z} \tag{3a}$$

$$D_{2x} = \frac{k_z}{\omega}(A e^{ik_{2x}^i x} + B e^{ik_{2x}^r x}) e^{ik_z z} \ ,$$

$$D_{2z} = -\frac{1}{\omega}(A k_{2x}^i e^{ik_{2x}^i x} + B k_{2x}^r e^{ik_{2x}^r x}) e^{ik_z z} \tag{3b}$$

$$D_{3x} = \frac{k_z}{\omega} A_3 e^{-\alpha x} e^{ik_z z} \ ,$$

$$D_{3z} = \frac{\alpha}{\omega i} A_3 e^{-\alpha x} e^{ik_z z} \tag{3c}$$

The permittivity tensor of the core can be expressed as

$$\overline{\overline{\varepsilon}} = \begin{pmatrix} \varepsilon_\perp \cos^2\theta + \varepsilon_\parallel \sin^2\theta & 0 & -\sin\theta\cos\theta(\varepsilon_\perp - \varepsilon_\parallel) \\ 0 & \varepsilon_\perp & 0 \\ -\sin\theta\cos\theta(\varepsilon_\perp - \varepsilon_\parallel) & 0 & \varepsilon_\perp \sin^2\theta + \varepsilon_\parallel \cos^2\theta \end{pmatrix} \tag{4}$$

Applying $\overline{D} = \overline{\overline{\varepsilon}} \overline{E}$, we can get the z component of electric field ($E_{2z}$) as bellow

$$E_{2z} = D_{2x} \sin\theta \cos\theta (\frac{1}{\varepsilon_\parallel} - \frac{1}{\varepsilon_\perp})$$
$$+ D_{2z}(\frac{\cos^2\theta}{\varepsilon_\parallel} + \frac{\sin^2\theta}{\varepsilon_\perp}) \tag{5}$$

To find the guidance condition, we match the boundary conditions at x=0 and x=d to obtain

$$A_1 = A + B \tag{6a}$$

$$A e^{ik_{2x}^i d} + B e^{ik_{2x}^r d} = A_3 e^{-\alpha d} \tag{6b}$$

$$(A + B)k_z \sin\theta \cos\theta (\frac{1}{\varepsilon_\parallel} - \frac{1}{\varepsilon_\perp})$$
$$-(k_{2x}^i A + k_{2x}^r B)(\frac{\cos^2\theta}{\varepsilon_\parallel} + \frac{\sin^2\theta}{\varepsilon_\perp}) = -\frac{\alpha}{i\varepsilon_1} A_1 \tag{6c}$$

$$(A e^{ik_{2x}^i x} + B e^{ik_{2x}^r x})k_z \sin\theta \cos\theta(\frac{1}{\varepsilon_\parallel} - \frac{1}{\varepsilon_\perp})$$
$$-(k_{2x}^i A e^{ik_{2x}^i x} + k_{2x}^r B e^{ik_{2x}^r x})(\frac{\cos^2\theta}{\varepsilon_\parallel} + \frac{\sin^2\theta}{\varepsilon_\perp}) = \frac{\alpha}{i\varepsilon_1} A_3 e^{-\alpha d} \tag{6d}$$

The dispersion equation for the core can be expressed as

$$k_x^2(\varepsilon_\perp \cos^2\theta + \varepsilon_\parallel \sin^2\theta) + k_z^2(\varepsilon_\parallel \cos^2\theta + \varepsilon_\perp \sin^2\theta)$$
$$+ 2\sin\theta\cos\theta k_x k_z(\varepsilon_\parallel - \varepsilon_\perp) - \varepsilon_\parallel \varepsilon_\perp (\omega/c)^2 = 0 \tag{7}$$

where c is the speed of light in vaccum. For a given $k_z$, there are two solution for $k_x$, which are $k_{2x}^i, k_{2x}^r$.

We normalize A=1. From Eq. (6a) and (6c), we can solve for A1 and B, and from Eq. (6b) we can solve for A3. The Eq. (6d) can be regarded as the guidance condition. For a given $k_z$, we can get $k_{2x}^i$ and $k_{2x}^r$ from Eq. (7). If $k_{2x}^i, k_{2x}^r$ satisfy Eq.(6d), the guided mode exist in the present waveguide. The dispersion relation of the guided modes will be discussed in detail soon.

**Dispersion relation of the guided modes**

Following Ref. [16], we assume $\varepsilon_\parallel$ obey a plasmonic behavior as below

$$\varepsilon_\parallel = 1 - (\frac{\omega_p}{\omega})^2 \tag{8}$$

where $\omega_p$ is the bulk plasmas frequency of the metal. This dielectric function takes into account the contribution of free electrons only. Despite its apparent simplicity, the plasmonic model has been the source of valuable insights into the general behavior of real metals since the additional Lorentzian resonance effect was not added at some optical frequency [19]. In the perpendicular direction, we assume simply an insulator with $\varepsilon_\perp = 1.6$ [13, 16].

The dispersion relation curves of different orientation of optical axis of the anisotropic plasmonic core are shown in Fig.2. The thickness d in our calculation is $0.1\lambda_p$ ( $\lambda_p = 2\pi c / \omega_p$ ). The forward modes exist when $\theta = 0°$ as shown in Fig.2 (a). In Fig.2 (d), the backward and forward modes coexist in some frequency region with $\theta = 90°$ ,so we can use a prism to sperate these two type of modes [15]. Some interesting phenomena happen when $\theta = 40°$ as shown in Fig.2 (b). As the frequency increases, the dispersion curves show a transition from the backward modes to the forward modes and thus we can controal the propagation direction of the wave by change the frequency. At the frequency of $\omega = 0.55\omega_p$, the slope of the dispersion curves which stand for the group velocity of the guided modes approaches zero. This slow propagation of light can increase the efficiency of nonlinear processes which are at the root of all-optical interaction [19].

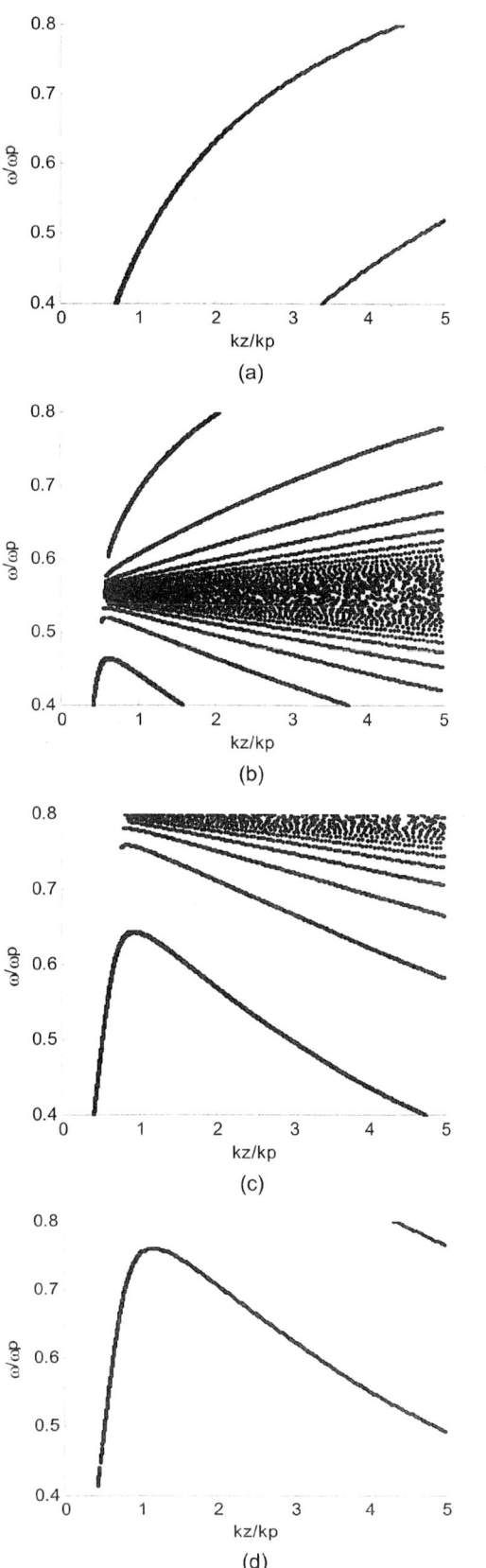

## Conclusions

The guided modes in the slab waveguide with an anisotropic dispersive plasmonic core are studied. It is shown that the different orientation of the optical axis of the core leads to different dispersion curves. The abnormal property of these dispersion relation curves is discussed. Since the modern advanced nanoscale fabrication techniques allow us to engineer very compact nanophotonic devices, fundamental research results here are expected to be applied to the design guidelines in other nanophotonic or subwavelength device applications.

## Acknowledgement

This work was supported by the National Basic Research Program (973) of China (No. 2004CB19800).

## References

1 Focus Issue, Opt. Express, 12, 3618-3706 (2004)

2 W. L. Branes et al, Nature, 424, 824-830 (2003)

3 C. Davis et al, J. Appl. Phys. 37, 461-462 (1966)

4 A.A.Oliner et al, J. Appl. Phys. 33, 231-233(1962)

5 E.N. Econnomou, Phys. Rev. 182, 539-554 (1969)

6 T.Takano et al, IEEE J. Quantum Elec. 8, 206-212 (1972)

7 K. Nosu et al, IEEE J. Quantum Elec. 12, 745-748 (1976)

8 D.K. Qing et al, Phys. Rev. B 71, 153107 (2005)

9 R.Zia et al J. Opt. Soc. Am. A 21, 2442-2446 (2004)

10 K. Tananka et al, Appl. Phys. Lett. 82, 1158-1160 (2003)

11 B. Wang et al, Appl. Phys. Lett. 85, 3599-3601 (2004)

12 Bae-lan Wu et al, Journal of Applied Physics, 93, No.11, June 2003.

13 Viktor.A. et al, Phys. Rev. B, 71, 201101 (2005)

14 Qiang Cheng et al, Appl. Phys. Lett. 87, 174102 (2005)

15 J.L.He et al, IEEE Microwave and Wireless Comp. Lett. vol.16, Feb 2006

16 O.Levy et al, Phys. Rev. B, 56, 8035 (1997)

17 D.R.Smith et al, Phys Rev. Lett. 84, 4184 (2000)

18 C.G.Parazzoli et al, Phys Rev. Lett. 90, 107401 (2003)

19 Ki Young Kim et al, Opt. Express, 14, 320-330 (2006)

*Fig.2 The dispersion curves with (a) $\theta = 0°$,(b) $\theta = 40°$, (c) $\theta = 60°$,(d) $\theta = 90°$*

# Transmission Property of the Nonmagnetic Media with a Hyperbolic Dispersion Relation

Guoan Zheng, Ke Chen

Department of Electronic and Optical Engineering, Zhejiang University, HangZhou, China, 310027

E-mail: guoanzheng@msn.com

**Abstract** *The transmission property of the nonmagnetic media with a hyperbolic dispersion relation is investigated. It is shown that the high directive transmission can be observed even in the presence of the material loss. Since the media we discussed can be fabricated in optical frequency, our result may find some application in novel optical devices.*

## Introduction

In the past years, left-handed material (LHM), which possesses simultaneous negative permittivity and permeability, has attracted much attention because of its exotic electromagnetic properties and the potential applications [1-4]. It is shown that the evanescent wave can be amplified inside the LHM and can be used to construct a "perfect" lens, theoretically capable to focus not only the propagating waves but also the near-field evanescent fields [4]. Recently, Ref. [5] proposed to use a waveguide system filled with an electrically anisotropic core whose permittivity tensor is partially negative to achieve "similar" properties of LHM, and the left-handed behaviors of some guided modes, in which including the enhancement of evanescent field, are observed in the waveguide system. Different from considering the guidance issue, however, in this paper, we investigate the transmission property to a similar structure also consisting of an electrically anisotropic slab with a partially negative permittivity tensor. We show that the high directive transmission property can be observed even in the presence of material loss. Since the anisotropic media discussed here can be fabricated in GHz, near- and mid-infrared frequencies, our result may have some potential applications in various novel optical devices for different frequency bands.

## Boundary conditions of the non-magnetic slab

We first consider an electrically uniaxial media with a scalar permeability $\mu = 1$ and a permittivity tensor

$$\overline{\overline{\varepsilon}} = \begin{pmatrix} \varepsilon_\perp & 0 & 0 \\ 0 & \varepsilon_\perp & 0 \\ 0 & 0 & \varepsilon_\parallel \end{pmatrix} \tag{1}$$

, which is given in the principle coordinates, say x-y-z system. The optical axis is assumed to along z direction with a dielectric constant $\varepsilon_\parallel$. For a conventional uniaxial media, $\varepsilon_\perp$ and $\varepsilon_\parallel$ are both

positive, but here we consider a special case that $\varepsilon_\parallel$ is negative while $\varepsilon_\perp$ is kept positive. In GHz frequencies, the idea to achieve $\varepsilon_\parallel < 0$ can be realized by fabricating a composite of periodically arranged metallic thin wires aligned along the optical axis [6, 7], and in optical, near- and mid-infrared frequencies, it can be respectively achieved by utilizing various techniques described in Ref. [5, 8], for example, for the composite of 10% of SiC nano-spheroids with an aspect ratio of 1/2, aligned with their shorter axis along the optical axis and embedded in quartz, we can obtain $\varepsilon_\parallel = -2.7 + 6\times10^{-4}i$, $\varepsilon_\perp = 1.6 + 1\times10^{-5}i$ for a wavelength of $CO_2$ laser of 12 $\mu$m.

First, we consider an anisotropic slab with thickness d shown in Fig.1, where the angle between the optical axis of the slab and the z-axis is $\theta$ (note that the optical axis is in the x-z plane).

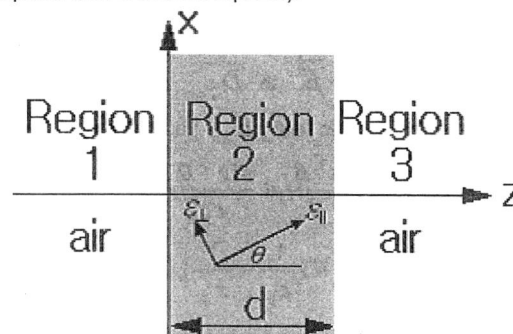

*Fig1. An anisotropic slab with $\varepsilon_\parallel$ <0 and $\varepsilon_\perp$ >0 .*

We consider an oblique propagation of monochromatic electromagnetic field (with the time dependence $e^{-i\omega t}$) in the structure with an oblique wave vector $k_x$ along the x-axis. For the H-polarization case, we have

$$H_{1y} = [e^{ik_1 z} + R e^{-ik_1 z}]e^{ik_x x} \tag{2a}$$

$$H_{2y} = [B e^{ik_2^1 z} + C e^{ik_2^2 z}]e^{ik_x x} \tag{2b}$$

$$H_{3y} = T e^{ik_1 z} e^{ik_x x} \tag{2c}$$

and

$$E_{1x} = \frac{k_1}{\omega \varepsilon_1}(e^{ik_1 z} - R e^{-ik_1 z})e^{ik_x x} \tag{3a}$$

$$E_{1z} = -\frac{k_x}{\omega \varepsilon_1}(e^{ik_1 z} + R e^{-ik_1 z})e^{ik_x x} \tag{3b}$$

$$D_{2x} = \frac{1}{\omega}(B k_2^i e^{ik_2^i z} + C k_2^r e^{ik_2^r z})e^{ik_x x} \tag{3c}$$

$$D_{2z} = -\frac{k_x}{\omega}(B e^{ik_2^i z} + C e^{ik_2^i z})e^{ik_x x} \tag{3d}$$

$$E_{1x} = \frac{k_1}{\omega \varepsilon_1}(T e^{ik_1 z})e^{ik_x x} \tag{3e}$$

$$E_{1z} = -\frac{k_x}{\omega \varepsilon_1}(T e^{ik_1 z})e^{ik_x x} \tag{3f}$$

The dispersion relations for the air and the anisotropic slab are

$$k_x^2 + k_{1z}^2 = (\omega/c)^2 \tag{4a}$$

and

$$k_x^2(\varepsilon_\perp \cos^2\theta + \varepsilon_\parallel \sin^2\theta) + k_{2z}^2(\varepsilon_\parallel \cos^2\theta + \varepsilon_\perp \sin^2\theta)$$
$$+2\sin\theta\cos\theta k_x k_{2z}(\varepsilon_\parallel - \varepsilon_\perp) - \varepsilon_\parallel \varepsilon_\perp(\omega/c)^2 = 0 \tag{4b}$$

, respectively, where c is the speed of light in air. For a given $k_x$, there are two solutions for $k_{1z}$, i.e., $k_1$, $-k_1$, and two solutions for $k_{2z}$, say $k_2^i$, $k_2^r$.

The permittivity tensor of the anisotropic media can be described by

$$\overline{\overline{\varepsilon}} = \begin{pmatrix} \varepsilon_\perp \cos^2\theta + \varepsilon_\parallel \sin^2\theta & 0 & -\sin\theta\cos\theta(\varepsilon_\perp - \varepsilon_\parallel) \\ 0 & \varepsilon_\perp & 0 \\ -\sin\theta\cos\theta(\varepsilon_\perp - \varepsilon_\parallel) & 0 & \varepsilon_\perp \sin^2\theta + \varepsilon_\parallel \cos^2\theta \end{pmatrix} \tag{5}$$

Applying $\overline{D} = \overline{\overline{\varepsilon}} \cdot \overline{E}$ to $\overline{D_2}$, we get the x component of electric field ($E_{2x}$) as bellow

$$E_{2x} = D_{2x}(\frac{\cos^2\theta}{\varepsilon_\perp} + \frac{\sin^2\theta}{\varepsilon_\parallel})$$
$$+ D_{2z}\sin\theta\cos\theta(\frac{1}{\varepsilon_\parallel} - \frac{1}{\varepsilon_\perp}) \tag{6}$$

The tangential electric and magnetic fields should be continuous at z=0 and z=d, i.e.,

$$H_{1y} = H_{2y}, E_{1x} = E_{2x} (z = 0) \tag{7a}$$

$$H_{2y} = H_{3y}, E_{2x} = E_{3x} (z = d) \tag{7b}$$

From Eq. (7), we have

$$R + 1 = B + C \tag{8a}$$

$$\frac{k_1}{\varepsilon_1}(1 - R) = (B k_2^i + C k_2^r)(\frac{\cos^2\theta}{\varepsilon_\perp} + \frac{\sin^2\theta}{\varepsilon_\parallel})$$
$$-k_x(B + C)(\frac{\sin\theta\cos\theta}{\varepsilon_\perp} - \frac{\sin\theta\cos\theta}{\varepsilon_\parallel}) \tag{8b}$$

$$B \cdot e^{ik_2^i d} + C \cdot e^{ik_2^r d} = T \cdot e^{ik_1 d} \tag{8c}$$

$$\frac{k_1}{\varepsilon_1}(T e^{ik_1 d}) = (B k_2^i e^{ik_2^i d} + C k_2^r e^{ik_2^r d})(\frac{\cos^2\theta}{\varepsilon_\perp} + \frac{\sin^2\theta}{\varepsilon_\parallel})$$
$$-k_x(B e^{ik_2^i d} + C e^{ik_2^r d})(\frac{\sin\theta\cos\theta}{\varepsilon_\perp} - \frac{\sin\theta\cos\theta}{\varepsilon_\parallel}) \tag{8d}$$

We can solve for R, T, B and C using Eq. (8a-8d).

**High directive transmission property**

Following Ref. [5, 8], we assume that $\varepsilon_\parallel$ obey the Drude model as below

$$\varepsilon_\parallel = 1 - \omega_p^2 / [\omega(\omega + i\gamma_e)] \tag{9}$$

where $\omega_p$ is the plasmas frequency, and simply assume $\varepsilon_\perp = 1.6 + 1 \times 10^{-5} i$ in accordance with the fabrication example we give before.

In Fig.2, we plot the magnitude of transmission coefficient T as a function of $k_x$ (normalized by $k_0 = \omega/c$) and $\omega$ (normalize by $\omega_p$) with different material loss $\gamma_e$. From this figure, we clearly see that the oblique incident wave can not propagate through the slab if $\omega$ is close to $\omega_p$, since the magnitude of transmission coefficient T is close to zero when $k_x \neq 0$, in addition, the smaller the material loss, the higher of the directivity of the transmission.

Magnitude of Transimission Coefficient

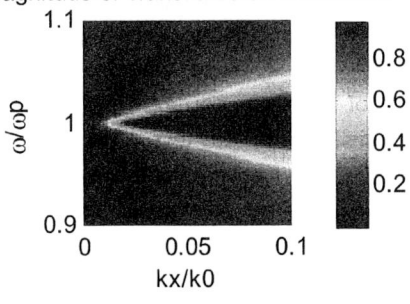

(a) $\gamma_e = 0.01$

Magnitude of Transimission Coefficient

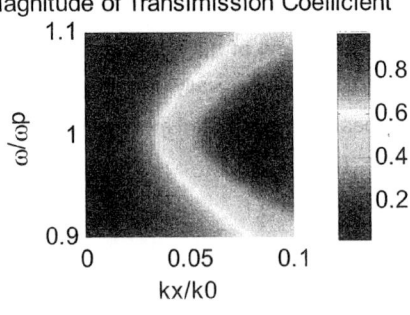

(b) $\gamma_e = 0.1$

*Fig.2 The transmission coefficient as a function of the transverse wave vector $k_x$ and* frequency $\omega$ .The thickness of the slab is $10\lambda$ (wavelength of the source) and $\theta = 0°$ in this calculation.

To give one more example, we also study the case of $\theta = 45°$, as shown in Fig.3.

Magnitude of Transimission Coefficient

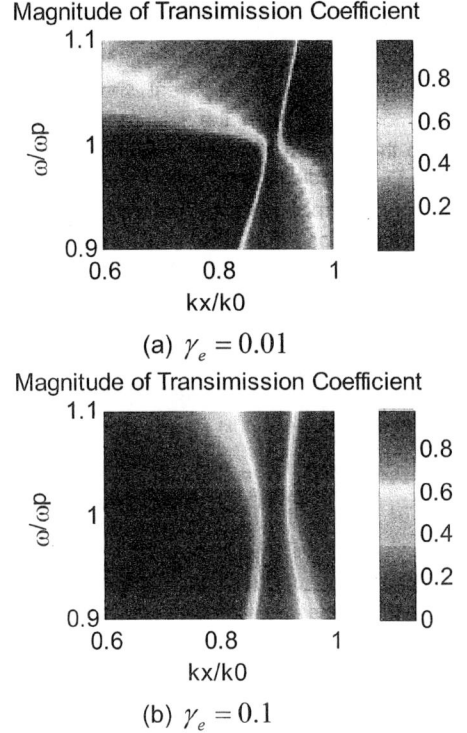

(a) $\gamma_e = 0.01$

Magnitude of Transimission Coefficient

(b) $\gamma_e = 0.1$

Fig.3 The transmission coefficient as a function of the transverse wave vector $k_x$ and frequency $\omega$ .The thickness of the slab is $10\lambda$ (wavelength of the source) and $\theta = 45°$ in this calculation.

## Conclusions

In this paper, we investigate the transmission property of an electrically anisotropic slab with a partially negative permittivity tensor. We show that the high directive transmission property can be observed even in the presence of material loss. In addition, by adjusting the orientation of the optical axis, we can get different high directivity for different $k_x$s. Since the electrically anisotropic slab discussed here can be fabricated in GHz, near- and mid- infrared frequencies, our result may find some application in developing and designing of some new microwave and optical devices.

## References

1. V. Veselago, "The electrodynamics of substances with simultaneously negative values of ", Soviet Phys. Usp. 10 (1968). 509-514.
2. R. Shelby, D.Smith and S. Schultz, "Experimental verification of a negative index of refraction", Science 292 (2001).77-79.
3. N. Engheta. "An idea for thin subwavelength cavity resonators using metamaterials with negative permittivity and permeability", IEEE Antennas Wireless Propagation Letter, 1 (2002), 10-13.
4. J.B Pendry, "Negative refraction makes a perfect lens", Phys. Rev. Lett. 85 (2000), 3966-3969.
5. Viktor.A.Podolskiv and Evgenii E. Narimanov, "Strongly anisotropic waveguide as a non-magnetic left-handed system", Phys. Rev. B, 71, 201101 (2005)
6. J.B. Pendry, A.J. Holden, W.J. Stewart and I. Youngs, "Extremely low frequency plasmas in metallic mesostructures", Phys. Rev. Lett. 76, 4773 (1996)
7. C. G. Parazzoli, R. B. Greegor, K. Li, B. E. C. Koltenbah, and M. Tanielian, " Experimental Verification and Simulation of Negative Index of Refraction Using Snell's Law", Phys. Rev. Lett. 90, 107401 (2003).
8. O. Levy and D. Stroud, "Maxwell Garnett theory for mixtures of anisotropic inclusions: Application to conducting polymers", Phys. Rev. B, 56, 8035 (1997)

# Slow Propagation of Light in a Dielectric-Metal-Dielectric Waveguide

Guoan Zheng (1), Mingwu Gao (1), Hongyu Chen (2)

1 : Electronic and Optical Engineering, Zhejiang University, 310027, HangZhou, China

2 : College of Science, Zhejiang University, 310027, HangZhou, China

Email: guoanzheng@msn.com

**Abstract** *The dispersion property of the guided modes propagating along a dielectric-metal-dielectric waveguide is studied. The energy flux is discussed and we propose a tapered waveguide based on it. Our result is verifed by using the FDTD simulation.*

## Introduction

The development of a simple solid-state-based technology to slow the propagation of light could prove the step in the realization of the high-bit-rate communication systems of the future and increase the efficiency of nonlinear processes which are at the root of all-optical interaction. There are several ways to reduce the group velocity of light. The simplest method is to let the light propagate in a medium with a high refraction index, however, its application is limited by the availability of optically transparent materials. Another approach is to exploit the dependence of the group index ng=n+dn/dω (the first term is the index of refraction; the second term is the dispersion of the medium). However, highly dispersive material is always accompanied by large absorption. A more promising approach is the band-gap property in the periodic structure. If the operating frequency approaches the band-gap, the group velocity approaches zero. In the last years the idea of slow-wave propagation has been investigated in the optical domain [1-7]. Recently, some authors propose to use a dielectric slab waveguide with a left-handed material (possess simultaneously negative values of the electric permittivity and magnetic permeability) substrate to slow down the propagation of electromagnetic wave [8].

In this paper, we proposed a dielectric-metal-dielectric waveguide to reduce the group velocity of the light (operating in the optical frequency). The energy flux as a function of the thickness of the metal is discussed and it is shown that the propagation speed of the guided modes can approach zero if the thickness is chosen appropriately. Our result is verified by using the FDTD simulation.

## Determination of the guided modes in a dielectric-metal-dielectric waveguide

We consider a dielectric-metal-dielectric waveguide, which is shown in Fig.1. The thickness of the metal is $d$. To highlight the essential physics in this structure, we begin by describing the dielectric function of the metal with a free-electron Drude mode [9]:

$$\varepsilon_0 = 1 - \frac{\omega_p^2}{\omega(\omega - i\Gamma_e)} \tag{1}$$

where ωp is the bulk plasmas frequency of the metal and $\Gamma_e$ is the collision frequency. This dielectric

function takes into account the contribution of free electrons only. Despite its apparent simplicity, the plasmonic model has been the source of valuable insights into the general behavior of real metals [10]. We choose $\Gamma_e = 0$ (indicating a lossless case) and the permittivity of dielectrics are 2.4 [8]. It was found from previous calculations that the real part of the dispersion relation is hardly affected by absorption for realistic amount of absorption [11]. We use $\omega_P = 1.36884 \times 10^{16} \, rad/s$, which is representative of metals (e.g., silver). We also assume that both the metal and dielectric are non-magnetic (i.e. $\mu = 1$).

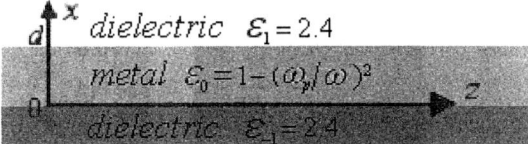

*Fig.1 A dielectric-metal-dielectric waveguide*

All the modes in Fig.1 have TM polarization with the magnetic field perpendicular to the wave propagation direction. In order to guide the field in the metal, the field in the dielectric must be evanescent in the x direction. We write the field solutions for the TM modes as follow [12]

$$\vec{H}_1 = \vec{y} A_1 e^{-\alpha x} e^{ik_z z} \quad x \geq d \tag{2a}$$

$$\vec{H}_0 = \vec{y}[A_0 e^{ik_x x} + B_0 e^{-ik_x x}] e^{ik_z z} \quad 0 \leq x \leq d \tag{2b}$$

$$\vec{H}_{-1} = \vec{y} A_{-1} e^{\alpha x} e^{ik_z z} \quad x \leq 0 \tag{2c}$$

and

$$\vec{E}_1 = \frac{1}{i\omega\varepsilon_1}(\vec{x} ik_z - \vec{z}\alpha) A_1 e^{-\alpha x} e^{ik_z z}$$

$$x \geq d \tag{3a}$$

$$\vec{E}_0 = \frac{1}{i\omega\varepsilon_0}(\vec{x} ik_z - \vec{z} ik_x) A_0 e^{ik_x x} e^{ik_z z}$$

$$+ \frac{1}{i\omega\varepsilon_0}(\vec{x} ik_z + \vec{z} ik_x) B_0 e^{-ik_x x} e^{ik_z z}$$

$$0 \leq x \leq d \tag{3b}$$

$$\vec{E}_{-1} = \frac{1}{i\omega\varepsilon_{-1}}(\vec{x} ik_z + \vec{z}\alpha) A_{-1} e^{\alpha x} e^{ik_z z}$$

$$x \leq 0 \tag{3c}$$

where *kz* is the wave vector in the z direction, *kx* is the transverse wave number in metal, which can be real or imaginary, and $\alpha$ is a positive real number that corresponds to the evanescent waves outside the metal.

To find the guidance condition, we match the boundary conditions at x=0 and x=d to obtain

$$A_{-1} = A_0 + B_0 \qquad (4a)$$

$$A_0 e^{ik_x d} + B_0 e^{-ik_x d} = A_1 e^{-\alpha d} \qquad (4b)$$

$$- p_{21} A_{-1} = A_0 - B_0 \qquad (4c)$$

$$A_0 e^{ik_x d} - B_0 e^{-ik_x d} = p_{21} A_1 e^{-\alpha d} \qquad (4d)$$

where we define $p_{21} = \varepsilon_0 i\alpha / \varepsilon_1 k_x$ and normalize $A_0 = 1$. Upon solving, we have

$$(\frac{1 - p_{21}}{1 + p_{21}}) e^{ik_x d} = e^{im\pi} \qquad (5)$$

So the guidance condition can be written as

$$\tan(\frac{k_x d}{2} - \frac{m\pi}{2}) = (\varepsilon_0 \alpha / \varepsilon_1 k_x) \qquad (6)$$

When $k_x$ is real, the allowable modes from the above dispersion equation are oscillating guided optical modes. The wave number $k_x$ will become purely imaginary ($\alpha_x i$) if the propagation constant $kz$ exceeds a critical value. For purely imaginary wave number, the dispersion equation determining the modes becomes [13]

$$\tanh^{\pm 1} \alpha_x d / 2 = -(\varepsilon_0 \alpha / \varepsilon_1 \alpha_x) \qquad (7)$$

The dispersion relations are

$$k_z^2 - \alpha^2 = \omega^2 \mu_1 \varepsilon_1 = k_1^2 \qquad (8a)$$

$$k_z^2 + k_x^2 = \omega^2 \mu_0 \varepsilon_0 = k_0^2 \qquad (8b)$$

For the wave to be guided, we must have $k_0$ larger than $k_1$. When $k_z$ becomes smaller than $k_1$, the wave in upper cladding will no longer be evanescent and begins to propagate in the x direction. Thus the wave in the core is guided only when $k_z > k_1$.

The dispersion curves for the surface modes are shown in Fig.2 (no oscillating modes exist in our case).

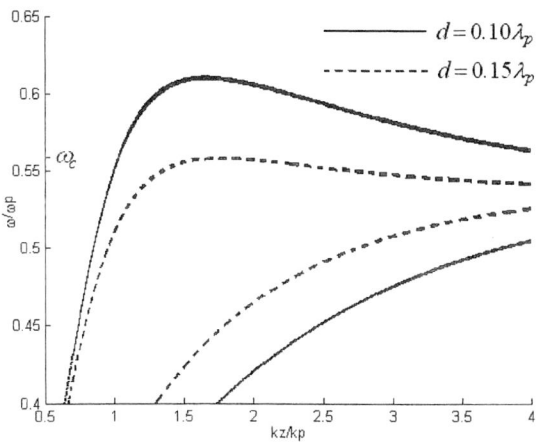

Fig.2 The dispersion curve for the surface modes when $d = 0.10\lambda_p$ and $d = 0.15\lambda_p$.

From Fig.2, one see that two surface modes with opposite signs of group velocity can co-exist in the present waveguide when the frequency is in some

special range. If a prism is used to excite the guided waves, the two modes propagate in opposite directions. When d=0.15λp (dashed curve), the two surface modes will slow down as the frequency approaches $\omega_c \approx 0.563 \omega_p$. In Fig.2, we also plot the dispersion curve for d=0.10λp (solid curve). When d=0.10λp, one forward mode exists at the frequency of $\omega_c$. The profile of the surface mode is shown in Fig.3.

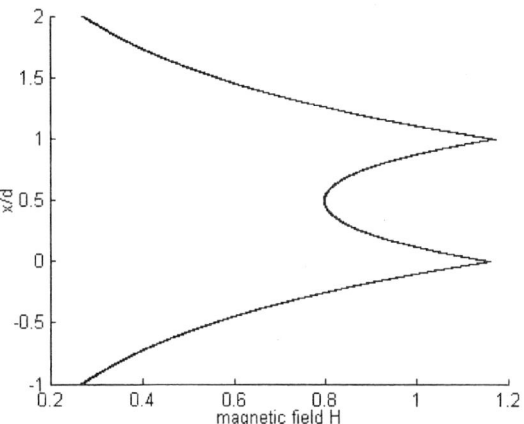

Fig.3 The surface mode profile (the magnitude of magnetic field H) for $\omega = 0.560\omega_p$ and $d = 0.15\lambda_p$

**Dispersion as a function of the thickness of the metal and its application in a tapered waveguide**

In order to study the slow propagation of the surface mode, we define the energy flux as below

$$EF = \int_{-\infty}^{+\infty} S_z dx \qquad (9)$$

where $Sz$ is the z component of the time average Poyting vector $S = \mathrm{Re}[E^* \times H]$. Fig.4 shows the instantaneous distribution of the Poyting vector of the guided mode for such a case (with $\omega = 0.560\omega_p$ and $d = 0.15\lambda_p$). The vortex structure of the energy flow in the present waveguide makes the guide mode flow much slower than the guided wave in the conventional waveguide. When the energy flux in the dielectrics is almost equal to the energy flux in the metal-core, the mode is almost stopped.

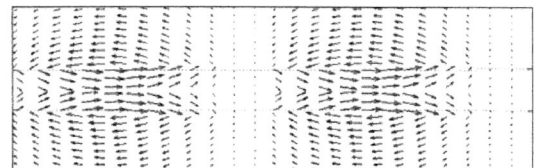

Fig.4 Snap-shot distribution of the Poynting vector for the mode near $\omega_c$ ($0.560\omega_p$) with $d = 0.15\lambda_p$.

As shown in Fig.5, the propagation of the guided modes slows down when the thickness d increases until reaching a critical thickness (0.15λp). Utilizing such a property, we design a tapered waveguide to trap the light. As shown in Fig.6, the thicknesses in region A and B are 0.10λp and 0.15λp respectively. The frequency we choose is 0.560 $\omega_p$ which is close to $\omega_c$. The wave can propagate through the region A with a large group velocity and be trapped in region B (propagate with a very small group velocity). Fig.6 shows the field distribution of the metal-core. As we see, the guided mode in the region B propagates very slowly.

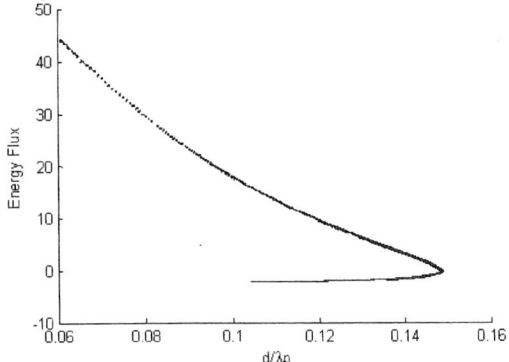

*Fig.5 Energy flux as the thickness of the guiding core layer varies.*

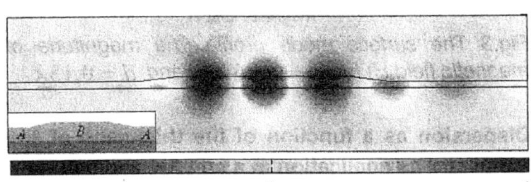

*Fig.6 The field distribution in a tapered waveguide at a moment of 20 periods after switching off the source (place left side of the region A). The thicknesses in region A and B are $0.10\lambda_p$ and $0.15\lambda_p$ respectively. The frequency we choose is $0.560\omega_p$ which is close to $\omega_c$. The computational domain is discretized by $1200 \times 600$ cells with spatial increment $0.002\lambda_p$, and the time step is about $0.001\lambda_p/c$ (c is speed of light in vacuum).*

## Conclusion

We have analyzed the dispersion and slow propagation property of the surface mode in a dielectric-metal-dielectric structure. Some metals such as silver and aluminum are natural candidates for the present structure, because negative permittivity is easily attainable in them as a consequence of the collective excitation of conduction electrons. Silver has a well-defined plasma frequency in the visible domain and the loss is small. Aluminum is also an ideal material for the present structure due to its small imaginary part of the permittivity in the UV [15]. Our conclusion in this paper may have some potential applications in large-scale integrated optical circuits and nanodevice.

## Acknowledgement

This work was supported by the National Basic Research Program (973) of China (No. 2004CB19800).

## References

1 L.V. Hau et al., "Light speed reduction to 17 metres per second in an ultracold atomic gas," Nature 397, 594 (1999).

2 A. Melloni et al., "Linear and nonlinear pulse propagation in coupled resonator slow-wave optical structures," *Opt.Quantum Electron.* 35,365 (2003).

3 H.F.Taylor, "Enhanced electro-optic modulation efficiency utilizing slow-wave optical propagation", J. Lightwave Technol. 17(10), 1875-1883 (1999).

4 N.Shaw, W. J. Stewart, J. Heaton and D.R. Wight, "Optical slow-wave resonant modulation in electro-optical GaAs/AlGaAs modulators", *Electron. Lett.* 35, 1557 (1999).

5 J.E. Heebner and R.W. Boyd, "Enhanced all-optical switching by use of a nonlinear fiber ring resonator", *Opt.Lett.* 24(12), 847-849(1999).

6 A. Melloni, F. Morichetti and M.Matinelli, "Pulse propagation I coupled resonator slow-wave structure", Proceeding of the 10th International Workshop on Optical Waveguide Theory and Numerical Modeling, Nottingham, UK, pag.33 (2002).

7 M. Soljactc, S.G. Johnson, S.Fan, M.Ibanescu, E.Ippen and J.D.Joannopoulos, "Photonic-crystal slowlight enhancement of nonlinear phase sensitivity", *J.Opt.Soc.Am.B* 19(9), 2052-2059 (2002).

8 J.L.He, S.L.He, "Slow propagation of electromagnetic waves in a dielectric slab waveguide with a left-handed material substrate," *IEEE Microwave and Wireless Components Letters, vol.16, Feb 2006*

9 Hocheol Shin et al, "All-angle nega-tive refraction for surface plasmon waves using a metal-dielectric-metal structure," *Phys.Rev.Lett.*, vol.96, Feb. 2006.

10 Peter B. Catrysse, et al, "Guided modes supported by plasmonic films with a periodic arrangement of subwavelength slits", *Appl.Phys.Lett.*,vol.88, 031101 (2006)

11 Han van der Iem, et al, "Band structure of absorptive two-dimensional photonic crystals", J. Opt. Soc. Am. B vol.20, 1334 (2003).

12 Bae-Ian Wu et al, "Guided modes with imaginary tran-sverse wave number in a slab waveguide with negat-ive permittivity and permeability", *Journal of Applied Physics*, vol.93, no.11, June 2003.

13 I. V. Shadrivov, A. A. Sukhorukov, and Y. S. Kivshar, "Guided modes in negative-refractive-index waveguides," *Phys. Rev. E* 67, 057602, 2003.

14 Tao Liu, Armis R. Zakharian et al. "Applications of photonic crystals in optical data storage", Proceeding of SPIE, vol.5380, 2004.

15 S Ioanid, M Bai, N García, A Pons, P Corredera, "Light Collimation and Focussing by a Thin Flat Metallic Slab", to be publish in *Optics Letter,*

16 Mehmet Fatih Yanik et al, "Stopping Light in a Waveguide with an All-Optical Analog of Electromagnetically Induced Transparency", *Phys.Rev.Lett.*, vol.93, Dec. 2004

# Performance and reliability issues of SC/APC plug style attenuators

**Andrzej Tymecki**

*Telekomunikacja Polska SA, Research and Development Branch, ul.Energetyków 23, 20-468 Lublin, Poland*
*Andrzej.Tymecki@telekomunikacja.pl*

**Abstract:** Tolerances of spectral attenuation of plug style SC/APC attenuators have been investigated. Results proved need for new attenuation specifications reflecting values expected in real life random mating conditions. Thorough analysis confirmed serious mechanical interface issues of SC/APC plug style attenuators.

## 1. Introduction

Fibre optic networks require power level control for different purposes or applications, e.g. equalizing power in DWDM channels, power balancing in Passive Optical Networks branches or adjusting link loss budget.

One of the most popular components used for such a purposes is plug style attenuator (or built-out-attenuator (BOA)). Commercially available BOAs are marketed with tolerance values of 0.5dB and 10 % (or even smaller) of nominal value for attenuators of nominal values ≤5dB and >5dB respectively. However field observations of BOAs mated in random configuration indicated the need for revision of attenuation tolerances. Moreover, feedbacks coming from users of damages to the plugs and attenuators interfaces (endfaces) raised concerns regarding possible mechanical damages and problems with these types of components.

This paper presents results of recently completed CENELEC round-robin test of built-out SC/APC attenuators that aimed at verification of reasonable tolerances for attenuators of tested style.

## 2. Round-robin tests on SC/APC built-out attenuators

Test plan comprised spectral attenuation measurements against common reference plugs, measurements against "Grade B" own plugs and PDL measurements.

To obtain a clear understanding of the accuracy and repeatability of the BOA spectral attenuation we selected for the tests three nominal attenuation values; 1dB, 5dB and 15dB (a sample from each of the manufacturers). In total we collected 18 samples from 6 different manufacturers. Five laboratories took part in round-robin tests. We set up tests plan with detailed description of setting requirements to decrease uncertainty between different laboratories.

Spectral attenuation measurements were conducted in wide wavelength range from 1260nm up to 1650nm in 5nm measurement steps. Test setup comprised unpolarized broadband light source and optical spectrum analyzer set to 2nm resolution. Spectral attenuation tests included measurements against common reference plugs and reference adapter and measurements against own grade B plugs.

When analyzing the measurement results we found out that the results obtained for common reference plugs were very consistent. The same attenuators measured against own "Grade B" plugs show higher attenuation distribution. Moreover, in case of measurements against reference plugs smaller numbers of attenuators did not pass the attenuation criteria.

Relaxed pass/fail criteria from the initial 0,5dB/10% to 0,75dB/15% (for nominal values ≤5dB and > 5dB respectively) decreased the number of attenuators failing to pass tests. Table 1 summarizes measurement results of CENELEC round-robin tests on plug style SC/APC attenuators.

It is clearly seen from Table 1 that in the case of measurements against "Grade B" own plugs, which represents random mating field conditions, we have very high percentage of attenuators failing to pass the criteria. Situation was much better in case of relaxed pass/fail criteria, however from user's point of view it's hard to accept e.g. 0,75 dB tolerance in case of 1 dB attenuators.

Table 1. Overview of plug style attenuators measurements results

| Attenuation criteria | Nominal attenuation value | Number of attenuators failing to meet criteria | |
|---|---|---|---|
| | | Common ref. plugs | Own grade B plugs |
| Tight | 1 dB | 2 of 30 | 5 of 30 |
| | 5 dB | 6 of 30 | 14 of 30 |
| | 15 dB | 6 of 30 | 13 of 30 |
| Relaxed | 1 dB | 0 of 30 | 1 of 30 |
| | 5 dB | 1 of 30 | 5 of 30 |
| | 15 dB | 0 of 30 | 0 of 30 |

During spectral attenuation measurements we noticed in some measurement cases strong oscillations in the attenuation in the 1310 nm region. The same kind of oscillations we noticed during previous round-robin tests of SC/PC plug style attenuators [1]. The occurrence of oscillations was not dependent on nominal value and we could have not indicate a specific manufacturer whose attenuators were attributed with such an oscillations. The phenomenon that caused these oscillations is known as modal interference noise. Two important observations were:

- oscillation appeared for given specimen both during measurements against common reference plugs and own "Grade B" plugs, however extremes in both cases were different,
- oscillations were much stronger in 1310nm region and decreased noticeably in 1550nm region. The example of spectral attenuation measurements for specimen selected from test sample is shown in Figure 1.

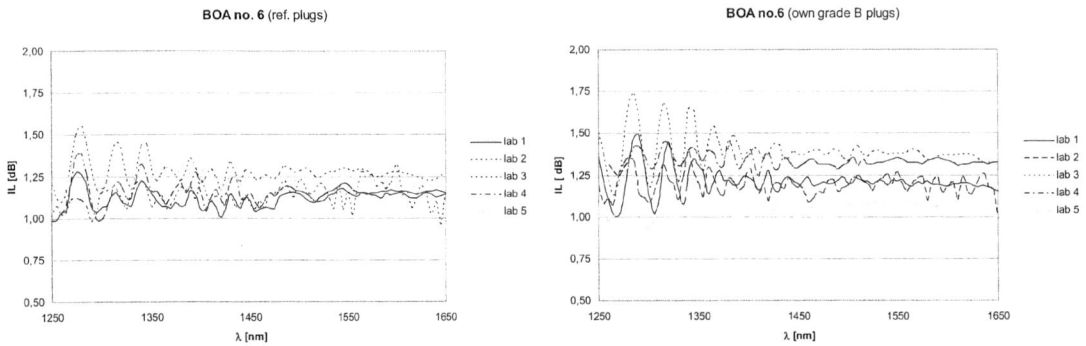

Fig. 1. Oscillations during spectral attenuation measurements for common reference plugs (left) and own grade B plugs (right)

Another important observation was that during measurements in the second laboratory, one of the reference plugs became mechanically damaged. Figure 2 presents reference plug A before (a) and after (b) measurements in second laboratory.

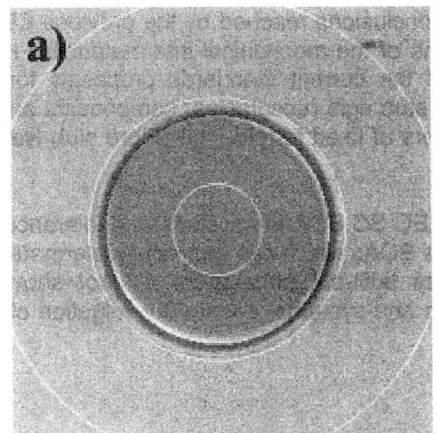 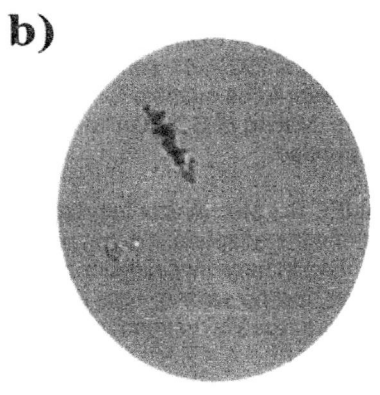

Fig. 2. Reference plug A before (a) and after (b) measurements in second laboratory

The accidental damage to the reference plug raised an even a larger interest in mechanical behaviour of specimens, which lead to launching a thorough investigation on the mechanical interoperability of BOAs in general (and the SC/PC and SC/APC BOAs in particular) with different components. An extensive investigation of the mechanical behavior of mechanical behaviour of SC/PC BOA was conducted by T.Bolhaar, D.Deams and J.Elenbaas [2]. The calculations were made for two main connecting configurations: an attenuator placed between 2 spring loaded ferrules (i.e. attenuator between two plugs) and an attenuator plugged into active component (e.g. transceiver) with plug on the second side. Here, the Transceiver is considered to be of a rigid (fixed) ferrule (fibre stub) construction type.

Table 2 presents summary of analysis included in [2].

Table 2. Overview of different scenarios of mechanical behaviour for various component configurations and attenuator constructions

| Standard | Configuration | Worst case* | Attenuator's ferrule fixture type | |
|---|---|---|---|---|
| | | | Fixed | Floating |
| EN 50377-4-4 [3] | Plug-Adapter-BOA-Plug | 1 | Damage by over-travel | Damage by over-travel |
| | | 2 | Correct physical contact | Correct physical contact |
| | Transceiver-BOA-Plug | 1 | Damage by over-travel | Damage by over-travel |
| | | 2 | Damage by over-travel | Physical contact |
| prEN 50378-2-1 [4] (withdrawn) | Plug-Adapter-BOA-Plug | 1 | Damage by over-travel | Damage by over-travel |
| | | 2 | No physical contact | No physical contact |
| | Transceiver-BOA-Plug | 1 | No physical contact | Damage by over-travel |
| | | 2 | Damage by over-travel | No physical contact |

* Worst case 1 means maximum possible cumulative tolerance while case 2 means minimum possible cumulative tolerance.

Overview presented in Table 1 shows that for each configuration and attenuator construction there is at least one case of tolerances accumulation that leads to possible lack of Physical Contact (PC) or endface damage. It's necessary to remember that the calculations were done for components manufactured according to commonly used international standards. The CENELEC TC86BXA round-robin tests proved that it is possible to have either cumulative ferrule over-travel leading to ferrule endface damage or lack of physical contact leading to spectral power oscillations in measurements.

## 4. Conclusion

Round-robin tests of plug style SC/APC attenuators conducted in CENELEC CLC/TC 86BXA standardization committee [5] proved that very tight attenuation tolerances marketed and specified by most of BOA manufacturers do not reflect random mating conditions.

The measurements results confirmed most of the above conclusions reached by the previous CENELEC round-robin tests of SC/PC BOAs. Extensive investigations of the mechanical intermateability issues of particularly SC/PC and SC/APC confirmed that none of the current standards proposals for SC/PC (SC/APC) BOAs seem to be suitable for both active (fibre stub rigid receptacles) components and spring loaded connectors. Mating of SC BOA to active transceivers of fixed ferrule type (Fibre stub Receptacle type) should be avoided.

With the current IEC SC Mechanical Interface and CENELEC SC BOA dimensions and tolerances of the connector plugs and the adapters, it is not possible to make BOAs which can guarantee intermateability in all applications. Preliminary investigations of other styles built-out attenuators did not show similar problems, however, for the sake of fibre optic components and systems, deeper investigation of various plug style attenuators types is recommended.

## 5. References

[1] *Connector sets and interconnect components to be used in optical fibre communication systems – Technical Report, Round robin test results on SC-PC simplex connector type fixed optical attenuator using EN 60793-2-50 type B1.1 single mode fibre* – CENELEC Working Document
[2] Ton Bolhaar, Daniel Daems, Jacco Elenbaas, *SC Plug style attenuator(BOA) mechanical interface issues*, Tyco Electronics
[3] *EN 50377-4-4:2006, Connector sets and interconnect components to be used in optical fibre communication systems - Product specifications - Part 4-4: Type SC-PC simplex terminated on IEC 60793-2 category B1.1 single mode fibre*
[4] *prEN 50378-2-1, Passive components to be used in optical fibre communication systems – Technical Report - Part 2-1: SC(SC2)-PC connector-type fixed optical attenuators using IEC 60793-2 Category B1.1 singlemode fibre (Note: document is withdrawn)*
[5] *prTR 50378-2-2, Round robin test results on SC-APC simplex connector type fixed optical attenuator using highly attenuating single mode fibre*

# BER performance analysis of PIN photodiode in 10Gbps fiber optical communication

Shang-Bin Li*

*Shanghai Research and Development Center of Amertron-global, Zhangjiang High-Tech Park, 299 Lane, Bisheng Road, No.3, Suite 202, Shanghai, 201204, P.R. China*

In this report, the performance analysis of 10Gbps optical receiver for fiber optical link is presented. Non-Return to Zero (NRZ) electrical signals from a Pseudo-Random Bit Sequence (PRBS) Generator are used as modulation input. Signals in the optical domain are detected at the PIN photodiode (PIN-PD) and are converted into electrical output signals. These electrical signals are filtered through a low pass Bessel filter. Effects of bandwidth and thermal noise of receiver on the link performance are studied in detail. For the 10Gbps fiber optical communication, a Q factor of 6.55 and the BER of $5.23E^{-11}$ are observed for the worst case, when the receiver is operated at $100^oC$. It is shown that BER heavily depends on thermal noise. The relation between the receiver sensitivity and the thermal noise is also investigated.

Key words: Bit error rate; Fiber optical communication; PIN photodiode; Thermal noise.

## I. INTRODUCTION

Recently, many efforts have been paid to the 10Gbps optical receiver due to the commercial demands for high speed transmitters or receivers used in fiber optic communication [1–5]. Many schemes of 10Gbps optical receiver have been proposed in past few years. Avalanche photodiode has been considered as an available candidate due to its significant gain capability [6, 7], though the shot noise issue has to be carefully dealt with. With the development of technology in PIN photodiode (PIN PD), the bandwidth of PIN PD has been achieved at 100GHz or higher, which open the way to consider the PIN PD in the design of 10Gbps optical receiver [8]. In PIN PD, the thermal noise plays the dominating role in the effect of noise on the performance of the receiver, if compared with the shot noise. In this report, we investigate the bit error rate (BER) performance and receiver sensitivity of 40GHz bandwidth PIN photodiode in 10Gbps fiber optic link. The influence of thermal noise on the BER and receiver sensitivity is analyzed.

Optical communication systems are very complex and difficult to analyze. It is often hard to predict the effect of various characteristics of the devices in a fiber optic link [9]. This work involves simulation of 10Gbps optic link performance. Such simulation can help in analyzing the device under development and predict the performance for a given link distance and the simulation output helps eliminating any possible performance degradation before implementing the actual hardware. The objective of the fiber optic link is to transport data or communication signals reliably over a longer distance. The desired Q factor is approximately 7 and the desired BER is approximately $10^{-12}$. The Q factor and BER obtained from either the eye diagram analyzer or the BER analyzer can be used

to analyze the degradation of the signal at the receiver components.

Here, simulation results of 10 Gbps optic link are presented. Non Return to Zero (NRZ) electrical signals from a Pseudo-Random Bit Sequence (PRBS) generator are used as modulation input. Signals in the optical domain are detected at the PIN photodetector and are converted into electrical output signals. These electrical signals are filtered through a second order low pass Bessel filter. Filtering removes distortion caused by noise in the signal. The simulation inputs include the Laser Diode optical and electrical properties, PIN PD and low pass filter, and the link optical attenuation. For the 10 Gbps fiber optic link with received optical power of -6dBm, a Q factor of 6.55 and the BER of $5.23E^{-11}$ are observed for the worst case, when the receiver is operated at 100C.

A schematic outline of optical link is shown in Fig.1. The main components of the optical link are PRBS generator, NRZ pulse generator, Laser diode, single mode fiber, optical attenuator, PIN PD, low pass filter, BER analyzer, and Eye diagram analyzer for analyzing the performance of the link. we have considered laser diode at wavelength 1550 nm with output power of 1mw (0dBm). Rise and fall time for laser signal are both equal to 0.4 ns with an overshoot and undershoot of 9 percent. Insertion loss of 1dB and depth 40dB are taken in consideration for 2 or 4 order low pass filter. PIN-photodiode has responsivity of 0.6A/W and dark current of 0.1nA. Thermal noise is assumed to be equal to $4k_BT$, where $k_B$ is the Boltzmann constant and $T$ is the absolute photodiode temperature. The binary data are produced by PRBS generator and converted into electrical pulses with exponential edge shapes by NRZ pulse generator, which is finally passed into a directly modulated laser. The electrical pulses generated have maximum amplitude 1 a.u and minimum amplitude 0 a.u. The rise times and fall times of the pulses are both 0.05 bit. A Low Pass Bessel Filter, with cutoff frequency of (0.75 Bits rate) and depth 40 dB, is included after the photodetector to

---

*Electronic address: Stephenli74@yahoo.com.cn

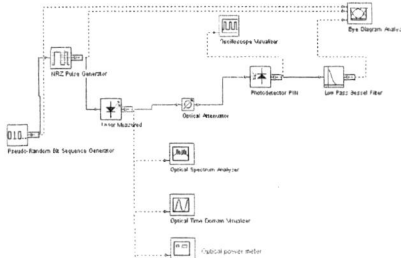

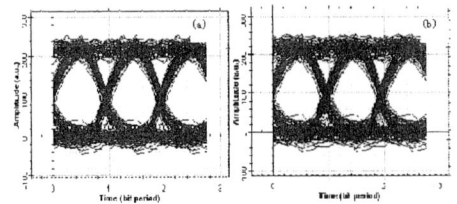

FIG. 1: The systematic layout of the simulated setup. In the simulation, we adopt the PRBS7 NRZ data pattern, and the data rate is 10 Gbps. The dominated wavelength of the transmitter is 1550nm with laser output power 0 dBm.

FIG. 3: The eye diagrams for two different photodiode temperatures are depicted with receiver optical power −6 dBm. (a) $T = 25^{\circ}$C; (b) $T = 100^{\circ}$C. The order of low pass Bessel filter is 4. The bandwidth of PIN PD is 40 GHz.

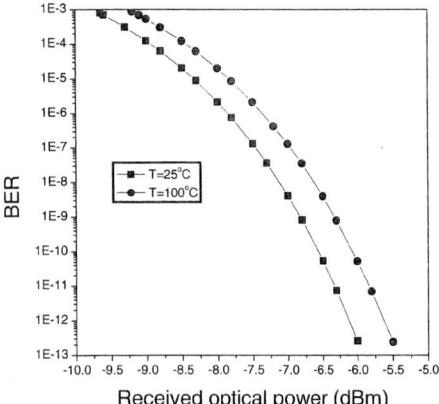

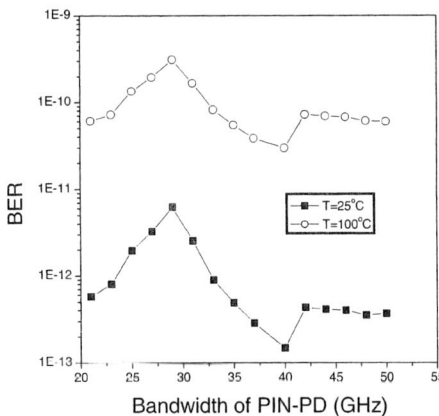

FIG. 2: The BER is plotted as the function of received optical power for two different device operating temperatures of $25^{\circ}$C and $100^{\circ}$C. The bandwidth of PIN-PD is 40 GHz. The bit rate is 10 Gbps. The order of low pass Bessel filter is 4. The bandwidth of PIN PD is 40 GHz.

remove distortion in the electrical signals.

## II. RESULTS AND DISCUSSION

Results of the simulation are observed by connecting the oscilloscope, BER and Eye diagram analyzer to PRBS generator, NRZ pulse generator and filter (see Fig.1).

In Fig.2, the BER is plotted as the function of received optical power for two different device operating temperatures of $25^{\circ}$C and $100^{\circ}$C. At temperature of $25^{\circ}$C, the BER increases from $2.57 \times 10^{-13}$ to $7.92 \times 10^{-4}$ as the received optical power decreases from −6dBm to −9.65dBm. At temperature of $100^{\circ}$C, the BER increases from $2.42 \times 10^{-13}$ to $8.84 \times 10^{-4}$ as the received optical power decreases from −5.5dBm to −9.2dBm. Comparing two cases, it is easy to understand the power penalty is about 0.5dB due to the increase of temperature from $25^{\circ}$C to $100^{\circ}$C.

FIG. 4: The BER is plotted as the function of bandwidth of PIN photodiode for two different levels of thermal noise. Received optical power is −6dBm. The order of low pass Bessel filter is 2.

Fig.3 shows the obtained eye diagrams by eye diagram analyzer for the 10Gbps optic link operating at two different temperatures. The Eye Diagram Analyzer is also used to measure the performance of a digital transmission system. For the room temperature $T = 25^{\circ}$C, the maximal Q factor is 7.31 and the minimal BER is $2.57 \times 10^{-13}$. For the high temperature $T = 100^{\circ}$C, the maximal Q factor is 6.55 and the minimal BER is $5.23 \times 10^{-11}$. The BER is the number of bits received with errors compared to the total number of bits received. A smaller BER has fewer errors. These values are acceptable according to the critical requirements for 10Gbps optic link standard. In high speed optical communication systems, in order to avoid undesirable effects, large modulation depth is required along with large modulation frequency and minimal power loss. A high modulation depth is desirable, because the optical signal will exhibit a greater margin to resist the noise. Extinction ratio (ER) is a measure of the modulation depth of a source transmitter and it expresses the proportional relationship between the power levels of the binary '1' and '0' signals, averaged while transmitting a pseudo-random binary-sequence signal. Effect of the laser diode extinction ratio variation on the Q factor and BER has also been investigated. The maximum

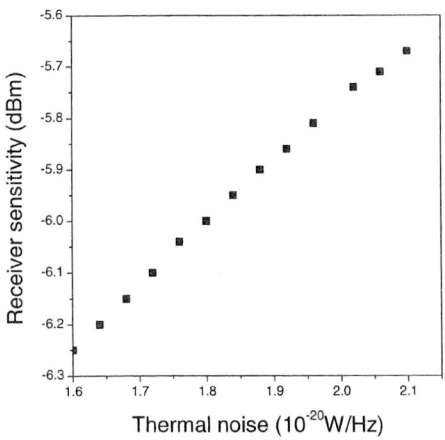

FIG. 5: The receiver sensitivity versus the thermal noise is depicted. The order of low pass Bessel filter is 2.

Q factor decreases from 7.31 to 6.62 for the decrease of extinction ratio from 45dB to 10dB for the case with temperature of 25°C and $-6$dBm received optical power. Similarly, BER is increased significantly from $2.57 \times 10^{-13}$ to $3.36 \times 10^{-11}$. Thus the signal has been degraded a lot by the decrease in the extinction ratio.

It is also very desirable to analyze the dependence of high speed performance on the bandwidth of PIN PD. In Fig.4, the dependence of BER on bandwidth of PIN PD is shown for two operating temperatures. With the increase of bandwidth from 21GHz to 50GHz, BER achieves a local maximal value at 29GHz and a local minimal value at 40GHz for both room temperature and high temperature. It implies that 40GHz may be the most suitable bandwidth for 10Gbps optic link in present situation.

As mentioned above, the increase of temperature causes the degradation of the performance. Therefore, we need to quantitatively discuss how the thermal noise affects the receiver sensitivity. In Fig.5, the receiver sensitivity (denoted by $P_{rec}$) is shown as a function of the thermal noise (denoted by $\eta$). We can see that the value (in unit of dBm) of receiver sensitivity nearly linearly increases with thermal noise in the range from $1.6 \times 10^{-20}$W/Hz to $2.1 \times 10^{-20}$W/Hz. By linear fitting, we obtain the approximate relation between the receiver sensitivity and thermal noise as follows: $P_{rec}=-8.12$dBm$+1.172 \times 10^{20}\eta$dB·Hz/W. In ideal situation, the thermal noise can be regarded as a linear function of absolute temperature in unit of Kelvin. Therefore, from the above equation, we can easily know how the receiver sensitivity varies with temperature.

### III. CONCLUSION

In summary, we have analyzed the performance of 10Gbps optical receiver for fiber optical link. NRZ electrical signals from a PRBS Generator were used as modulation input. Signals in the optical domain were detected at the PIN PD and were converted into electrical output signals. These electrical signals were filtered through a low pass Bessel filter. Filtering removes distortion caused by noise or interference in the signal. The simulation inputs include the Laser Diode optical and electrical properties, PIN-PD and attenuation of laser output power, etc.. Effects of bandwidth and thermal noise of the optical receiver have been discussed in details. As the received optical power equals -6dBm, a Q factor of 7.31 and the BER of $2.57E^{-13}$ were observed for the case of room temperature, and a Q factor of 6.55 and the BER of $5.23E^{-11}$ were observed for the worst case, when the optical receiver is operated at 100°C. It has been shown that BER heavily depends on thermal noise and received optical power. The relation between the receiver sensitivity and the thermal noise has been also investigated. The simulation results also revealed that 40GHz bandwidth of PIN PD might be very suitable for 10Gbps optic link. The present work may have potential applications in the design and manufacture of 10Gbps TOSA and ROSA.

[1] R. S. Tucker *et al.*, Electron. Lett. **22**, 917 (1986).
[2] C. C. Barron *et al.*, Electron. Lett. **30**, 1796 (1994).
[3] I.-H. Tan *et al.*, IEEE Photon. Technol. Lett. **7**, 1477 (1995).
[4] K. Kato, and Y. Akatsu, Opt. Quantum Electon. **28**, 557 (1996).
[5] T. Takeuchi *et al.*, IEICE Trans. Electron. E **82C**, 1502 (1999).
[6] R. S. Fyath, and J. J. O'Reilly, J. Lightwave Techol. **7**, 62 (1989).
[7] H. Ishikawa *et al.*, Electron. Lett. **29**, 1874 (1993).
[8] Y.-G. Wey *et al.*, J. Lightwave Techol. **13**, 1490 (1995).
[9] G. P. Agrawal, *Fiber-optic communication systems* (John Wiley and Sons, Inc. 2002).

# Optimum Isotropic Dimension Phenomenal For Nematic Liquid Crystal
# Display By Twisted Nematic Cell Gap

**No,949,Da Wan Rd,Yung-Kang City.Tainan Hsien,710,Taiwan,R.O.C.**
**Telephone: +  886(6)2727-175; anderson880@yahoo.com.tw**
**Chia -Fu, Chang   Zou-ni, Wan   Chia-Hi, Chen**

The transition can be approached by changing the impurity concen trationor, indirectly, by tuning the tempera
ture since the pinning strengths of the random and crystal potential havein general a different temperature depend ence.The nematic liquid crystal is clear only when a long range order exists, in the whole medium using the Jones matrix method. Director can change from point to point and is,in general, a function of space.These two technol ogies have ca used more and more product designers to turn to liquid crystal (LC) displays, which have conseq uently exper ienced phenomenal growth. The resist profile simulation is carried out using thecombined data thus obtained. Details of the lens structure and of the devices fabrication and performance are described.

Simulateion.Light from conventional light source or laser is passed through a polarizer and then incident on the specimen. Liquid crystal displays has fostered continued de velopment, to the point where full color video displays have been realized which can rival.At the nematic isotropic, transition temperature, the medium becomes isotropic and looks clear and transparent.

# A Critical Study of the Soot Deposition Rate in ACVD Process

G.C.Mishra, Dattatray Pasare

Sterlite Optical Technologies Ltd

E-2, MIDC, Waluj, Aurangabad, Maharashtra, 431136, India

E-mail: 1:Gopal.Mishra@sterlite.com 2:datta_pasare@yahoo.com

**Abstract** *Atmospheric Chemical Vapor Deposition (ACVD) process is widely used for silica soot deposition in the manufacture of optical fiber. In the present work a series of experiments were done on the ACVD lathe to study the deposition rate. Target core rod diameter, number of burners, distance between the burners and the distance between the target core rod and the burners were found the key parameters affecting the deposition rate. Average deposition rate of 30.2 gm/min with a material efficiency of 34% were achieved in three-burner experiment.*

## Introduction

In recent years optical fiber is being extensively used in telecommunication and sensor applications. Reduction in prices and increase in demand of optical fiber has induced the need to develop new techniques to increase the productivity of the machine. For manufacturing of optical fiber using the ACVD process, sub micrometer porous soot particles are deposited on the rotating and traversing target rod by passing the vapor stream of chloride and fuel gases through a burner. Soot deposition rate decides the capacity of the deposition machine. Target core rod diameter, chloride vapors and fuel gases were found to be the key process parameters affecting the deposition rate. Similarly no. of burners, distance between the burners and the distance between target core rod and the burners were the key hardware parameters affecting the deposition rate. The equipments used in this study consist of a lathe that is used for deposition of soot particles onto the surface of target core rod (Fig 1).

## Experiments and Findings

The first step in the ACVD process is soot deposition on the target rod. In this step, a hot stream of soot particles of desired composition is generated by passing the vapor stream through a fuel gas i.e. oxy hydrogen flame directed towards a rotating and traversing target rod. The raw materials used were $SiCl_4$, $GeCl_4$ and $O_2$ and the burner gases were $H_2$ and $O_2$. The dominant mechanism of soot deposition in the ACVD process is thermophoresis, which is the tendency of particles to migrate down with the local gas temperature gradient. During deposition, the burner traverses back and forth parallel to the target rod axis so that one layer of the soot is deposited per pass. Soot particles are built up on this rod layer by layer. The soot for the core material is made of mixture of $SiO_2$ and $GeO_2$ with an appropriate proportion and deposited on the target rod depending on the refractive index profile of the desired optical fiber. The soot for the cladding material is silicon dioxide ($SiO_2$). When enough soot particles are deposited for both the core and the cladding of the optical fiber, deposition is stopped and the porous preform is slipped off the target rod. A weight measurement system was built to calculate the instantaneous deposition rate during the deposition process.

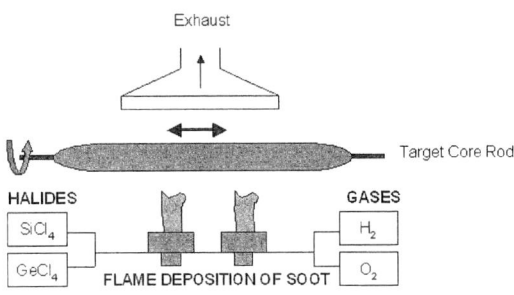

Fig: 1 Schematic diagram of over cladding process

The chemical reactions involved in the formation of the soot are given in equation (1) and (2). The chlorides react with Oxygen ($O_2$) in the flame at above 1500 $^0$C.

$$SiCl_4 + O_2 \rightarrow SiO_2 + 2Cl_2 \qquad (1)$$
$$GeCl_4 + O_2 \rightleftharpoons GeO_2 + 2Cl_2 \qquad (2)$$

Equation (2) is shown as equilibrium because $GeO_2$ is not significantly more stable than $GeCl_4$ at this temperature. The water from the combustion of the fuel gas reacts with chlorine to produce hydrogen chloride.

$$2H_2O + 2Cl_2 \rightleftharpoons 4HCl + O_2 \qquad (3)$$

The effect of this reaction is to consume a large portion of the chlorine gas produced by the formation of $SiO_2$ and $GeO_2$.

In the next step, called sintering, a hollow porous preform is dehydrated and collapsed in a furnace at a temperature of 1550 $^0$C in the presence of helium to form the desired mother preform. Core rods were drawn from mother preform for further over cladding. Normally 8 to 12 core rods were drawn from 10 kg of mother preform.

In the soot over cladding process, the desired amount of porous soot was deposited on the core rod (Fig 1). Soot preforms up to 1200 layers have been fabricated. Soot preforms were about 1300 mm in length, 180 mm in diameter and had a weight of around 15 kg. The average deposition rates were around 24 gm/min.

The porous preform is then taken to sintering stage. In this process again, the porous soot preform is sintered or consolidated to a dense glass rod by passing it vertically through the hot zone of a special annular furnace. During this step, the temperature is raised to 1550 °C. By using an atmosphere of helium with a few percent of chlorine within this furnace, OH- ions was removed very effectively resulting in very low OH- content fiber preforms. The preforms are then as usual drawn to fibers in the next process step.

The soot particle begins within about 10 mm from the burner face and is nearly complete within 100 mm. The first particles formed are of the order of 0.1μm. These grow by collision and coalescence to produce particles of up to 0.25 μm. Soot particles vary in composition, depending on the part of the flame in which they are formed and are collected on the target rod due to thermophoresis. Average deposition rate is defined as total deposited weight on the target rod divided by total deposition time, whereas the instantaneous deposition rate is the ratio of increase in weight deposited on the target core rod and time required for depositing the same.

Following experiments were conducted to study the effect of various parameters on deposition rate in over cladding process of target core rod.

### I) Target core rod diameter versus deposition rate

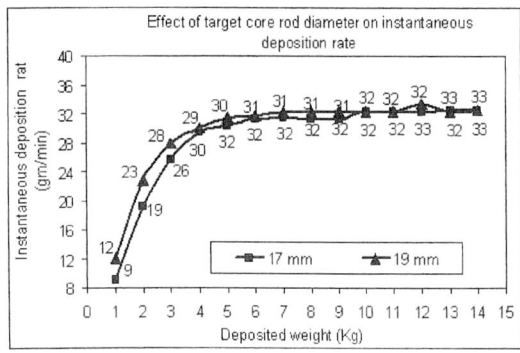

Fig: 2 Instantaneous deposition rate as a function of the target rod diameter.

Core rod diameter is found to be the key process parameter affecting the deposition rate (Fig 2). At the beginning of soot deposition, larger target diameters result in high deposition rate. For the first one kg of

deposition, deposition rate was 14 and 10 gm/min for 19 and 17 mm target core rod diameter respectively.

High circumference and surface area of the preform allows more time and area during which the particles are close enough to the surface to be collected. Also, the average preform temperature is reduced, thereby increasing the thermophoretic force. Deposition rate and material efficiency both were found high with the higher target core rod diameter. Average deposition rate was found 26 gm/min in case of 19 mm target rod diameter and 24 gm/min in case of 17 mm target core rod diameter. The material efficiency of the process was 44 % and 40% in case the core rod diameter as 19 and 17 mm respectively.

### II) Number of burners versus deposition rate

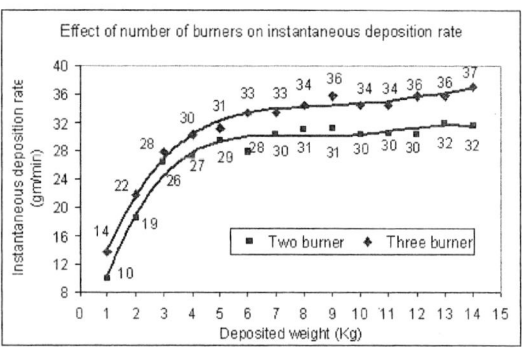

Fig: 3 Instantaneous deposition rate as a function of no. of burners

Fig.3 presents the instantaneous deposition rate for the two burner and three burner experiments. The instantaneous deposition rate went to 37 gm/min at the end of process in three-burner experiment and 32 gm/min in two-burner experiment. The deposition rate saturates during the soot deposition process when a critical soot diameter is achieved. Due to separation between the burners there was a formation of conical shape at the ends of the target core rod. A good quality optical fiber cannot be drawn from these conical parts due to undesired proportion of core and cladding. Lengths of cones were longer in case of three-burner than two-burner. Three-burner can be cost effective in increasing the capacity in case of longer target rod length. Again if the burners are too far apart, the length of the end regions increases too much. Average deposition rate in three-burner was 30.2 gm/min whereas in two-burner it was 25 gm/min. The material efficiency in three-burner was 34 % where as in two-burner it was 44 %.

### III) Distance between target core rod and burner versus deposition rate
Increasing the distance of the target core rod from the burners increases the density of the soot layer. This is

due to coagulation of the particles in the flame before they touch the target core rod. It is also possible that when the burner is moved farther from the target core rod, a hotter part of the flame touches the target core rod and partially sinters the formed soot layers. The increasing density again restricts the growing diameter of the target core rod. This explains why increasing the burner distance from the substrate decreases the deposition rate. Initial deposition rate was 13 and 11 gm/min in case of distance between burner and target rod as 36 and 30 cm respectively. However in case of 36 cm deposition rate was high towards end of the process (Fig 4).

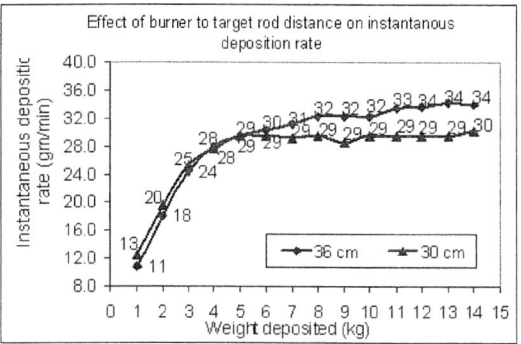

Fig: 4 Instantaneous deposition rate as a function of distance between burner and target core rod

Average deposition rate was 27 gm/min and 25 gm/min and material efficiency was 41 % and 44 % in case of burner to core rod distance as 36 cm and 30 cm respectively.

## IV) Distance between two burners versus average deposition rate

Effect of distance between the two burners on average deposition rate has been given in fig 5.

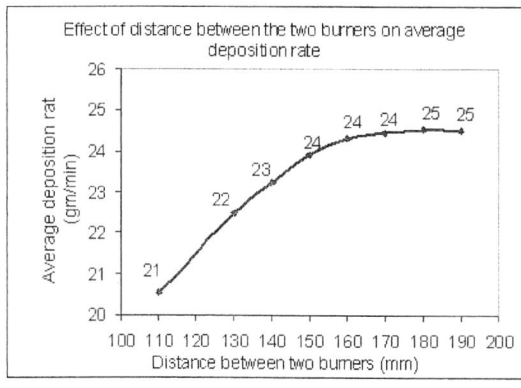

Fig: 5 Average deposition rate as a function of distance between two burners

Experiments were performed to find the optimum distance between the two burners. The trials were done for the distances between 100 to 200 mm. There was interaction between the burner flames when the distance between the burner was small. This interaction guides the soot particles past the target and in turn decreases the deposition rate and material efficiency. If the burners are too far apart, the length of conical part at the ends region of target core rod increases too much and a good quality optical fiber cannot be drawn from these parts. The optimum distance was found to be around 150 mm.

Fiber drawn from the above experiments had an attenuation of less than 0.320 dB/km at 1310 nm and less than 0.210 dB/km at 1550 nm wavelengths. All the drawn fiber had very good geometrical parameters particularly core cladding concentricity, which was 0.22 µm, and cladding non-circularity, which was 0.3 % as measured using the PK 2400 geometry analyzer. Typical tensile strength was 4.4 Gpa and corrosion susceptibility factor n was found 20 as measured by dynamic tensile mode. All the optical, geometrical and reliability parameters were comparable with currently commercially available optical fibers.

## Conclusions
The thermophoretic force has been found the main force pulling the particles towards the target core rod. By optimizing the process and hardware parameters one can increase the thermophoretic forces and in turn deposition rate. The average deposition rate was found 30.2 gm/min in three-burner experiment with material efficiency of 34%. There is trade-off between deposition rate and material efficiency as in case of two-burner material efficiency is 44% but deposition rate is 25 gm/min. The future potential lies in the larger preform size for which higher deposition rates and material yield can be achieved.

## References
1. Tingye Li, Optical Fiber Communications Volume 1 Fiber Fabrication, Page 65-92, Academic Press Inc (London) Ltd, 1985.
2. C. K. Wu, R. Greif, "Thermophoretic deposition including an application to the outside vapor deposition process", International journal of heat and mass transfer, Vol. 39, No 7, 1429-1438, 1996
3. T. Izawa and S. Sudo, Optical Fibers: Materials and Fabrication, Page 64-69, KTK Scientific Publishers Tokyo, 1987.
4. Heikki Ihalainen, Jouko Kurki, "Soot-over cladding process for enlarging modified chemical vapor deposition preforms", Optical Engineering 34 (9), 2538-2542 (September 1995).
5. Michael G. Blankenship, Charles W. Deneka, "The Outside Vapor Deposition Method of Fabricating Optical Waveguide Fibers", IEE journal of Quantum Electronics, Vol. QE-18, 1418-1424, October 1982.

# Upgrades of Submarine Systems using Higher Bitrates and Advanced Modulation Formats

Jörg Schwartz (1), Ronald Freund (2), Lutz Molle (2), Christoph Caspar (2),
Steve Webb (1) and Stuart Barnes (1)

1 : Azea Networks Ltd., Bates House, Harold Wood, Romford RM3 0SD, United Kingdom.
Email: joerg@azea.net
2: Fraunhofer-Institute for Telecommunications, Heinrich-Hertz-Institut, Einsteinufer 37,
10587 Berlin, Germany. Email: Ronald.Freund@hhi.fraunhofer.de

**Abstract** *Upgrading undersea communication systems beyond their design capacity is an attractive alternative to deploying new cables and increasing the bitrate is an obvious way to facilitate this. Nevertheless, this comes with tightened performance requirements, such as reduced noise and dispersion tolerance, which will reduce any available margins. Applying new modulation formats is a route to mitigate this effect. This paper discusses the technical challenges for carrying out upgrades, either as overlays or on dark fibres, and also presents recent research results on how to add capacity to systems currently operating with Nx10.7 Gbit/s RZ-ASK signals over transoceanic distances. Experiments and simulations are used to analyse performance and limitations, with 20 Gbit/s RZ-DPSK being a strong candidate for upgrades.*

## Introduction

Upgrades of telecommunication systems using existing fibres has become a very interesting proposition for carriers since it allows stepwise, low cost, and small granularity increases of the network capacity. Although the transmission design of long-haul undersea systems is far more demanding than for terrestrial links, the concept of upgrades has attracted a lot of attention recently. Previously, upgrades were mainly limited to adding WDM channels to an under-equipped system up to the original design capacity. However, since first reports on how today's technology can be utilised to upgrade legacy systems, e.g. in [1], the concept of extending the capacity beyond the design limits has been adapted in commercial deployments for submarine systems.

First and second generation systems, designed for single channel, 2.5 or 5 Gbit/s, and N x 2.5 Gbit/s WDM operation, respectively, are upgraded by increasing the bitrate to 10 Gbit/s and/or increasing the channel number. However, systems of the most recent, third generation have been designed for N x 10 Gbit/s utilising a narrow spaced DWDM grid. Therefore taking those systems beyond their design capacity by adding channels often represents a

challenge and options to upgrade them – similar to Generation 1 and 2 systems – by increasing the bitrate are studied in this paper.

## Dark fibre vs. overlay upgrades

Another degree of freedom is the way the upgrade is performed. The most straight-forward approach is the so-called dark fibre upgrade (see Figure 1). However, this requires an unused fibre pair to be available or existing channels to be removed and re-routed before the upgrade can be performed. Dark fibre upgrade scenarios using N x 40 Gbit/s using Differential Phase Shift Keying (DPSK) modulation on non-slope matched undersea fibre spans, have already been demonstrated [2,3]. However, most recently a side-by-side comparison of 10 Gbit/s and 40 Gbit/s RZ-DPSK using the same spectral efficiency over transoceanic distances showed 1 dB less margin for 40 Gbit/s, with further reduction due to significantly worse performance fluctuations [4].

Alternatively, overlay upgrades are an option, adding new channels by a (new or already existing) coupler. This has the benefit that existing traffic remains largely unaffected. However, usually this approach limits the maximum achievable capacity since the power of newly added channels is limited by the margins available on the existing ones. Furthermore, mixing bitrates and modulation formats on the same fibre can also lead to penalties by channel interactions [5].

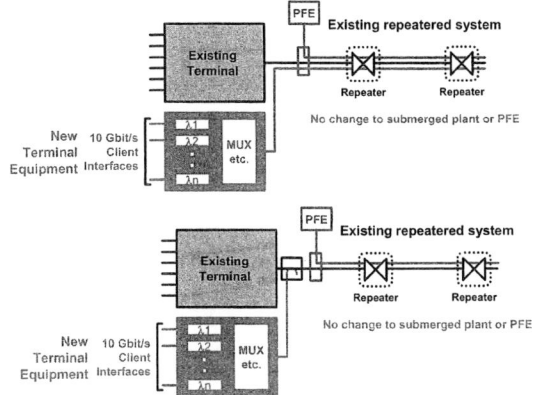

Figure 1 Dark fibre (upper) vs. overlay upgrades (lower).

## 40 Gbit/s DPSK for overlay upgrades

For experimental studies of both 10 Gbit/s RZ-ASK and 20 or 40 Gbit/s RZ-DPSK a re-circulating fibre twin-loop testbed (see Figure 2) using non-slope matched submarine fibres was used. This offers excellent flexibility and allows replication of the dispersion and OSNR maps of typical transoceanic submarine links.

Four Erbium Doped Fibre Amplifiers (EDFA) with characteristics similar to submarine line amplifiers were used to compensate for the fibre attenuation and insertion loss of the loop components. In addition, the loop contained two Fibre Bragg Grating (FBG) based notch filters to suppress Amplified Spontaneous Emission (ASE) noise peaks. The output power of the EDFAs was adjusted to about +9 dBm total launch power into both the Non-Zero Dispersion Shifted Fibre (NZDSF) section and the standard Single-Mode Fibre (SMF) sections. Different NZDSF fibre variants, including large effective area fibre (LEAF) and low-slope fibre were selected to give a good representation of a Generation 3 system.

The power of the WDM channels was

equalised using a channel-based dynamic Gain Equaliser (GEQ) in the inner as well as a two-stage Mach-Zehnder based gain Equaliser (MZ-EQ) in the outer loop. Fast (~ 700 kHz) polarisation scrambling (PS) was applied to mitigate polarisation effects. The sixteen launched WDM channels (1546.12... 1558.17 nm) were co-polarised (representing the worst case) and modulated with PRBS of length $2^{23}$-1.

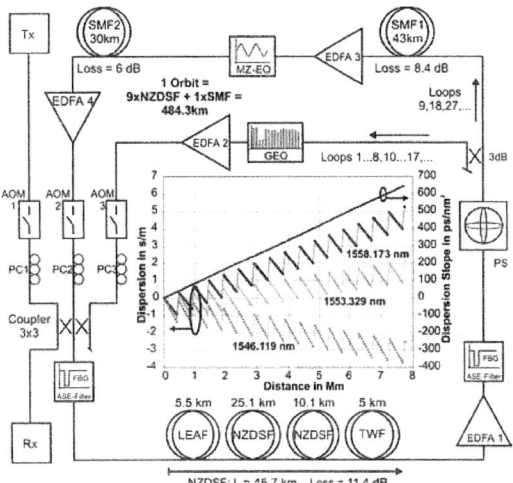

Figure 2 Twin-loop configuration and dispersion map.

As a reference point, the transmission performance of 16x10.7 Gbit/s RZ-ASK was evaluated using a proprietary submarine transponder. As shown in Figure 3 the signal was transmitted over 6000 km with a BER ~$10^{-8}$ without use of additional phase modulation (chirp).

To emulate an upgrade using commercial 40 Gbit/s technology, the 1553.33 nm channel was then replaced by a 42.8 Gbit/s RZ-DPSK modulated signal. From the OSNR of ~7 dB/nm after 6000 km and the noise-loaded back-to-back performance at 42.8 Gbit/s one would have expected a BER of about $10^{-7}$, however only a value of ~$5x10^{-4}$ was observed (see Figure 3) – which is above the level of $1x10^{-6}$ (Q=13.5) required for operating a system using enhanced Forward Error Correction (FEC) with sufficient margins.

Additional computer simulations [5] confirmed that this difference can be somewhat reduced, but this would require in-band dispersion slope compensation, which would have to be applied at the terminals since the installed submarine fibre plant cannot be altered.

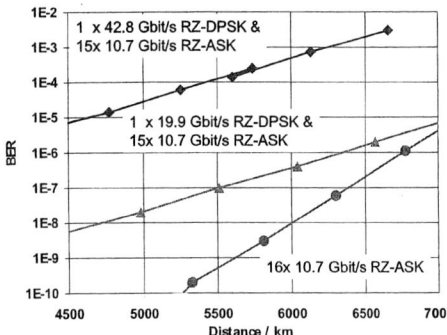

Figure 3 Measured BER vs. distance for different upgrade scenarios for 100 GHz channel spacing.

### 20 Gbit/s vs. 40 Gbit/s
As an alternative approach to enhance the design capacity, 10 Gbit/s channels were upgraded to 20 Gbit/s RZ-DPSK (Figure 3). Without the need for in-band slope compensation of chromatic dispersion and even with a non-optimised receiver the 20 Gbit/s channels showed a much less severe degradation and sufficient margins at 6000 km. Furthermore, the 20 Gbit/s performance converged towards the 10 Gbit/s BER at longer distances.

The maximum spectral efficiency achievable with 20 Gbit/s RZ-DPSK was investigated further by reducing the channel spacing. In contrast to the 40 Gbit/s experiments, 3 channels were replaced by 20 Gbit/s RZ-DPSK channels with variable spacing.

The results show that the channel separation can be reduced to 50 GHz before the onset of significant performance degradations, which is directly relevant for dark-fibre upgrades and most promising for overlay scenarios. A channel spacing reduction from 100 to 50 GHz relates to a BER degradation of less than 1 decade (Q degradation of < 0.4 dB) at 6000 km. For 40 Gbit/s however, the same spectral efficiency of 0.4 bit/s/Hz was achieved using 50 GHz spaced 20 Gbit/s RZ-DPSK at 6000 km – but with about 3 orders of magnitude lower BER.

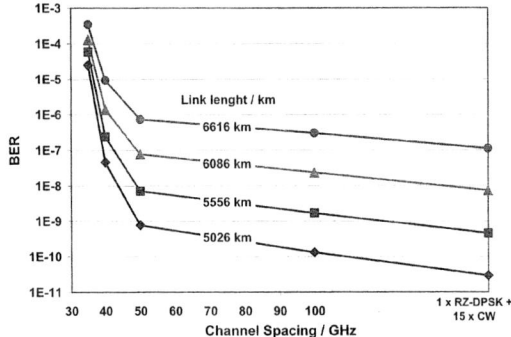

Figure 4 Measured BER vs. channel spacing at different transmission distances.

### Conclusions
Upgrading existing long-haul submarine links beyond their design capacity using 40 Gbit/s technology has been found to be difficult and expensive task and only results in marginal Q performance. On the other hand, 20 Gbit/s RZ-DPSK is a promising candidate for upgrading non-slope matched transoceanic transmission systems. It does not require in-band dispersion slope compensation and for longer distances it offers performance comparable to 10 Gbit/s RZ-ASK at 100 GHz channel spacing – but with twice the spectral efficiency.

### References
1 J. Schwartz et al, Proc. Suboptic 2004, We9.3.
2 J.-X. Cai et al, Proc. OFC 2005, PDP 26.
3 J.-X. Cai et al, Proc. ECOC 2005, Th1.2.2.
4 J.-X. Cai et al, Proc. OFC 2006, OFD3.
5 R. Freund et al, Proc. ECOC 2006, Th1.6.5.
6 M. Rohde et al, El. Lett. 36 (2000), 1483 – 1484.

**Prospective Optical Communication Industry in China**
**Feng Wang, Infostone Communication Consultant**
**Yixin Chen, Shanghai Jiao Tong University**

The growth of optical fiber communication manufacturing industry and market recent years in China are analyzed and reviewed in the presentation. It is prospect that the global market share of optical transmission equipment and optical devices in China will increase with steady steps.

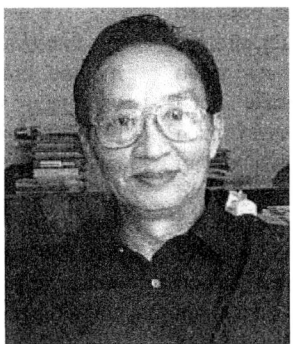

**Professor: Yi Xin Chen**
**Bookham Technology**
**Shanghai Jiao Tong University, China**
**AOE 2006 Steering committee**

**Yi-Xin Chen** is the chief engineer of Bookham Technology (Shenzhen) Co., Ltd and the consultant professor in Shanghai Jiao Tong University (SJTU) physics department. After he graduated from SJTU, department of electrical engineering in 1956, then became the faculty member of this university. He was the first batch of visiting scholars from China to US at University of California, San Diego, in 1980 to study integrated optics. He was the professor, PhD student advisor and chairman of applied physics department in SJTU during 1981 to 1985. He was continuously as a professor there in the optoelectronics, integrated optics, fiber optics, magnetic recording, microfabrication and MEMS etc areas in late 1980s and most of 1990s. He trained more than fifty PhD, Master, and Post Doctor students and published 5 technical books, and more than180 papers. He has also organized many domestic and international technical conferences in these areas. He joined ETEK Dynamics Inc in US from 1998 to 2000, and later JDS Uniphase from 2002 to 2004.

# A Novel Bidirectional Link of Millimetre-Wave Radio-over-Fiber System[*]

Minglei Xiu(1), Rujian Lin(2)

1 : School of Communication and Information Engineering, Shanghai University,
No.149, Yanchang Road, Shanghai 200072, China, minglei_xiu@163.com
2 : School of Communication and Information Engineering, Shanghai University,
No.149, Yanchang Road, Shanghai 200072, China, rujianlin@vip.sina.com

**Abstract** *In the paper, a novel bidirectional link of millimetre-wave ROF system is brought forward for the first time, which is based upon the improved OFM technology. The characteristic of the new bidirectional link structure is millimetre-wave referenced source can be gain at the same time when millimetre-wave carrier is generated.*

## Introduction

The sustaining and rapid growth of the wireless communication traffic has resulted in that the frequency spectrum source in use becomes more and more congested. In addition, the demands for wireless broad bandwidth services, such as wireless video services, mobile multimedia services, and so on, need data rate supporting up to 100Mbps and even much higher [1][2]. In order to satisfy the requirement of great capacity communication, millimetre-wave band is counted as a promising solution due to its characteristics of ultra-broad bandwidth, high efficient reuse of the same frequency, compact base station equipment and small size terminal device. But it is not feasible to directly distribute millimetre-wave wireless signal over long haul in the air space owing to high levels of atmospheric radio frequency (RF) energy absorption by oxygen molecules. Radio-over-Fiber (ROF) technology provides a practical scheme to realize the delivery of millimetre-wave, due to its broad bandwidth, high cost-effective and completely immunity to EMI/EMR.

The central issue of millimeter-wave ROF transmission system is how to deliver millimetre-wave signal over fiber link. It includes two aspects: one is to find an economical method to generate millimetre-wave carrier modulated by data information in downlink; and another is to provide a cost-effective millimetre-wave referenced source avoiding expensive millimetre-wave oscillator employed at base stations (BSs) in uplink.

At present, researchers paid more attention on downlink design than on uplink design. The schemes of uplink structure are approximately divided to two ways: one is lightwave carrier is directly modulated by microwave signal, which has advantage of quite simple BS equipment and disadvantages of high rate external modulator and severe dispersion; another is employing a local oscillator at BS to downconvert the microwave frequency into intermediate frequency, which has advantage of depressed dispersion and disadvantages of increased cost and complexity at BS.

We improved the optical frequency multiplying (OFM) method present by T. Koonen [3] to extend the technology to be applicable for millimetre-wave band [4][5]. In this paper, a novel bidirectional link of millimetre-wave ROF system is brought forward for the first time, which is based upon the improved OFM technology. The characteristic of the new bidirectional link structure is millimetre-wave referenced source can be gain at the same time when millimetre-wave carrier is generated.

## Principle Description of the System

The system is composed of center station (CS), base station and bidirectional optical fiber link. The structure sketch is shown as Figure 1.

At the sending branch of CS, the baseband data is joined with a sinusoidal sweeping signal at frequency $f_{sw}$. The synthetical signal is put on the control drive port of LiNbO3 phase modulator which makes the lightwave output by LD phase modulated by it. the electric field of output lightwave at CS can be denoted as:

$$E(t) = E_c \exp[j\omega_c t + j\beta \sin \omega_{sw} t + j\pi g(t)] \quad (1)$$

Where, $\omega_c$ and $E_c$ are centre radial frequency, electromagnetic amplitude of LD output lightwave, respectively. $\omega_{sw}$ is radial frequency of sweeping signal. $\beta$ is phase modulation index. $g(t)$ is baseband data denoted by NRZ code.

The lightwave is first divided into two branches at BS: one is the transmitting branch of downlink signal, which is applied to generate a millimetre-wave carrier modulated by baseband data; another is millimetre-

---

[*]This project is supported by National Nature Science Foundation of China (60377024) and Nature Science Foundation of Shanghai (04ZR14055).

wave referenced source branch, which is provided to uplink as oscillator of down-conversion.

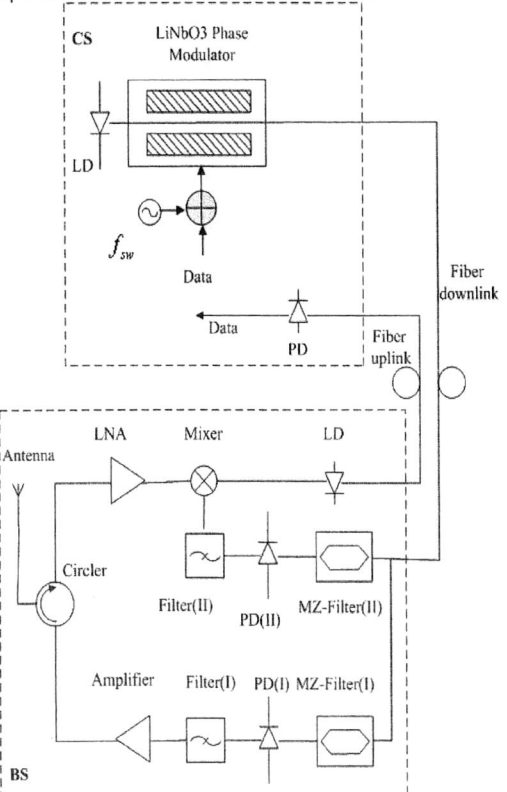

Figure 1. bidirectional link structure od millimeter-wave RoF system

The difference between the two branches consists in different parameters of comb-like optical filters (MZ filter I and II) and band-pass filters (Filter I and II). The output lightwave of MZ filters can be uniformly expressed as:

$$E(t) = \frac{1}{2} E_c \{ \exp[j\omega_c t + j\beta \sin \omega_{sw} t + j\pi g(t)]$$
$$+ \exp[j\omega_c(t - \tau_{mz}) + j\beta \sin \omega_{sw}(t - \tau_{mz})$$
$$+ j\pi g(t - \tau_{mz})] \} \quad (2)$$

Where, $\tau_{MZ}$ is the time delay difference between the two Mach-Zehnder branches, which is $\tau_{MZ1}$ for MZ filter I and $\tau_{MZ2}$ for MZ filter II.

The output photocurrent of the photodiode can be expanded using Bessel function into:

$$i_d(t) = i_0 + \cos[\omega_c \tau_{mz} + \pi g(t) - \pi g(t - \tau_{mz})]$$

$$\cdot J_0(2\beta \sin \frac{\omega_{sw}\tau_{mz}}{2}) + 2\cos[\omega_c \tau_{mz} + \pi g(t) - \pi g(t - \tau_{mz})]$$

$$\cdot \sum_1^\infty (-1)^n J_{2n}(2\beta \sin \frac{\omega_{sw}\tau_{mz}}{2}) \cos 2n(\omega_{sw}t - \frac{\omega_{sw}\tau_{mz}}{2})]$$

$$+ 2\sin[\omega_c \tau_{mz} + \pi g(t) - \pi g(t - \tau_{mz})]$$

$$\cdot \sum_1^\infty (-1)^n J_{2n+1}(2\beta \sin \frac{\omega_{sw}\tau_{mz}}{2}) \cos(2n+1)(\omega_{sw}t - \frac{\omega_{sw}\tau_{mz}}{2})$$
(3)

In order to maximize the power of millimetre-wave frequency component desired, let $\omega_c \tau_{mz} = k\pi$ to eliminate the odd harmonic components, and let $\sin \frac{\omega_{sw}\tau_{mz}}{2} = 1$ to maximize the Bessel function.

Then the millimetre-wave signal at output port of filter can be denoted as:

$$i_m(t) = \cos[\pi g(t) - \pi g(t - \tau_{mz})]J_{2n}(2\beta)\cos(2n\omega_{sw}t)$$
(4)

The equation (4) indicates that the base-band data $g(t)$ originally modulated at phase of lightwave becomes to modulating amplitude of millimetre-wave carrier after filter, and the amplitude $g(t) - g(t - \tau_{mz})$ of millimetre-wave carrier turns into RZ code. When $\tau_{mz}$ value is small, the duty ratio of the signal is quite low and the power of signal distributes at very broad spectrum range. Contrarily, when $\tau_{mz}$ value is large, the duty ratio of the signal is quite high and the power of signal is convergent at low frequency spectrum region. The analyse is show as Figure 2.

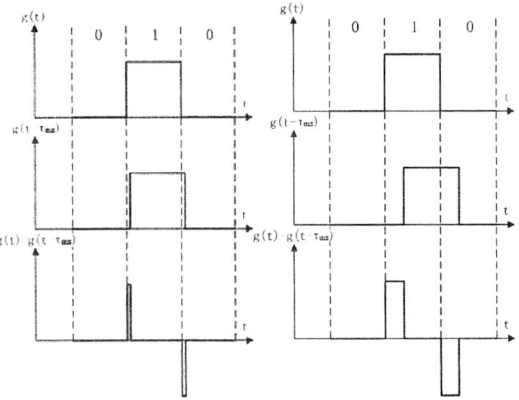

Figure 2. (a)baseband signal    (a)baseband signal wave-shape at small $\tau_{MZ}$    wave-shape at large $\tau_{MZ}$

That is to say, if $\tau_{MZ1}$ is designed at large value connected with a broader passband filter and $\tau_{MZ2}$ is designed at small value connected with a more narrow passband filter, then a millimetre-wave referenced source can be gain at the output of Filter II at the same time when millimetre-wave carrier is generated at the output of Filter I.

**Simulation Result**

The system parameters are set as follows:

At CS, the center wavelength $\lambda_0$ of LD is 1550nm with the full width at half maximum (FWHM) linewidth 1MHz, while the fiber-coupled power is kept as a constant 10mW. The sweeping voltage source is with frequency $f_{sw} = 5$ GHz, and phase modulation index $\beta = 6.3$, and data rate is 100Mbps.

At BS, the MZ filter I is designed with $\tau_{MZ1} = 1$ ns (FSR=1GHz), the centre frequency and the passband of the Filter I are 60GHz, 2GHz respectively. The MZ filter II is designed with $\tau_{MZ2} = 0.1$ ns (FSR=10GHz), the centre frequency and the passband of the Filter II are 60GHz, 1GHz respectively.

Based on the configuration mentioned above, the duty ratios of baseband signal are 0.1 at transmitting branch, 0.01 at referenced source branch.

The spectrum of photocurrent of PD I at transmitting branch is shown as Figure 3. The wave-shape of baseband data after correlative demodulation is shown as Figure 4.

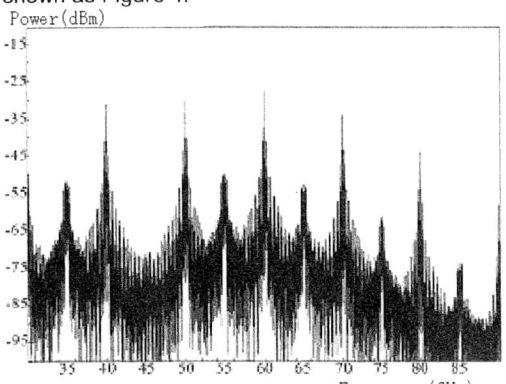

Figure 3.spectrum of 60GHz carrier modulated by data at transmittin branch

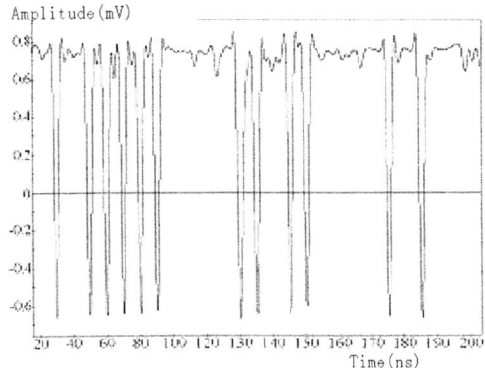

Figure 4. wave-shape of baseband data after

correlative demodulation

Figure 3. and Figure 4. indicate that the power of baseband data is relative high and can be demodulated correctly when the duty ratio is 0.1. Upon this situation baseband data is not very easy to be effected by noise, but is impressible to timing signal.

The spectrum and the wave-shape of 60GHz referenced source at transmitting branch are depicted as Figure 5. and Figure 6. respectively.

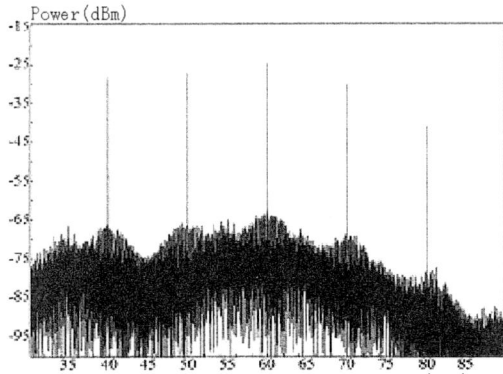

Figure 5.spectrum of 60GHz referenced source

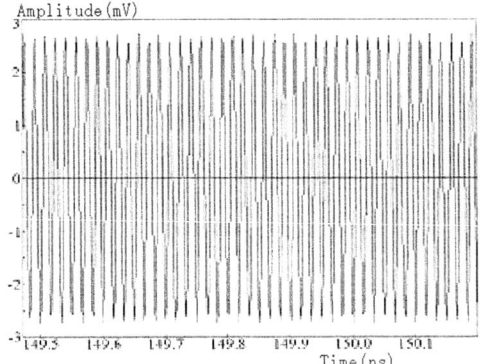

Figure 6.wave-shap of 60GHz referenced source

Figure 3. and Figure 4. indicate that the power of 60GHz millimeter-wave referenced source is around -24dBm, and the power of baseband data is well restrained.

## Conclusions

We present a novel bidirectional link of millimetre-wave ROF system for the first time, which is based upon the improved OFM technology. The characteristic of the new bidirectional link structure is millimetre-wave referenced source can be gain at the same time when millimetre-wave carrier is generated. The bidirectional link structure is proved to be available by the simulation results agreeing with the analysis.

## References

[1] Wake, D., *Trends and prospects for Radio over Fiber picocells*, International Topical Meeting on Microwave Photonics, 2002. pp 21-24.

[2] Ken-Ichi, Kitayama, *Architectural Considerations of Fiber-Radio Millimeter-Wave Wireless Access Systems*, Fiber and Integrated Optics (2000), pp: 167-186.

[3 Ton Koonen, Anthony Ng'oma,, *In house network using multimode polymer optical fiber for broadband wireless services*, Photonic Network Communications,5:2,177-187,2003.

[4] Hailin Qin, Minglei Xiu, Xinqiao Chen, Rujian Lin, *A novel millimetre-wave optical fibre transmission system*, Communication Technology, 2005

Supplement, pp: 43-44.

[5] Minglei Xiu, Hailin Qin, Rujian Lin, *Study on Methods to Yield a Periodically Wavelength-swept Lightwave Signal Based on 60GHz Radio-Over-Fiber System*, Proc. SPIE Vol. 6025, p. 294-300, ICO20,2005.

# Research on the light source of 60GHz millimeter wave ROF system

Yinghua Pan, Minglei Xiu and Rujian Lin

1. School of Communication and Information Engineering, Shanghai University, Shanghai
200072 ,China;Email:shiguangzhidao@163.com
2. minglei_xiu@163.com
3. rujianlin@vip.sina.com

**Abstract:** *A novel light source is proposed and applied in a 60GHz millimeter wave Radio-over-Fiber(ROF) system. The Fabry-Perot semiconductor laser diode（F-P SLD）which shows a multimode output can be converted to a single-mode laser source by employing a F-P SLD locked to the externally injected narrow-band amplified spontaneous emission(ASE). We successfully apply this source in the BS (base station) of 60GHz millimeter wave ROF system.*

### Introduction

As the one of the mayor options for next generation wireless communication, the technology of radio-over-fiber system has been drawing increasing attention. It has been recognized as flexible, bandwidth-efficient and cost-effective option for wireless communication. They process signal in a centralized station (CS) and produce modulated 60GHz millimeter wave signal which is delivered by antennas in the BS. At the same time, received signal must be modulated onto a downstream laser and sent to CS after frequency conversion. Thus a light source to achieve intensity modulation is necessary in the BS. Traditional F-P SLD is cost-effective, but it shows a multimode output and is not suitable for transmission. The method which converts F-P SLD to single-mode laser by injection locking with a stable single-mode light source is not cost-effective.

To solve this problem, a new light source is proposed in the 60GHz millimeter wave system which achieves single-mode output by injecting a narrow band ASE source into an F-P SLD.

### 1. Principle of narrow band ASE-injected F-P SLD

Usually the F-P SLD shows a multimode output. The origin of this feature is the randomness of the amplified spontaneous emission coupled to each longitudinal mode of the laser. The RoF system requires that the optical spectrum of the laser should be narrow enough to minimize the fiber dispersion affection in a high bit rate optical communication system. So we hope the F-P SLD can radiate in a single-mode with the mode partition depressed.

The mode power of F-P SLD is proportional to the spontaneous emission which is coupled to the lasing mode. So we can achieve a single-mode output from

an F-P SLD by injecting a narrow-band ASE source. The mode nearest to the peak wavelength of the injected ASE will be locked to the injected light and the other modes will be suppressed a lot. Then we can expect a sigle-mode output source without partition noise.

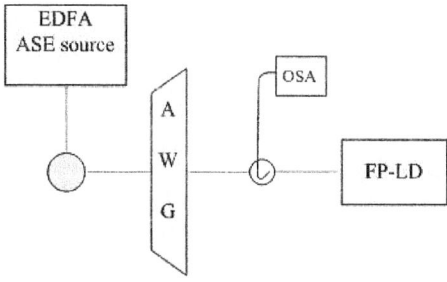

Fig 1 the experimental setup of ASE-injected F-P SLD

Fig 1 shows the experimental setup of ASE-injected F-P SLD.

ASE broadband source can be implemented by amplified spontaneous emission of EDFA. The basic concept of the two-staged ASE source is to generate the long-wavelength ASE in the first stage and inject it into the second stage .It is amplified in the second stage. The pump power for the first stage of EDFA is 50mW,and for the second stage is 100mW.

The broadband ASE is sliced spectrally by an AWG and becomes a narrow band ASE. We can get good injection locking if the mode spacing is larger than the AWG channel spacing. The mode spacing of the F-P SLD is chosen to be 0.6nm.

---

This project is supported by National Science Foundation of China (60377024) and Natural Science Foundation of Shanghai Science and Technic committee (04ZR14055).

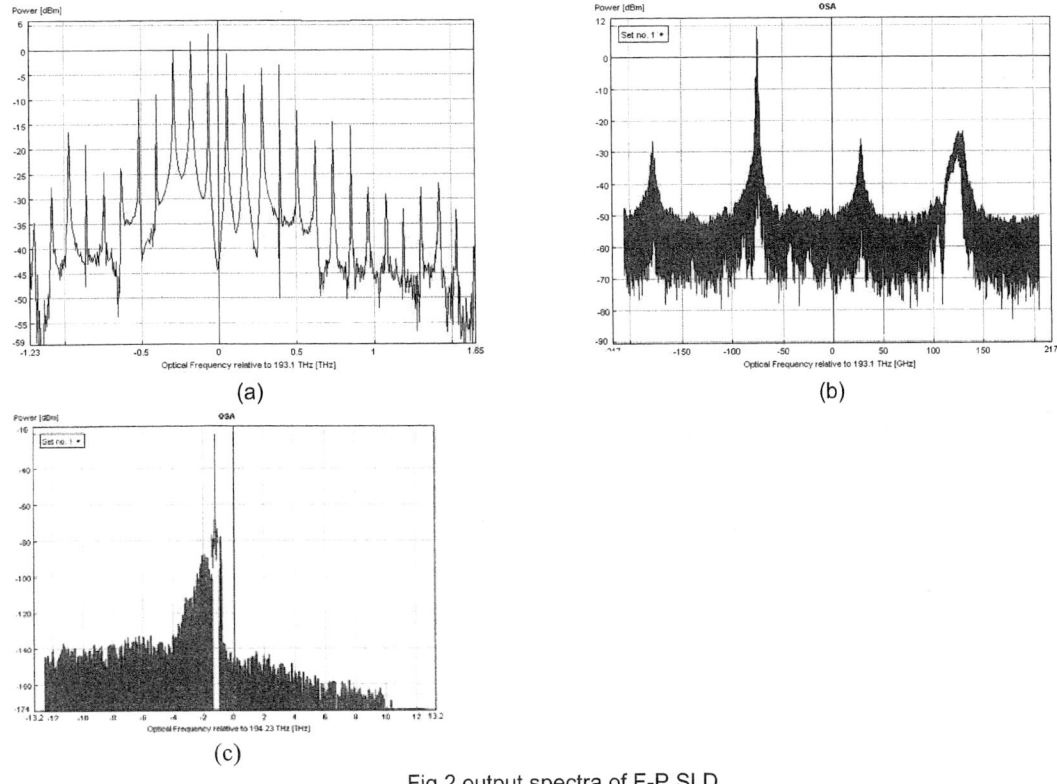

(a)

(b)

(c)

Fig 2 output spectra of F-P SLD

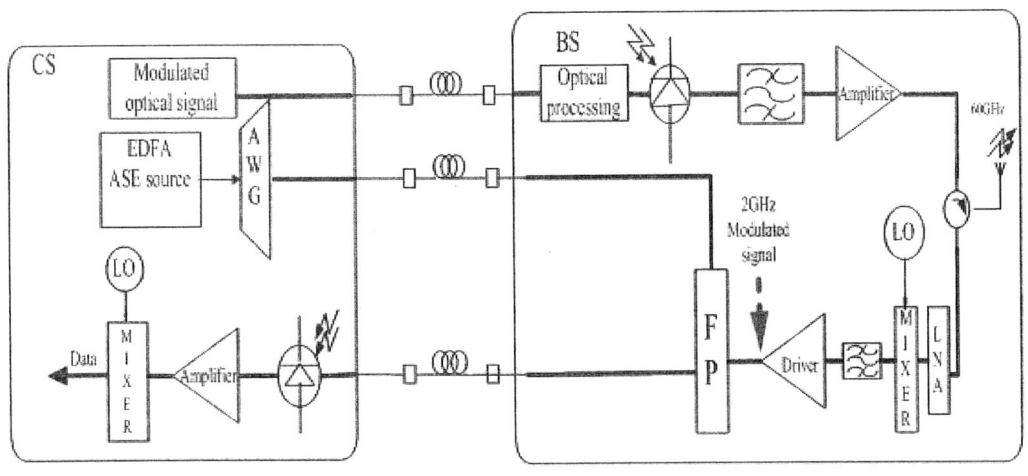

Fig 3  setup of 60GHz millimeter wave system

Fig 2 shows the spectra of F-P SLD with and without narrow band ASE injection.

Fig 2(a) shows the multimode output of F-P SLD without ASE injection and (b) describes the spectra of narrow-band ASE source. Fig 2(c) shows the output of F-P SLD with ASE injection. We can see that the output wavelength of the F-P SLD is locked to the

peak wavelength of the injection ASE and other modes are well suppressed.

As shown in Fig 2(b), the output power of narrow–band ASE is -16dBm. After injecting -16dBm ASE into the F-P SLD, its multimode output turns into single-mode output as in Fig 2(c).The side mode suppression ratio (SMSR) increases from less than

2dB to 30dB, which is enough for system to transmit data. We can increase SMSR by enhancing the narrow band ASE injection power.

## 2. Application of injection locked F-P SLD in 60GHz millimeter wave system

Fig 3 shows the setup of 60GHz millimeter wave system which includes injection locked F-P SLD. The broadband ASE source is located at CS. Then the Fig 3 shows the setup of 60GHz millimeter wave system which includes injection locking F-P SLD. The broadband ASE source is located at CS. Then the spectrally sliced ASE is transmitted to the BS and injected into the F-P SLD which is used in uplink transmission. Every BS will be equipped with one F-P SLD.

To minimize the cost of system, the AWG should be located in the CS together with the broad-band ASE light source .When the narrow-band ASE source is transmitted to BS, the modulated lightwave is sent to BS at the same time. The signal-modulated 60GHz carrier is delivered from BS by an antenna.

In uplink of BS, a 60GHZ signal from the antenna is received and down-converted to a 2GHz IF signal which then intensity-modulates the F-P SLD. The modulated lightwave is transmitted back to the CS.

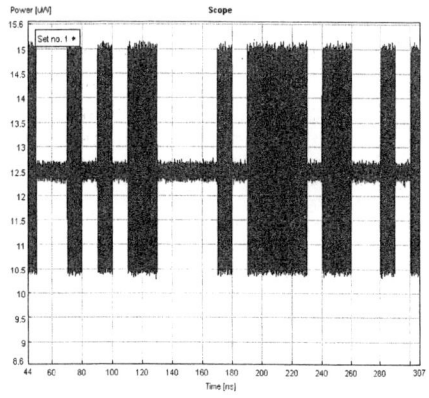

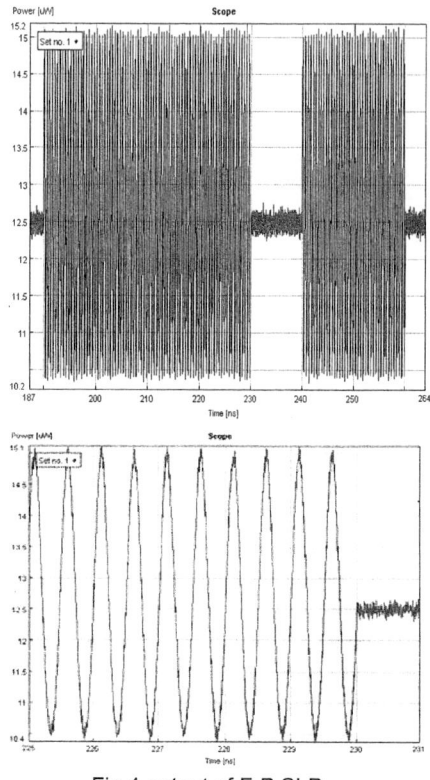

Fig 4 output of F-P SLD

Fig 4 shows the modulated optical output of F-P SLD.

Because the side mode can not be suppressed absolutely, there still is some mode partition noise for level "1" and level "0".We can decrease the affection by increasing injection ASE power properly.

## 4. Conclusion

We propose and demonstrate a novel cost-effective light source employing an Fabry-Perot semiconductor laser diode(F-P SLD ) of which output wavelength is locked to the externally injected narrow-band amplified spontaneous emission(ASE).The F-P SLD with multimode output can be converted to a single-mode laser. This source can be successfully applied in the BS of 60GHz millimeter wave ROF systems.

## References

1. H.D. Kim. et-al. "A low-cost WDM some with an ASE injected Fabry-Perot semiconductor laser," IEEE Photon. Technol. Len. 12.1067-1069 (2000).
2. R. P. Espindola, G. Ales, J. Park, and T. A. Strasser, "80 nm spectrally flattened, high power erbium amplified spontaneous emission fiber source," Electron. Lett., vol. 36, no. 15, pp. 1263–1265, 2000.
3. P.Healey, el-al, "Spectral slicing WDM-PON using wavelength-seeded reflective SOAS," Electron. Lett 37. l18 1-1 182 (2001).

# Research on Wireless Indoor Channel in 60GHz Radio over Fiber System[*]

Qi Zhang (1), Rujian Lin (2), Minglei Xiu (3)

1: School of Communication and Information Engineering of Shanghai University; No.149, Yanchang Road,
Zhabei District, Shanghai, China; zhangqigalford@yahoo.com.cn
2: School of Communication and Information Engineering of Shanghai University; rujianlin@vip.sina.com
3: School of Communication and Information Engineering of Shanghai University; minglei_xiu@163.com

**Abstract** *A kind of framework of next generation wireless communication system based on RoF technology is first introduced in this paper. The research of the wireless indoor propagation channel is mainly presented to serve this new structure system. A new 60GHz model is also developed to simulate the impulse response and makes some necessary statistical analyses by using Matlab. The results include the temporal-spatial impulse response, TOA/AOA statistics and RMS delay.*

## Introduction

Today more and more people have been paying attention to the high-quality communication. Excluding the method of modulation-demodulation, the now existing structure of communication system, the material of system component, our RoF R&D unit has developed a new RoF system structure, using fiber as the connection media between central station (CS) and base station (BS), making BS electrical passive and optical simple.

On the other side, a successful wireless communication system must have reliable wireless channel part, which plays an important role of the whole system. To serve the RoF development, this paper has modelled statistically-geometrically mixed indoor radio propagation channel. The aim of this model is to estimate the shape of impulse response in temporal and spatial field, to calculate the RMS delay, to describe the Angle of Arrival (AOA) of rays, to analyse statistics, such as joint PDF of AOA and TOA (Time of Arrival) and finally to give the conclusion of the general characters of indoor wave behaviour.

This model can match the actual measurement data well and can give a good prediction of system data rate, the longest propagation distance and the mean power level of received signals.

Here is the structure of this article. Section I introduces the RoF technology. Section II starts modelling the indoor channel statistically and geometrically. Section III displays the simulation results and explains them. Section IV conclude the whole article and give some expectation.

## Section I: RoF system introduction

Radio over fiber technique uses fiber as the connection media between central station (CS) and base station (BS). Some RoF techniques have active components to create or to assist to yield 60GHz radio frequency. Some RoF system uses unique

components, such as EAM, EAT and etc. [1] to achieve the similar aim. Others use different approaches, like optical self-heterodyning technology, external modulation technology, optical transceiver, etc.

Our Radio over Fiber (RoF) system uses Optical Frequency Multiplying technique to create 60GHz carrier and our optical and electrical components are also regular and available in market. The typical 60GHz RoF communication system structure is shown in Figure 1. In this figure, CS can connect many BSs, which have their own effective radiation ranges. If BS can be achieved passively, the cost of the whole system will decrease much comparing the existing one.

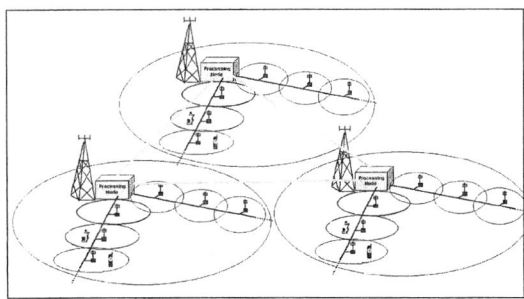

Figure 1. The typical RoF communication system structure

According to the international rule about free-radio license, the 60GHz RoF system's radio power is limited under 10mW. Considering the strong air attenuation of 60GHz millimetre-wave (see Figure 2[2]), the construction structure of BSs is neither micro-cellular nor pico-celluar. It maybe called as "femto-celluar".

Regard as the main topic of this paper, article [3], [4] is recommended for more detail information about our 60GHz RoF system.

---

[*]This project is supported by National Nature Science Foundation of China (60377024), Nature Science Foundation of Shanghai (04ZR14055).

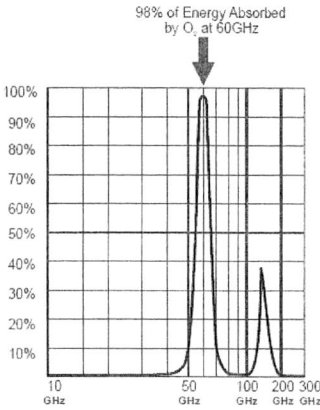

Figure 2. The oxygen absorption of different frequencies after 1000M transmission distance

## Section II: Model characteristics

This indoor 60GHz wave propagation model can be separated into two parts. In Part A, a statistical method is used to describe the impulse response in temporal field. This part shows the relationship between excess delay and amplitude (or power level). This kind of model was first suggested by Turin[5] and lately developed by Saleh and R.A.Valenzuela[6]. The second part describes the spatial response using geometrical approach. This method originates from Richard B. Ertel and Jeffrey H.Reed[7]and is modified to satisfy the circumstance of our building. The whole 60GHz wireless indoor radio impulse response is described as

$$h(t) = \sum_{N(t)} \underbrace{\beta_k(t) \cdot e^{j\theta_k(t)} \cdot \delta(t - \tau_k(t))}_{\text{Temporal Part}} \cdot \underbrace{\delta(\varphi - \psi_k(t))}_{\text{Spatial Part}}$$

(1)

### Part A: Temporal part

In this function, the temporal part uses four-variable model to describe the time-variant channel:

- $N$ : number of muti-path components (MPCs),
- $\{\theta_k\}$ : associated phase shifts,
- $\{\beta_k\}$ : real positive gains,
- $\{\tau_k\}$ : propagation delays.

These four variables are also time-variant. To simplify, all $\{\tau_k\}$ in this article is excess delay. This means $\tau_0 = 0$, $\tau_1 = t_2 - t_1$, $\tau_2 = t_3 - t_2 \cdots$ and all the data can be compared at the same level.

In the temporal field, from observation of real measurement data, the rays arrive in the form of cluster. So "inter-cluster" and "intra-cluster" can be used to describe this phenomenon clearly. Figure 3 is the typical impulse responses with inter-/intra-clusters.

The cause of it is that many rays will arrive at the receiver at different times, different powers, different angles and different phases. These rays will be added vectorially. Among many arrival rays, there

must be several rays with similar arrival delays, similar phases and similar AOAs. So these ones will be added to enhance the total power and on the contrary, the total power will decrease if the coming rays have very different characteristics. This effect may lead to the "cluster".

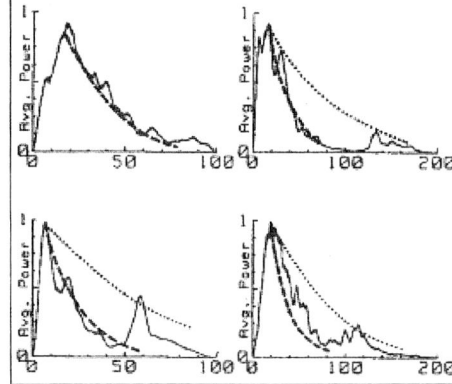

Figure 3. The phenomenon of inter-/intra-cluster

Upon the cluster suppose, the temporal part of (1) can be changed into:

$$h_{temporal}(t) = \sum_{l=0}^{\infty} \sum_{k=0}^{\infty} \beta_{kl} e^{j\theta_{kl}} \delta(t - T_l - \tau_{kl})$$

(2)

where $l, k$ represent the $l^{th}$ cluster and $k^{th}$ ray in this cluster and $T_l, \tau_{kl}$ are excess delay of the $l^{th}$ cluster and $k^{th}$ ray in this cluster respectively. It should be noticed that here $l, k$ are different from the ones in (1). The superposition schematic is shown in Figure 4.

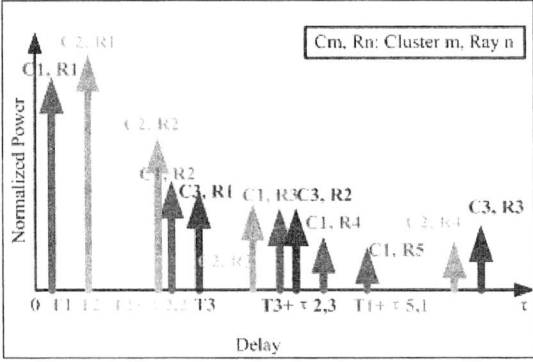

Figure 4. The schematic of vector superposition

So if one can describe four variables in (1) or (2), the exact temporal function can be achieved. The four variables will be discussed below.

Generally, associated phase shifts $\theta_k$ are supposed to satisfy the uniform distribution within the range $[0, 2\pi)$.

In TOA aspect, it is suggested that excess delays fluctuate around the mean arrival rates $\Lambda$ and $\lambda$. Each of them forms the Poisson arrival time sequences, called as double-Poisson distribution. Using the discrete time distribution, the following two Poisson PDFs describe these two delays respectively:

$$P\{T = T_k\} = \frac{\Lambda^{T_k} \cdot e^{-\Lambda}}{T_k !} \qquad (3)$$

$$P\{\tau = \tau_{kl}\} = \frac{\Lambda^{\tau_{kl}} \cdot e^{-\lambda}}{\tau_{kl} !} \qquad (4)$$

where $\Lambda, \lambda$ stand for the mean rate of clusters and rays. Note that $T_0$ is 0 while $\tau_{1,l}(l = 1, 2, \cdots)$ is 0 too.

From Figure 4, my model describes the impulse response power as the Exponential process. This means the amplitude of each ray in one cluster (C1-R1, C1-R2, C1-R3, C1-R4, C1-R5, ...) is Exponential distribution and the first rays of each cluster (C1-R1, C2-R1, C3-R1, ...) also belong to that form (of course with other different parameters). Based on this hypothesis, the power gain can be written as

$$\overline{\beta_{kl}^2} \equiv \overline{\beta^2(T_l, \tau_{kl})} = \overline{\beta^2(0,0)} e^{-T_l/\Gamma} e^{-\tau_{kl}/\gamma} \qquad (5)$$

where $\overline{\beta^2(0,0)}$ is the average power gain of the 1st ray of the 1st cluster and $\Gamma, \gamma$ are power-delay time constants for the clusters and rays, which describe how fast the clusters or rays attenuate, respectively. Based on the average gain, each specific power can be created using exponential PDF model, which describes below:

$$p(\beta_{kl}^2) = \frac{1}{\overline{\beta_{kl}^2}} \cdot \exp(-\frac{\beta_{kl}^2}{\overline{\beta_{kl}^2}}) \qquad (6)$$

Finally, $\overline{\beta^2(0,0)}$ can be solved as

$$\overline{\beta^2(0,0)} \approx \frac{G(1m)}{\lambda \gamma r^\alpha} \qquad (7)$$

where $G(1m)$ is the power gain obtained at 1M from the source transmitter.

In the simulation, we can set the threshold to stop the program. This happens when $e^{-T_l/\Gamma} \ll 1$ and $e^{-\tau_{kl}/\gamma} \ll 1$ in (5). Every time the computer simulates, the number of MPC will be added. So at the end of simulation, one can count the simulation time as $N$, the number of MPC.

**Part B: Spatial part**
Spatial part of (1) mainly computes the AOA at the site of receiver. The approach is quite different from [8] but the aim is similar to get the PDF of AOA, TOA and joint PDF of TOA/AOA.

This model only considers the circular situation to

meet with the actual circumstances of our building. To begin, some necessary hypothesises should be made.
- All the scatters will distribute in the circle uniformly,
- There should be one direct propagation channel between Tx and Rx.
- Tx and Rx should be set in the same high level height.

The joint PDF of TOA/AOA can be written as[7] function (8). The marginal PDF-TOA and AOA are (9) and (10) (See Appendix), where $D, R, c$ are the radius of circle (or room), distance between Tx-Rx and light velocity, respectively. The assistance functions are listed below which will be used in (10).

$$A_\tau(\tau) = R^2\alpha + \int_\alpha^\pi (\frac{c^2\tau^2 - D^2}{2c\tau - 2D\cos(\psi)})^2 d\psi \qquad (11)$$

$$\alpha = \arccos(\frac{D^2 + 2R\tau c - c^2\tau^2}{2RD}) \qquad (12)$$

From (8)~(10), the AOA and TOA relationship can also be obtained. You can see simulation results in the next section.

To complete (1), some special random numbers, which follow (8) ~ (11), must be created. The approach is to first integrate PDF to get the CDF and then create uniform random numbers in [0,1), finally use these numbers to get the numerical results of reversed CDFs.

**Section III: Simulation results**
Based on the Section II, some simulations can be obtained.

**Part A: Temporal part**
Figure 5 shows a 2D simulation result of one-time computation. You can clearly see the cluster and ray.

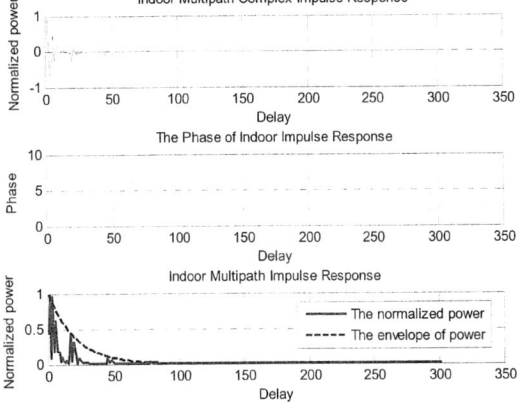

Figure 5. 2D simulation result of one-time computation

Figure 6 shows 3D simulation result. From the simulation times, you can see that the indoor channel is time-variant and Figure 7 shows the RMS Delay

---

\* Function (11) can be solved by numerical approach, using MATLAB.

result.

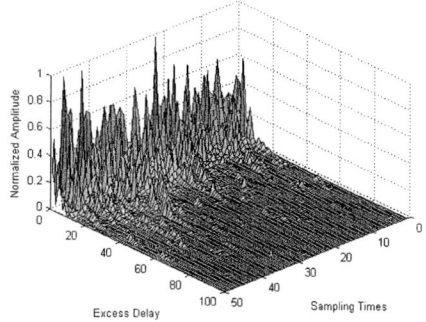

Figure 6. 3D simulation result

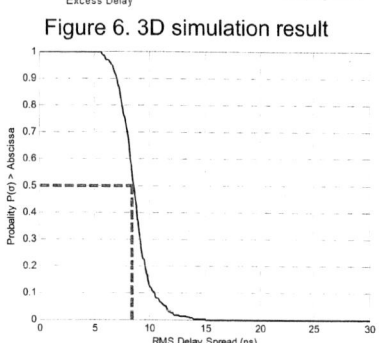

Figure 7. RMS Delay

From this figure, it can be found that one half of excess delays are less than about 8 ns (LOS). The maximum excess delay is no more than 15 ns. This can determine the mean data rates of digital system.

**Part B: Spatial part**

The joint PDF of TOA/AOA is shown in Figure 8.

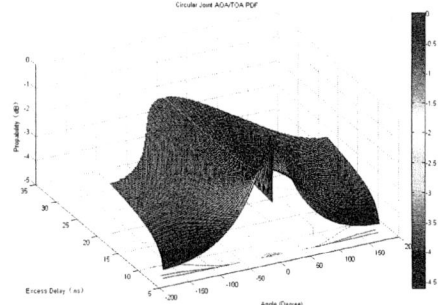

Figure 8. The Joint PDF of TOA/AOA.

Figure 9 and 10 are AOA PDF and TOA PDF respectively.

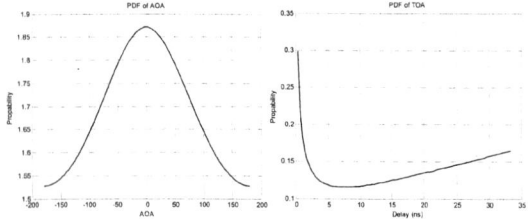

Figure 9. TOA PDF          Figure 10. AOA PDF

Figure 8~10 show that one half of AOA occurs between ±80° and within 8ns most rays arrive at the receiver. This fits to the result of Part A (See Figure 7.).

**Section IV: Conclusions**

We use MATLAB to simulate the 60GHz indoor radio propagation channel. The model combines traditional temporal statistical approach and novel spatial geometrical approach. Two different methods can give the similar results, such as TOA PDF, etc., which proves that they can work compatibly. From the simulation results, this model can match the real 60GHz wave behaviour well and testify that our RoF system has good performances in wireless aspect. This model can be also transferred into other measurement circumstances. It should be noticed that because of the randomness of this model, it can only provide the general characteristics of special indoor channel by changing some program input parameters. Up to date, the exact model presenting the channel perfectly doesn't exist in the world. This model is the one that compromises the complication and preciseness.

**References**

[1] KEN-ICHI KITAYAMA, *Architectural Considerations of Fiber-Radio Millimeter-Wave Wireless Access Systems*, Fiber and Integrated Optics (2000), pp: 167-186.

[2] Terabeam Corporation, *Performance Characteristics of 60GHz Communication Systems*, Sep. 3rd, 2002, p. 2.

[3] Minglei Xiu, Hailin Qin, Rujian Lin, *Study on Methods to Yield a Periodically Wavelength-swept Lightwave Signal Based on 60GHz Radio-Over-Fiber System*, Proceedings of SPIE-The International Society for Optical Engineering (2005).

[4] Hailin Qin, Minglei Xiu, Xinqiao Chen, Rujian Lin, *A Novel fiber millimeterwave communication system (2005)*, techniques of communication, pp: 43-44.

[5] G. L. Turin, *Communication through noisy, random-multipath channels*, IRE Convention Record, part 4 (1956), pp. 154-166.

[6] A.A.M. Saleh, R.A.Valenzuela, *A statistical model for indoor multipath propagation*, IEEE Journal on selected areas of communication. Vol. SAC-5 (February 1987), pp.128-137.

[7] Richard B. Ertel and Jeffrey H.Reed, *Angle and Time of Arrival Statistics for Circular and Elliptical Scattering Mode*, IEEE Journal on selected areas in comm., Vol. 17, No. 11(Nov. 1999), pp: 1829-1840.

[8] Joseph C.Liberti and Theodore S. Rappaport, *A geometrically based model for Line-of-sight Multipath radio channels*, IEEE (1996), pp:844-848.

**Appendix:**

$$f_{\tau,\psi}(\tau,\psi) = \begin{cases} \dfrac{(D^2-\tau^2 c^2)(D^2 c+\tau^2 c^3-2\tau c^2 D\cos(\psi))}{4\pi R^2}, & \dfrac{D^2-2\tau cD\cos(\psi)+\tau^2 c^2}{\tau c-D\cos(\theta)} \le 2R, \ \psi \ne 0 \\[2ex] \dfrac{c(D+\tau c)}{4\pi R^2}, & \dfrac{D}{c} \le \tau \le \dfrac{D+2R}{c}, \psi = 0 \\[2ex] 0, & \text{else} \end{cases} \tag{8}$$

$$f_{\psi}(\psi) = \frac{1}{2\pi R^2}(D^2\cos(2\psi)+R^2+2D\cos(\psi)\sqrt{R^2-D^2\sin^2(\psi)}), -\pi \le \psi \le \pi \tag{9}$$

$$f_{\tau}(\tau) = \begin{cases} \dfrac{c}{4R^2} \cdot \dfrac{2\tau^2 c^2-D^2}{\sqrt{\tau^2 c^2-D^2}}, & \dfrac{D}{c} \le \tau \le \dfrac{2R-D}{c} \\[2ex] \dfrac{1}{\pi R^2}\dfrac{dA_{\tau}(\tau)}{d\tau}, & \dfrac{2R-D}{c} < \tau \le \dfrac{2R+D}{c} \end{cases} \tag{10}$$

# Model of PPM Receiver used in Deep Space Communication Systems

Xiaoyan Wang, Sami Fadali, and Moncef B. Tayahi
University of Nevada, Reno
Advanced Photonics Research Lab
Reno Nevada 89557- USA
wangx4@unr.nevada.edu

**Abstract** – A receiver for optical Deep Space Communication System (DSCS) is designed and modeled. A detailed architecture of the Pulse Position Modulation (PPM) and demodulation is addressed and simulated. The optical link part and the free space channel are also designed and simulated. Physical characteristics of the free space communication channels are modeled, including path attenuation, noise, distortions, time delay, etc. Preliminary results show that external modulation is adequate for characterizing complex and costly DSCS links.

**Key Words -** Deep Space Communication Systems (DSCS), Pulse Position Modulation (PPM), slot synchronization, frame synchronization.

## I. Introduction

Pulse Position Modulation is an attractive modulation technique for the DSCS. With this coding technique, M bits of information are encoded onto one of $L = 2^M$ PPM words by establishing a one-to-one correspondence between the possible states of M binary digits and the location of an optical pulse among L possible slots [1-2]. Currently, NASA had adopted this coding technique for the deep space communications systems. Unlike the binary modulation, one does not require a prior knowledge of the signal or background noise radiation levels to implement an optimum PPM receiver. The other key requirement of systems considered for space applications is the peak laser power level that must be large enough to survive huge deep-space losses. For this reason, Q-switched lasers typically are employed for such applications.

The DSCS link model needs to be analyzed and simulated accurately, because the choice of a suitable architecture, the optimal algorithms and the performance of the demodulation part directly depend on the characteristics of the DSCS channel itself and the properties of the optical components to be used. The receiver is one of the crucial parts of such systems [3]. Many types of the demodulation design were discussed and analyzed [4]. The discrete-time demodulation architecture, which combines the post detection filtering and slot synchronization,

is investigated in the reference articles [1-3]. The simulation setup of one kind of the demodulation architecture is implemented in detail.

This work builds on a large body of prior work in deep space optical communication [6-8]. The primary task is to model and simulate a discrete-time pulse position modulation and demodulation link setup. Section III describes the simulation of the electrical transmitter part of pulse position modulation in detail. The preliminary characteristics of the deep space channel are also analyzed in this section and the special compound component for this channel model is built. Section IV shows the preliminarily demodulation of PPM.

## II. Proposed deep space communication system

The Deep Space Communication System (DSCS) is developed for digital transmission between spacecrafts and ground receivers. An end-to-end system validation relies on optical links between transmitters flying onboard interplanetary spacecrafts and earth-based receiving stations. The deep space communication link diagram is illustrated in Figure 1. A data source component generates the digital- data that will be transmitted. The information can have data, audio or video format. The encoder generates

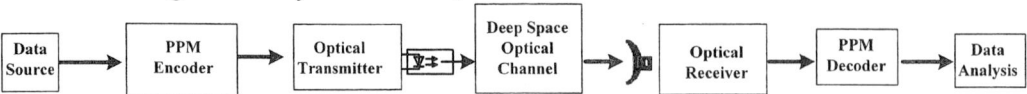

Figure 1: Diagram of the Deep Space Communication link

the PPM code. The optical transmitter converts the electrical PPM signal using an external modulator. The ground optical signal is detected by an optical detector and converted into a PPM electrical signal which is fed to the decoder stage.

The high-speed PPM decoder section performs slot synchronization, symbol synchronization, and pulse position demodulation with other digital signal processing functions to recover an error free signal. A 16-PPM symbol set and 16 silent slots are used as a proof of concept through the simulation with the high speed signal processing frequencies up to 1GHz. In this work, the parts of PPM generator, PPM encoder and PPM decoder are carried out. The parts of optical transmitter, deep space channel and optical detector, are modeled in optical simulation software. An approach which is fit for the field programmable gate arrays (FPGAs) design for the PPM encoding and decoding is adopted. The method of using the CW laser and Mach Zhender modulator for the PPM transmission is confirmed. Although the simulation is only preliminary and basic, this approach provides very useful insights into the PPM transceiver design.

## III. Simulation of pulse position modulation transmission link

The architecture of the PPM encoder is modeled and shown in Figure 2. The PPM encoder includes the slot clock source, the frequency divider, the serial-to-parallel converter, the 4-to-16 line decoder for the parallel to the serial conversion, the clock counter, the selector, and other secondary components.

In the architecture, the slot clock is generated by a pulse generator with a frequency of 1GHz. Lower frequencies are achieved by frequency dividers. The frequency dividers are made of sets of D flip-flops and their output frequencies are slower than the frequency of the input signal generated by the pulse generator. The tributaries are used to trigger a pseudorandom generator to generate a pseudorandom digital sequence. Every rising edge of the generator's input square pulses will trigger the pseudorandom generator to output one in the pseudorandom output sequence to the serial-to-parallel converter adjacent to it. The serial-to-parallel converter does the conversion of the pseudorandom sequence from the serial sequence into the

parallel sequence. The output parallel sequence is sent to a 4-to-16 line decoder which is comprised of a combinatorial logic block. The 4-to-16 line decoder changes every four bits of the input sequence into an array of $L = 2^m$ elements ( $m = 4, L = 16$ ) and sends them to the selector. According to the slot number obtained from counter 1 block, the selector selects the exact slots and organizes them into the signal frames. There is also the counter 2 block which counts the total slots of one frame and sends the exact number of the slots in one frame to the middle input of a switch block. The switch block compares the number of the slots with the switch's threshold. The threshold presents the total number of the data slots in one frame and here it is chosen as "16". If the input slot number is smaller than or equal to the switch's threshold, the switch block will choose the upper signal slot into the frame, so this signal slot occupies the data slot in the considered frame. If the input slot number of the switch block is bigger than the switch's threshold, the switch will choose the lower silent slot into the frame. The "zero" value is used as the lower silent slot and occupies the silent slot in the frame. The above procedure repeats for the 16 signal slots and 16 silent slots in one frame. The generated and modulated data is sent to the "To File" module. The "To File" module receives the modulated data and saves it into a data file according to a special data format of the simulation software.

For illustration purposes and to explain how the PPM encoder works, we use the following example: a digital sequence "100111001101...... " is generated by the pseudorandom generator and sent into the serial-to-parallel converter. The serial-to-parallel converter will convert every four serial digits into four parallel digits, i.e., from "1001" serial digits to "1001" parallel digits. The output parallel sequence is then sent to a 4-to-16 line decoder. The 4-to-16 line decoder converts the "1001" parallel digits into a sequence of "0000000010000000". The two counters will control the selector and the switch to select this "0000000010000000" sequence as the signal data slots and select zeros as the silent slots. The output sequence from the switch is "00000000100000000000000000000000" for the first frame which presents the first four digits "1001". The diagrams of PPM waveform, frame clock, slot clock, counters output and the

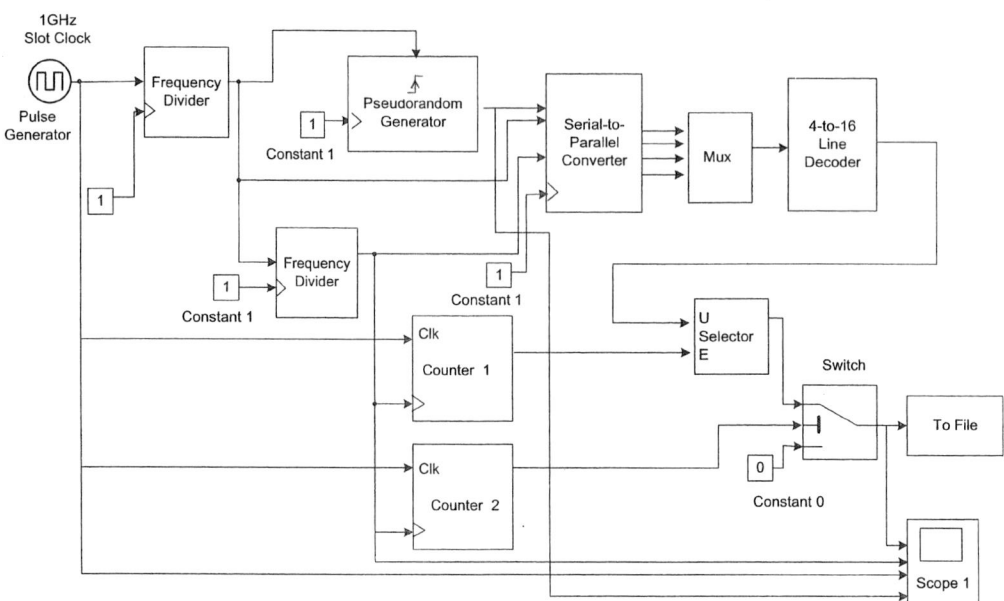

Figure 2: Simulation architecture for the PPM encoder

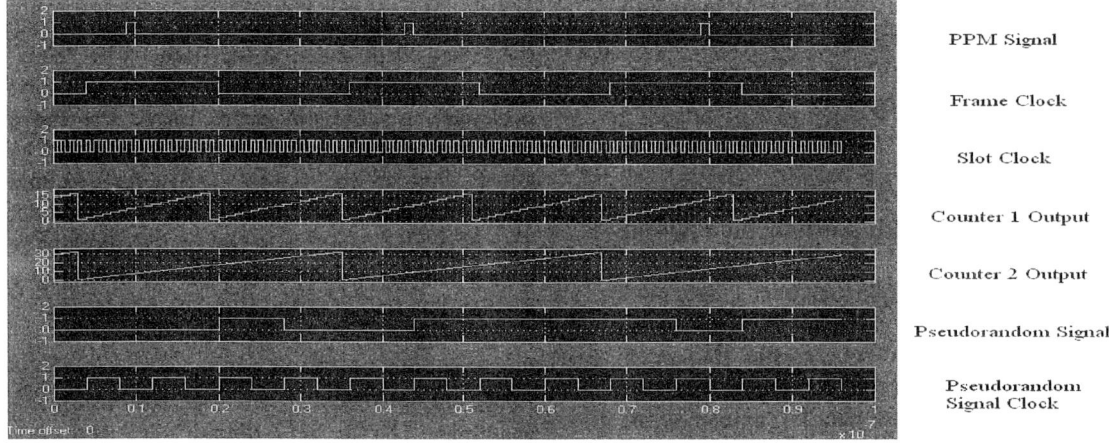

Figure 3: Diagram of the output stage for PPM encoder

pseudorandom signal and signal clock of the illustrated example are shown in Figure 3.

In order to achieve optimum data rates for PPM based communication systems, and considering the physical characteristics of the transmitter, a CW laser (1064 nm) and Mach-Zehneder (MZ) modulator were chosen. Because the physical characteristics of the deep space channel change under different atmospheric conditions, the channel model with variable parameters, such as path attenuation, noise, distortions, and time delay, were modeled using optical simulation software.

The PPM optical channel setup is shown in Figure 4. The encoded signal is received from a data file by the electrical playback component in

OptSim, and then is sent to the optical transmitter which performs the functions of converting the electrical signal from the data files into the optical channel module, and sending the output signal into an OptSim data file for further processing.

In order to simulate a realistic setting, noise is injected with the input PPM signal before the MZ modulator. The signal is sent to the first gain controlled amplifier (Amp 1) to be adjusted to the suitable range according to the input signal amplitude. This amplifier serves as driver for the single arm Mach-Zehnder with $Sin^2$ response. The laser source is a continuous wave (CW) laser with a center wavelength of 1064nm. The output optical signal from the modulator is

144

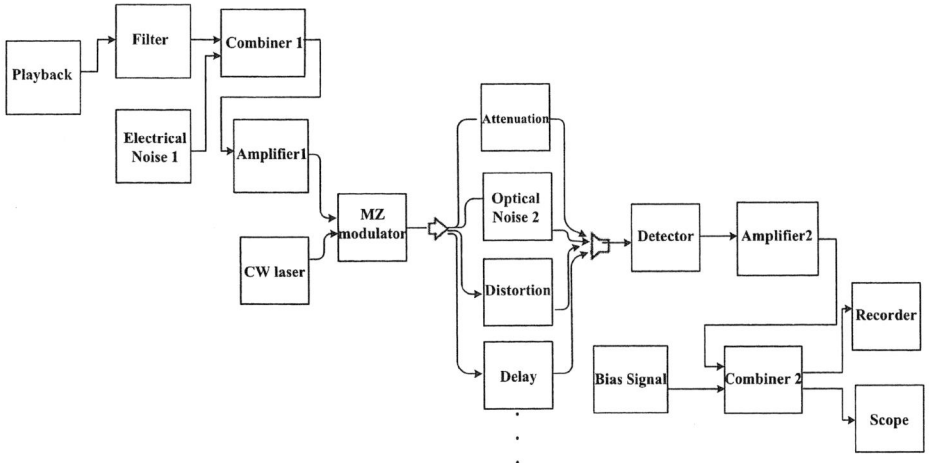

Figure 4: Optical simulation setup

fed into the optical channel module. The module has variable parameters such as attenuation, noise, distortions, delay, etc. At the end of the optical channel module, the optical signal is detected by a detector which is the high sensitivity avalanche photodiode (APD). Since the output signal has a DC offset, a negative bias is used to bring down the PPM signal to zero offset. The output signal is sent to the electrical recorder component, which saves the signal into an OptSim data file in the data format of OptSim. The data can be read later by Simulink for further signal processing.

The multi pulse diagram and the single pulse diagram of the output signal of the combiner 1 block are shown in Figure 5. The multi-pulse diagram and the single pulse diagram of the output signal of the combiner 2 are shown in Figure 6. It is shown that under realistic conditions and because of the noise in the optical components such as CW laser, MZ modulator, and the APD, the signal is distorted.

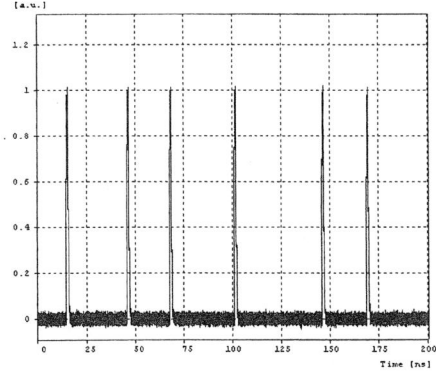

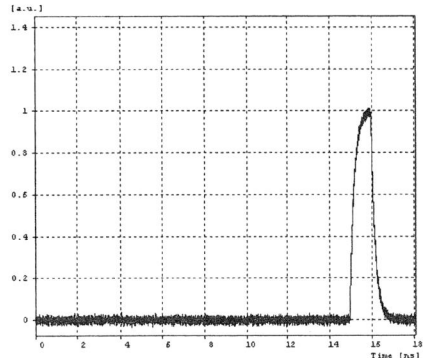

Figure 5: The multi pulse diagram and single pulse diagram of the input signal

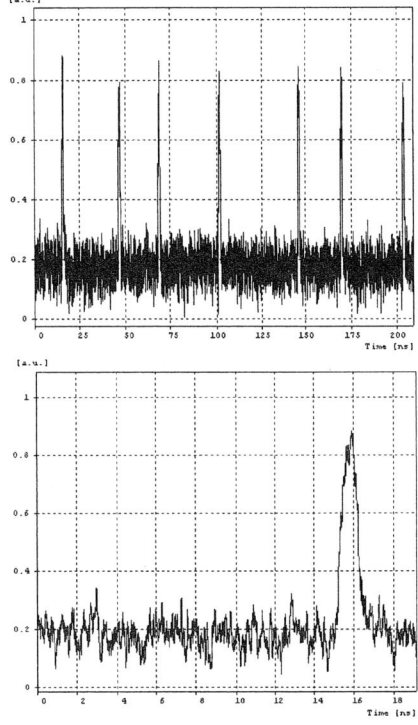

Figure 6: The multi pulse diagram and single pulse diagram of the output signal

145

Throughout the modeling, the signal data is transferred back and forth by data files from Simulink to OptSim. Because the data formats of Simulink and OptSim are format incompatible, a subroutine is used to convert the data format of the Simulink data file to the OptSim recognizable data file. The data transfer diagram is shown in Figure 7. The Simulink subroutine A1 generates the data file F1. Then subroutine A2 changes the data format and produces data file F2. File F2 can be received and interpreted by the OptSim program. After the simulation in the OptSim, the subroutine B2 changes the data format of a data file F3 and produces a new data file, F4. The subroutine B1 reads the data from the data file F4 and uses it for further simulation. In this module, the signal can also be plotted and checked. The plotted results given by the subroutine A1 are shown in Figure 8 and the plotted results given by the subroutine B1 are shown in Figure 9.

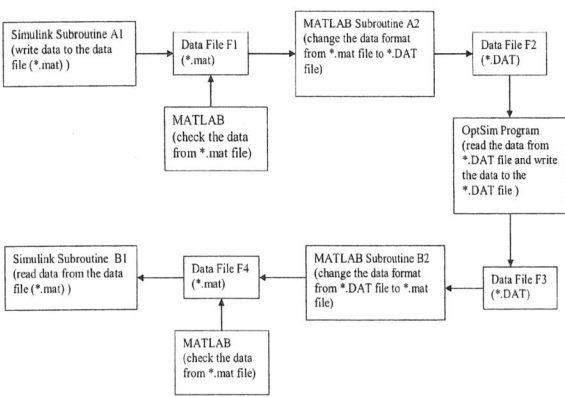

Figure 7: Diagram of the data transferring between Simulink and OptSim

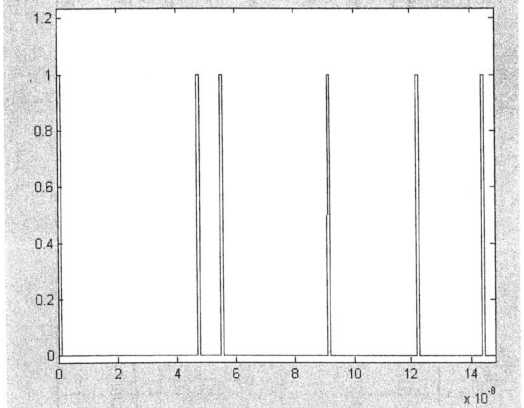

Figure 8: Diagram of the plotted results given by the program A1

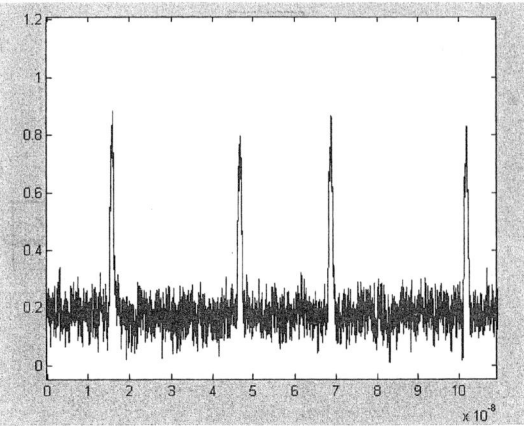

Figure 9: Diagram of the plotted results given by the program B1

## IV. Simulation of pulse position demodulation

The functions of the PPM receiver include signal detection, signal data reading, synchronization recovery (slot recovery, symbol recovery and frame recovery), serial-to-parallel conversion and the 16-to-4 line decoding. The PPM demodulator architecture is shown in Figure 10 and the received signal is read from a file block module. The electrical signal is sent to a signal detection block, which performs the data synchronization and recovery functions. The output signal of the data detection, which is sent to the synchronization recovery block and the serial-to-parallel converter block is shown in Figure 11. The synchronization recovery block recovers the slot rate, the symbol rate and the frame rate, respectively. The three types of recovered synchronization information are sent to the serial-to-parallel converter and the counter. The serial-to-parallel converter, which is made of a set of D flip-flops, converts the serial sequence obtained from signal detection block into the parallel sequence. The counter counts the number of the slots in one frame. The mux block receives the parallel sequence from the serial-to-parallel converter and changes it into the arrays of $L = 2^m$ elements. There is a 16-to-4 line decoder receiving these arrays and decoding each array into 4 digits of one frame. This 16-to-4 line decoder is made of a functional component. Through this component, the decoding subroutine can be used in the Simulink file. According to the recovered frame rate and the slot rate, every frame with $L = 2^m$ slots is changed to an array of 4 digits known as an element. The data signal from the synchronization recovery block is sent to the clock input of the counter. Another frame rate

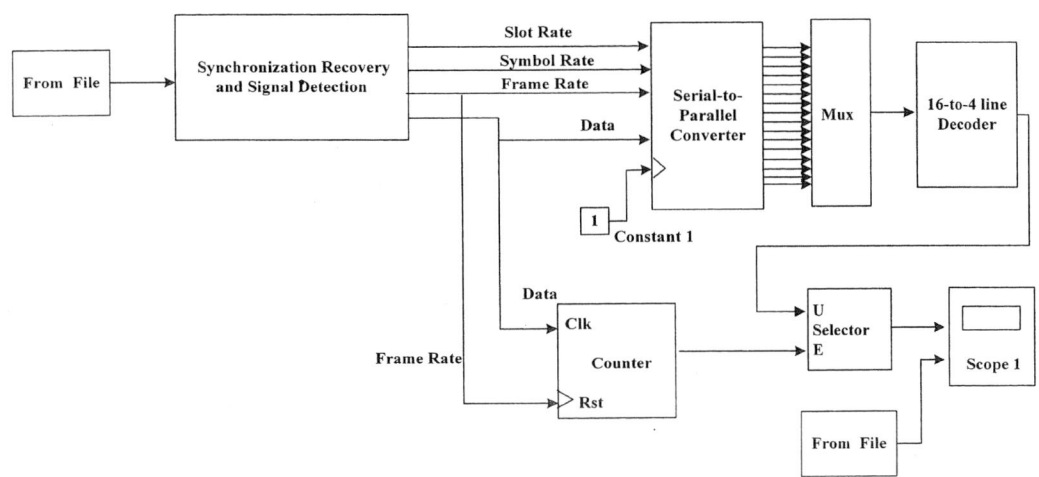

Figure 10: Diagram of the PPM demodulation architecture

signal from the synchronization recovery block is sent to the reset input of the counter. The counter calculates the number of slots in one frame and resets itself every frame. The output signal of the counter controls a selector to select the exact four digits in one frame from the output signal of the 16-to-4 line decoder. With the counter and the selector, the position in one frame of every pulse is calculated and decoded into a digital sequence.

For example, the following is a digital sequence received by the discrete-time demodulation architecture. With the help of the synchronization recovery and signal detection component, the slot and frame clocks are recovered and the value of every slot is detected as "0000000010000000000000000000000". According to the slot clock and the frame clock, the serial output sequence "0000000010000000000000000000000" from the synchronization recovery and signal detection component is sent to the serial-to-parallel converter and changed to the parallel sequence "0000000010000000". The 16-to-4 line decoder receives the parallel sequence and output "1001" according to the position of "1" in the frame. The counter and the selector will select the output "1001" into the exact slots for the final output sequence. An example of the output signal diagram is shown in figure 12.

Figure 11: Diagram of the output signal after the detection component

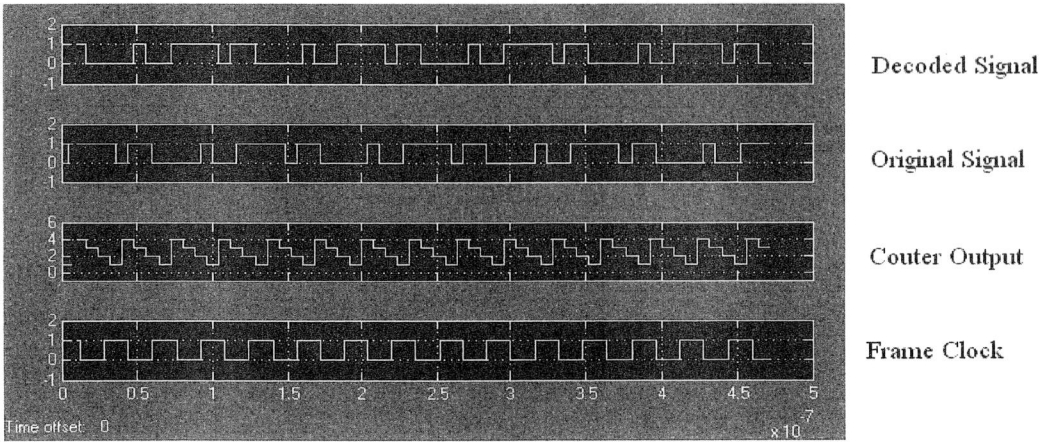

Figure 12: Diagram of the final output signal in the Simulink

## V. Conclusions

A Deep Space Communication Systems is designed and simulated. A discrete-time pulse position modulation and demodulation techniques are modeled. Realistic characteristics of the deep space channel were used in the model to gain insights to the deep space optical communication link. The pulse position modulated signal is transmitted through the optical communication channel, and demodulated successfully using our architecture. It is confirmed that the pulse position modulation and demodulation can be implemented using external modulation. A Mach Zehnder modulator was used in the PPM deep space communication test bed for a proof of concept. The simulation of PPM modulation and demodulation can be implemented into a DSP code or VHSIC Hardware Description Language (VHDL) using FPGAs. Although various models were investigated in this area, there are still many challenges into modeling and designing optical deep-space communication.

**Acknowledgement:** This project was supported in part by Nevada Space Grant EPSCoR Program.

## VI. References

[1] Andrew A. Gray, Clement Lee "Discrete-Time Demodulator Architecture for Free Space Broadband Optical PPM," *NASA JPL internal Document.*

[2] V. Vilnrotter, A. Biswas, W. Farr, D. Fort, and E. Sigman, "Design and Analysis of a First-Generation Optical Pulse-Position Modulation Receiver," *The Interplanetary Network Progress Report* 42-148, October–December 2001.

[3] M. Gebhart, E. Leitgeb, J. Bregenzer "Atmospheric Effects on Optical Wireless Link," *The 7th International Conference on Telecommunications – ConTEL,* Zagreb, Croatia, June 11-13(2003)

[4] T.-Y. Yan and C. and C. Chan, "Design and Development of Deep Space Baseline Optical Transceiver," *Proc. SPIE,* vol. 3615, San Jose, California, January 1999.

[5] J. Proakis, *Digital Communications,* New York: McGraw-Hill, Inc., 1995.

[6] M. Simon, S. Hinedi, and W. Lindsey, *Digital Communication Techniques,* Englewood Cliffs, New Jersey: PTR Prentice Hall, 1995.

[7] Heinrich Meyr, Gerd Ascheid, *Synchronization in Digital Communications,* Wiley Series in Telecommunications, John Wiley and Sons, New York, 1990.

[8] M. Srinivasan, V. Vilnrotter and C. Lee, "Decision-Directed Slot Synchronization for Pulse-Position-Modulated Optical Signals," *The Interplanetary Network Progress Report,* pp. 42-161, May15, 2005.

# All-wave Non-Zero Dispersion-Flattened Single Mode Fibers

Y.H.Wang

On leave from Shanghai Transmission Lines Research Institute

Room 301, No. 1, Lane 85,Yun Guang Road, 200437, Shanghai

yhwang 1940 @sina.com

**Abstract** *Why is the dispersion –flattened fiber incapable of practical applications? Is the PCF the only choice for all-wave fibers? All-wave non-zero dispersion-flattened single mode fibers have following features: The maximum absolute value of dispersion between 1300nm and 1625nm is smaller than the maximum dispersion of the G.656 fiber between 1460nm and 1625nm. Higher-order modes are negligible. The effective area becomes larger. Depressed triple-clad fibers could satisfy both present and future commercial needs and approach the ultimate in performance until the practical usage of the PBG type of PCF.*

## 1. Introduction

During the past thirty years, all-wave dispersion-flattened fibers have not been in practical applications, due to problems of tolerance control, possible higher order modes and effective area. [1]-[6] Thus it is necessary to study the concept of the dispersion-flattened fiber and the feature of the depressed cladding fiber. Perhaps the later is related to the endlessy single-mode feature of photonic crystal fibers (PCF). For this reason, we have investigated dispersion stability theory, leaky mode attenuation and simplified PCF during some recent years. [7]-[9]

## 2.Dispersion Characteristics

Before the discovery of the four-wave mixing (FWM), zero-dispersion was considered as one of ultimate design goals. In this background the low dispersion fiber over a range of wavelengths was proposed in 1974. [1] Then it has been considered as the

dispersion-flattened fiber and various depressed cladding fibers were studied. The depressed triple-fiber reported in [4] (briefly as the discussed fiber, below) is a representative of them.

As known from analyzing the discussed fiber, it is zero dispersion as one of ultimate design goals and large variation of dispersion slope that results in a strict tolerance control.

The non-zero dispersion-flattened fiber can avoid this trouble. For this purpose, let us modify the discussed fiber in such a way: only the thickness of the outer cladding is optimally reduced (e.g. 3.0 μ m), while other construction parameters keep constant. This is a modified discussed fiber.

## 3 single mode features

About twenty years ago, there was a disputation on the single mode feature of the depressed cladding fiber. [2]

Francois et al. pointed out that QC

fibers could support higher order modes, according to their bending loss measurement. [3] Etzkom et al. claimed that the discussed fiber is " a truly single mode fiber" based on their measurement results. [4] Two zero dispersion wavelengths of the discussed fiber are about 1300nm and 1550nm, while the cutoff wavelength for $LP_{02}$ mode and $LP_{11}$ mode is 2260nm and 3330nm, respectively. We have proposed a method to calculate the leaky mode attenuation and obtained following conclusions:

Francois' result can be interpreted by the additional material attenuation due to the absorption in the substrate tube cladding and the coatings. Etzkom 's report can be explained by the proper leaky mode attenuation.

Since higher order leaky modes are negligible due to attenuation, a multi-mode depressed cladding fiber could demonstrate single-mod feature in a wide range of wavelengths. This is the wide-band single-mode feature.

The wide-band single-mode feature can be considered as a general concept, while the endlessy single-mode feature a special limit.

Without taking the leaky mode attenuation into account, it is possible to underestimate the single mode operation window and the effective area of the depressed cladding fiber.

By means of the leaky mode attenuation calculation with some necessary measurements, one can solve the problems of possible higher order modes and the effective area.

There are in the modified discussed fiber not only non-zero flattened dispersion characteristics,

but also large effective areas with low dispersion slopes.

## 4.Design

Since the outer diameter of the modified discussed fiber is far smaller than 125 $\mu$ m, it is necessary to supply a layer of holes in the outer cladding. This is a simplified PCF to realize the all-wave non-zero dispersion-flattened single mode fiber.

When the outer cladding is thick enough, the dispersion characteristic is almost constant for different thickness of the outer cladding. It depends on the other construction parameters. The non-zero flattened dispersion characteristics can be designed according to our dispersion stability theory, [7] and the wide-band single-mode feature can be estimated by means of our leaky mode attenuation calculation method. [8] This is a depressed cladding fiber to realize the all-wave non-zero dispersion-flattened single mode fiber.

We have designed depressed triple-clad fibers having both positive and negative dispersions. See Tab.1 and 2.

The maximum absolute value of dispersion of these fibers between 1300nm and 1625nm is smaller than the maximum dispersion of G.656 fiber between 1460nm and 1625nm.

The effective area of the designed fiber having positive dispersion is equivalent to that of LEAF fiber.

It is possible to design the fibers so that a positive dispersion and a negative dispersion are compensated one by another over 1460-1625nm.

150

## 5.Conclusion

Some past understandings make the dispersion-flattened fiber incapable of practical applications during the past thirty years. It is the wide-band low dispersion fiber that requires a too strict tolerance control. Without taking the leaky mode attenuation into account, it is possible to underestimate the single mode operation window and the effective area of the depressed cladding fiber. The proposed all-wave non-zero dispersion-flattened single mode fiber can be realized by the depressed triple-clad fiber. Its maximum absolute value of dispersion between 1300nm and 1625nm is smaller than the maximum dispersion of G.656 fiber between 1460nm and 1625nm. The effective area becomes larger. The positive dispersion and the negative dispersion can be compensated one by another over 1460-1625nm. We have the simple and popular fibers satisfying both present and future commercial needs and approaching the ultimate in performance until the practical usage of the PBG type of PCF.

Tab.1

| Wavelength nm | 1300 | 1460 | 1550 | 1600 | 1625 |
|---|---|---|---|---|---|
| Dispersion ps/nm-km | 2.7850 | 9.2700 | 9.2676 | 8.6451 | 8.3791 |
| All dispersion error ps/nm-km | 0.5881 | 1.7000 | 2.4751 | 2.5681 | 2.4314 |
| Dispersion slope ps/nm$^2$-km | 0.0676 | 0.0132 | -0.0105 | -0.0123 | -0.0083 |
| Effective are $\mu m^2$ | 52.8 | 60.3 | 68.9 | 76.4 | 81.2 |

Tab. 2

| Wavelength nm | 1300 | 1460 | 1550 | 1600 | 1625 |
|---|---|---|---|---|---|
| Dispersion ps/nm-km | -6.3802 | -5.2600 | -6.9135 | -5.9471 | -4.5867 |
| All dispersion error ps/nm-km | 1.0520 | 2.4725 | 2.2054 | 1.2127 | 0.7411 |
| Dispersion slope ps/nm$^2$-km | 0.0440 | -0.0227 | -0.0010 | 0.0421 | 0.0666 |
| Effective are $\mu m^2$ | 41.7 | 53.9 | 70.0 | 85.1 | 94.9 |

### References

1 S.Kawakami et al., Electron. Lette., V.10 (1974), pp.38-40.

2 B.J.Ainsile et al., J.Lightwave Technol. V.LT-4(1986),pp.966-979.

3 P.L.Francois et al., Electron. Lett., V.20(1984), pp.37-38.

4 H.Etzkom et al., Electron. Lett., V.20,pp.423-424.

5 K. Ohsono et al., Proc. the 51st Inter.Wire and Cable Symp. (2002),12-5.pp.475-481.

6 T.A.Birks et al., Opt. Lett.,V.22 (1997),

pp.961~963.
7 Y.H.Wang, to be published.
8 Y.H.Wang, to be published.
9 Y.H.Wang, J. Otpoelectron. Laser (in chinese). V.17 (2006), Suppl., pp.48-49.


附言：

1、所讨论主题涉及光纤发展三十年来的一个关键问题和光子晶体光纤的研究方向，鉴于其重要性和普遍性，如有可能请安排特邀报告。

2、请先将本文转发给邬贺铨先生。我想尽快与他联系，能否提供电话号码和 **e-mail** 地址。

3、请先将本文转发给康克明先生。我想尽快与他联系，能否提供电话号码和 **e-mail** 地址。

4、我的电话号码：**021-55541534**。

谢谢！

汪业衡

# Design & Usage of Non-zero Dispersion Wideband Transport (NZDWT) Fiber

Saurav Dutta , Shashi kant

1: E-2, Sterlite Optical Technologies Ltd, MIDC, Aurangabad, India and dutta_s@sterlite.com

2: shashi.kant@sterlite.com

**Abstract** *NZDS fiber with unique profile design with low dispersion slope and high effective area that mitigates non-linear effects and compatible to future S band is presented in this paper. Lowering of the dispersion slope is accompanied by a contraction in the effective area of the core (Aeff), but what is needed, on the contrary, is an expansion of Aeff to overcome non-linear phenomena that hamper WDM transmission.*

## Introduction

The telecommunications industry has rapidly adopted DWDM technology so as to accomodate more number of channels in the same fiber. This results in high signal power density in the fiber leading to non-linear effects such as Four-wave mixing (FWM) and cross-phase modulation (XPM) that are determintal in signal transmission over longer distances. This calls for fiber with higher effective area $A_{eff}$, since they reduces non linear effects along with low dispersion slope so as to minimize dispersion variation among adjacent channels. Other properties that need to be addressed involves (a) fiber with low chromatic dispersion so as to reduce dispersion compensation in the network (b) low polarization mode dispersion (PMD) link design value (<0.20 ps/√km), for higher bandwidth (c) cable cut-off wavelength and cable attenuation coefficients in the C and L bands have to be same as commercially available G.655C. Fibre that qualifies all the above mentioned desired properties are denoted as G.656 & G.655E as per ITU specifications. These fibres overcomes the defects of G.652 and G.655C grade of ITU specification. To reach 40 Gb/s, we need to get a very tight grip on chromatic dispersion. Ironically, some chromatic dispersion is necessary because it helps to prevent the effect called FWM, which occurs when wavelengths spread out enough to begin interfering with one another. New non-zero dispersion shifted fiber (NZDWT) has been manufactured to produce less noise and less dispersion to take signals longer distances.

## Fiber characterstics

In order to obtain fibers with above characteristics, proper care has to be taken while designing the profile shape of the core rod. This include designing the profile for high effective area, low dispersion slope and positive but low chromatic dispersion. Taking care of these parameters help to get the required fiber properties.

Optical fibre effective area $A_{eff}$ (1) is defined as:

$$A_{eff} = \frac{2\pi \left( \int_{0}^{\infty} I(r) r \, dr \right)^2}{\left( \int_{0}^{\infty} I^2(r) r \, dr \right)} \qquad [1]$$

Higher effective area reduces the growth of non-linear effects. A typical value of effective area in the range between 60 to 65 $\mu m^2$ is desired.

The dispersion properties involve absolute chromatic dispersion and the dispersion slope. A positive and non zero dispersion value is desired since non linear effect like FWM causes interaction between signals i.e. optic channels in DWDM systems. The dispersion value needs to be positive over the entire range of operation. To achieve both long-distance and high-speed transmission with easy dispersion compensation for a wide wavelength band, the dispersion slope at 1550-nm region of 0.05 to 0.06 ps/km²/km is required (2).

The PMD coefficient of the fibre affects the link length of the fiber with lower PMD coefficient, the greater is the transmission capacity of the fiber. For bit rate of 2.5 Gbs, PMD coefficient has to be less than 2 ps/√km and for bit rate of 40 Gbs, PMD coefficient has to be less than 0.1 ps/√km.

## NZDWT fiber Design & Manufacturing

The refractive index profile was made by any of the perform manufacturing methods. The refractive index profile of the preform and its parameters are shown in Fig. 1. The profile was designed considering dispersion, MFD, and bending loss with core radius and was obtained through the simulation software. Enlarging the MFD and reducing the slope of the dispersion curve to a sufficient level developed non-zero dispersion-shifted fiber. The final profile design was obtained by considering large profile shapes and obtaining the required fiber characteristics through simulation.

An optical wave-guide fiber, which includes the central first core region surrounded by, inner depressed clad region, which again is surrounded by the ring core region. The ring core region is finally surrounded by outer depressed clad region. Each of the central core region, inner clad region & outer clad region are divided into two regions, which has got a delta% difference of 0.010 to 0.030 in between them. The central core region has got refractive index 'D1', ranging between 0.4% to 0.6% whereas the ring core has got a refractive index of 'D3', ranging between 0.1% to 0.2%. The inner clad region has got refractive index 'D4', ranging between -0.035% to -0.055% whereas the outer clad has got a refractive index of 'D2', ranging between -0.035% to -0.050%. In this case D1 > D3 & |D4|>|D2|. The core diameter lies between the limits of 4 to 6.5 μm.

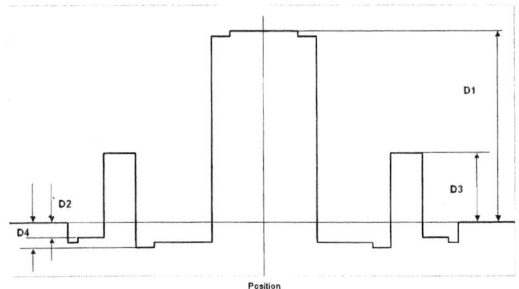

Figure 1.Profile design of NZDWT fiber.

The typical characteristic of NZDWT fiber is shown in table no1. The dispersion value was in the range of 2 to 5 ps/nm/km at the 1460 nm, 6 to 9 ps/nm/km at 1550 nm, and 7 to 14 ps/nm/km at the 1625 nm wavelength. The dispersion slope was computed from the dispersion of linearity rule in the 1550-nm region. It is in the range of 0.05 to 0.06 ps/km$^2$/km. The slope is important for usage of fiber in wide wavelength.

The attenuation at a wavelength of 1550 nm is less than 0.21dB/km. The PMD was about 0.1-ps/√km demonstrating a lower PMD. The effective area is quite high which is about 64 μm$^2$. The relation between dispersion slope and MFD is shown in fig 2.

Table 1: Typical characteristics of NZDWT fiber made out of the process.

| TYPICAL CHARACERISTICS OF NZDWT FIBER | | |
|---|---|---|
| Attenuation | @ 1460 dB/km | 0.25 |
| | @ 1550 | 0.21 |
| | @ 1625 | 0.22 |
| Dispersion | @ 1460 ps/nm/km | 2.5 |
| | @ 1550 | 6 |
| | @ 1625 | 10.2 |
| Dispersion slope | ps/nm$^2$/km | 0.05 |
| Cable cut off | nm | 1300 |
| MFD | um | 9.2 |
| Aeff | um$^2$ | 64 |
| PMD | ps/√km | 0.1 |

The fiber made with reduced dispersion slope to cover a wider bandwidth and with link length of about 2666 kms & 150 kms, that is greater than standard single mode fiber having link length about 941 kms & 60 kms transmitting 2.5 Gbps & 10 Gbps respectively. NZDWT fiber can be used for a wide band of wavelength i.e. C-L-S band as compared to G.655 type of fibres which can only be used for C-L band. The dispersion is high enough at 1550 nm for NZDWT fiber to counter non liner effects as compared to G.655 fiber.

Key advantages of NZDWT fiber are:
  a. It is "S" band compatible, efficiently supports 8 (1460 – 1625nm) channels CWDM transmission system.
  b. Optimised for operation without dispersion compensation in Metropolitan Area Network for 2.5Gb/s to 10Gb/s.
  c. 40Gb/s transmission possible with commercially available dispersion compensation devices.
  d. High MFD means high effective area mitigates non-linear Effects.
and thus fully compatible with exiting NZDSF Fiber network.

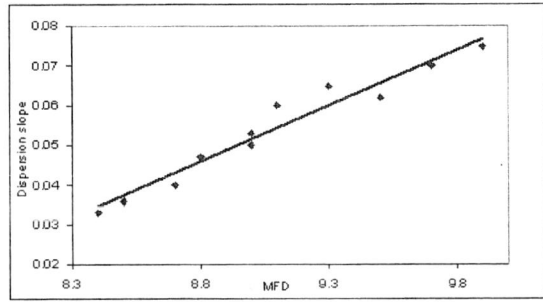

Figure 2: MFD vs. dispersion of fiber under test

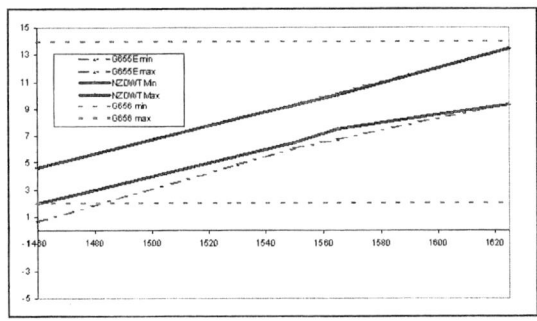

Figure 3: Band of dispersion Vs wavelength

**Result and discussion**

NZDWT fiber is a significant step in the evolution of optical networks. The development of standards in the area of optical fiber is important as network operators wants to reduce costs and provide more innovative services to their customers. In order to make easier for network operators to deploy bandwidth to maximize technology in their core networks, ITU has set a global standard for a new optical fibre. NZDWT fiber complies to G.655E & G.656 specification. Wave Division Multiplexing increases the data carrying capacity of an optical fibre by allowing simultaneous operation at more than one wavelength. NZDWT fiber permits easier deployment of Coarse Wave Division Multiplexing in metropolitan areas, and increases the capacity of fibre in Dense Wave Division Multiplexing systems. Chromatic dispersion is crucial as the number of wavelengths used in WDM systems increases.

NZDWT fiber of the present work is an unique optimization of effective area, chromatic dispersion and dispersion slope enables excellent distortion management, cost effective operation at 10 and 40 Gbps, tight channel spacing in C- and L-bands, compatibility with the future S-band and provides full compliance with the new ITU G.656 & ITU G.655E recommendation. NZDWT fiber is further optimized for long haul and ultra long haul applications along with Metro application. Its characteristic chromatic dispersion of 8 ps/nm/km at 1550 nm is optimized to be half that of standard single mode fiber resulting in lower costs for dispersion compensation, but high enough to counter cross-channel non-linearities.

The fiber developed permits easier deployment of Coarse Wave Division Multiplexing in metropolitan areas, and increases the capacity of fibre in Dense Wave Division Multiplexing systems.Chromatic dispersion compensatation is not required if NZDWT fiber is used by the operator for CWDM application. By using NZDWT fiber, at least 40 more channels can be added to Dense Wave length Division Multiplexing systems in 1460-1490 nm band,compared to G655 fiber (3). This low value of the chromatic dispersion coefficient in the S-C-L bands is the real novelty of NZDWT fiber. It allows the utilization of a larger wavelength band.

Such fiber has low attenuation at 1400nm wavelength region optimising performance for metropolitan backbone applications. Its optimisation of effective area, Chromatic Dispersion, and dispersion slope

enables less dispersion compensation, tight channel spacing in C and L bands, Cost effective operation at "S" band using CWDM system, supports operation at 10Gb/s and 40Gb/s. CWDM systems have channels at wavelengths spaced 20 nanometres apart, compared with 0.4 nm spacing for DWDM. It provides longest uncompensated reach (~200km) transmitting 10 Gbps in metropolitan network results in cost saving compared to conventional single mode fiber.

## Conclusions

Unique preform profile having large effective area, which mitigates non-linear effects and low dispersion slope to reduce dispersion variation along the channels, was obtained through the simulation software. The unique profile gives a band of chromatic dispersion that exceeds the requirement of ITU G.656 & G.655E specification. Such fiber is capable of transmission over a wider range of wavelengths from 1460 to 1625 nm making it useful in S-C-L Band.

## References

1. Effective area of optical fiber: definition and measurement tech. By R.Billington, page-3.
2. Low NZDSF for DWDM transmission, by kazumasa Ohsono, Page-2 Hitachi cable review, no-19, Aug00.
3. International Telecommunication Union, Press Release, Geneva, 13 May 2004.

# Single-Frequency Operation of a Widely Tunable SOA-Based Fiber Ring Laser

CHEN Hongxin, HE Gang, François BABIN* and Gregory W. SCHINN

EXFO Electro-Optical Engineering Inc., Quebec City, Quebec, G1M 2K2, CANADA

E-mail: hongxin.chen@exfo.com; Tel.: +1(418) 683-0211

**Abstract** *In a SOA-based fiber ring laser, tunable over the range 1450 nm to 1640 nm, a "wide" intra-cavity filter leads, counter-intuitively, to singlemode oscillation, whereas a "narrow" filter results in multimode oscillation. However, multimode oscillation is always observed in the red part of the gain curve, above about 1615 nm.*

## Introduction

Recently widely tunable, semiconductor-optical-amplifier (SOA)-based fiber ring lasers (SFRL) have elicited considerable research interest on account of their potential applications in the fields of optical communications, optical test and measurement, optical sensors and biomedical imaging [1-4]. When lasing occurs on multiple, closely-spaced cavity modes extending over a few GHz of optical frequency, the SFRL exhibits a "medium-coherence" output. Such an output is well suited for many test-and-measurement applications [3]. Nevertheless, much effort has also been undertaken to induce single-longitudinal-mode (SLM) oscillation in a SFRL, such as by means of a coupled cavity filter [2]. Alternatively, an SOA has also been used to suppress the beat noise of erbium-doped fiber ring lasers [5,6]. Suppression of mode competition in an SFRL by SOA has also been theoretically analyzed and experimentally demonstrated [7,8].

In this paper we present a study of lasing behavior for very widely tunable SOA-based fiber ring lasers. By using an intra-cavity tunable bandpass filter (TBF) having a 3-dB bandwidth of >200 pm, our SFRL exhibits SLM lasing, with a side mode suppression ratio of 30 dB, for a tuning range extending over 160 nm from 1450 nm to 1615 nm. The SOA high-pass filtering due to its relatively fast carrier recovery rate and gain saturation effect is believed to be responsible for this SLM operation by suppressing the other laser modes [5-7]. However, we also observe that when the TBF central wavelength is >1615 nm, the SFRL exhibits multi-longitudinal mode (MLM) lasing oscillation. The exact wavelength where the laser behavior changes from SLM to MLM emission depends upon the SOA gain profile, laser operating conditions, cavity design and loss.

On the other hand, when a TBF bandwidth of ~16 pm was used in our SFRL, significant suppression of the laser mode competition was no longer observed and the laser output exhibited MLM oscillation behavior over its entire tunable wavelength range from 1445 nm to 1620 nm.

## Single Longitudinal Mode Oscillation of SFRL

*1. Set-up*

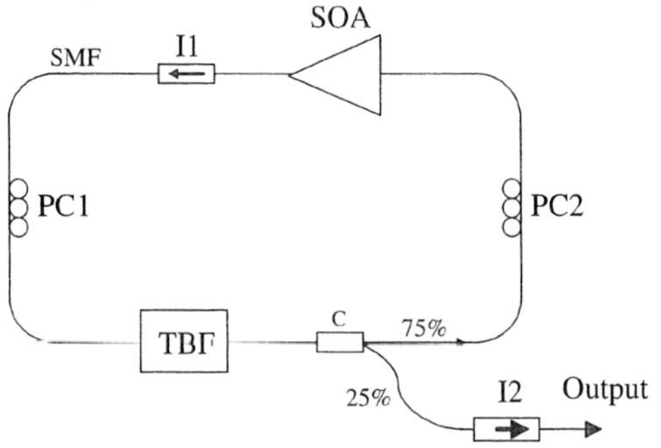

*Fig. 1 Set-up of widely tunable SFRL under either SLM or MLM oscillation. I1 and I2 – isolators, SMF – single mode fiber and C – 75/25 coupler.*

The widely tunable SFRL is depicted schematically in Fig. 1. It is a traveling-wave, fiber ring cavity containing an SOA and a TBF. The SOA serves not only as a gain medium but also to suppress the beat noise between the longitudinal laser modes or to suppress the mode competition due to the SOA gain saturation effect [5-8]. Therefore, SLM lasing oscillation occurs and a very narrow laser linewidth is also observed. The SOA used in the experiment has a maximum gain of ~23 dB near 1550 nm and its typical operating current is 850 mA. The TBF serves to select and scan the laser wavelength. In order to obtain SLM oscillation under our laser operating conditions, TBF bandwidth needs to be >200 pm. In this experiment, two TBF bandwidths are used: ~300 pm and ~1 nm. The state of polarization of light is controlled with two polarization controllers (PC1 & 2). The cavity length is ~13 m. The round-trip cavity loss using this "wide" TBF is ~6 dB, and does not vary significantly with wavelength. The laser output is recorded using an optical spectrum analyzer (OSA) or a power meter (PM). SLM or MLM operation is verified by detecting the light with a fast photo-detector (PD) and an *rf* spectrum analyzer. The laser linewidth is measured with a scanning Fabry-Perot interferometer having a free spectral range (FSR) of 8 GHz and a finesse of ≥300.

## 2. Results

### A: Laser spectrum and side mode suppression

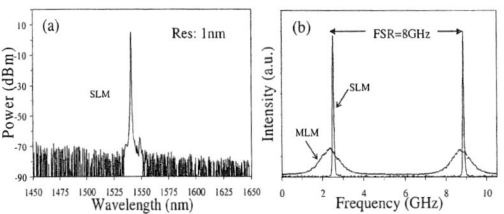

Fig. 2 (a) Laser spectrum measured with an OSA having a resolution of 1 nm. (b) Laser spectra recorded with a scanning Fabry-Perot interferometer having a resolution of approximately 25 MHz.

Fig. 2(b) shows output spectra when the laser operates under SLM (solid line) or MLM conditions (dashed line). SLM or MLM operation is deduced from the beat noise spectrum. The absence of tones at frequencies inversely proportional to the cavity length in the rf spectrum (Fig. 3(b)) indicates SLM oscillation, while a high beat noise (Fig. 3(a)) indicates a MLM oscillation. The laser emits on a single cavity mode for wavelengths falling within most of the SOA gain spectrum, from 1450 nm to 1615 nm (see Figs 5 and 6). However, MLM lasing was observed when TBF is tuned to wavelengths at the "red" end of the gain curve, such as at wavelengths >1615 nm under our experimental conditions (see Fig. 6). Fig. 4 shows that SLM oscillation dominates from 1440 nm to 1615 nm when the laser is tuned from shorter wavelengths to longer wavelengths, even though some residual beat noise is still seen. However, when the laser is tuned from longer wavelengths to shorter wavelengths, for example from 1640 nm to 1615 nm, MLM oscillation is maintained even at a wavelength of 1615 nm (see Fig. 3(a)). Exact physics of this hysteresis phenomenon remains to be elucidated. Fig. 2(b) shows the observed SLM linewidth, whose measurement is limited by the instrument resolution of the scanning Fabry-Perot interferometer, i.e. less than approximately 25 MHz.

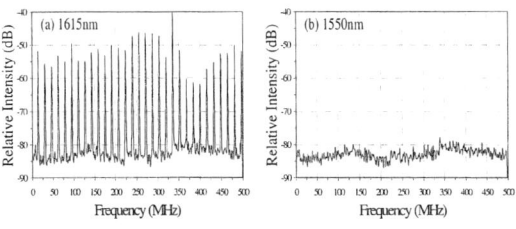

Fig. 3 Electrical baseband spectra for (a) MLM operation at 1615 nm and (b) SLM operation at 1550 nm.

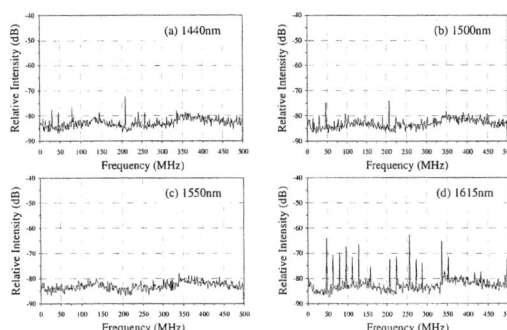

Fig. 4 Electrical baseband spectra for operating wavelengths of: (a) 1440 nm, (b) 1500 nm, (c) 1550 nm and (d) 1615 nm.

### B: Tuning range

The SFRL typically emits on a single cavity mode in the short- and "medium"-wavelength region of the SOA gain curve, such as from 1450 nm to 1615 nm. However, the laser operates on multiple cavity modes at longer operating wavelengths. This MLM oscillation may arise from a reduced level of laser-mode-competition suppression by the SOA at these wavelengths and/or be related to the linewidth enhancement factor [9,10]. Fig. 6 shows the SFRL wavelength tuning range where a grating-based TBF (bandwidth ~1 nm) is used to select laser wavelength while Fig. 5 shows the corresponding SOA gain and ASE power density profiles. Also indicated are the wavelength ranges over which SLM and MLM emission occurs.

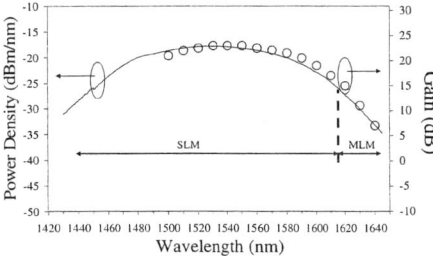

Fig. 5 Measured SOA gain and ASE power density with an OSA having a resolution of 1 nm.

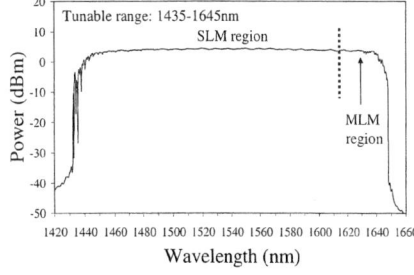

Fig. 6 Measured SFRL tuning range. The separation of SLM and MLM regions is indicated by the vertical dashed line along 1615 nm.

## Multi-Longitudinal Mode Oscillation of the SFRL

### 1. Set-up

In order to achieve a MLM operation across the entire SOA gain curve, we built a very narrow bandwidth tunable filter by using a four-pass grating-based TBF design with an optic circulator (see Fig. 7). The total cavity loss is ~10 dB. When the TBF bandwidth is small, four-wave mixing (FWM), arising from the high nonlinear coefficient of the SOA [11], may play an important role in the SFRL, thereby inducing lasing on multiple longitudinal cavity modes.

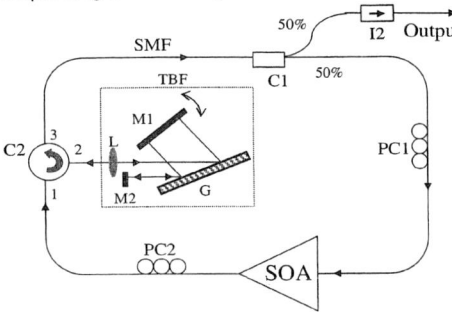

Fig. 7 Set-up of tunable SFRL that exhibits MLM oscillation where the TBF bandwidth of ~16 pm was used. C1 – 50/50 coupler, C2 – circulator, M1 and M2 – roof prism mirrors, G – grating.

### 2. Results

Fig. 8(a) shows SFRL tuning range for the fiber laser cavity design in Fig. 7 where the four-pass grating based TBF bandwidth is ~16 pm. Fig. 8(b) presents measured effective linewidths (defined by the time-averaged envelope of the multiple modes) of the SFRL using the same scanning Fabry-Perot interferometer from 1450 nm to 1624 nm. Typically, an effective linewidth of ~1.5 GHz is observed. With the cavity configuration shown in Fig. 7 with a TBF bandwidth of ~16 pm, MLM operation is indeed observed over the entire tuning wavelength range (as indicated in Fig. 8(b)).

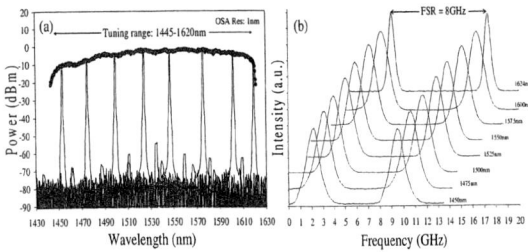

Fig. 8 (a) Measured SFRL tuning range. (b) Measured linewidths for a SFRL cavity design shown in Fig. 7.

### Discussions and Conclusions

We have reported an investigation of lasing behavior in very widely tunable narrow linewidth SFRLs, where the SOA acts as a gain medium and the TBF is used to select the laser wavelength. Somewhat counter-intuitively, when a "wide" TBF bandwidth of >200 pm is used in the SFRL, the laser emits on a single longitudinal mode for a tunable wavelength range extending over 160 nm from 1450 nm to 1615 nm, with a side mode suppression ratio of ~30 dB. We posit that the transition from SLM to MLM emissions in different wavelength regions is probably due to wavelength-dependent SOA gain saturation behavior and linewidth enhancement factor [9,10]: i.e. for longer wavelengths, the linewidth enhancement factor is larger for the SFRL. When a "narrow" TBF bandwidth of ~16 pm is used, the SFRL exhibits MLM operation across its entire tunable wavelength range. In this case, significant suppression of laser mode competition in this SFRL was not observed.

Even through SLM oscillation in an SFRL has been modeled previously [7], we recommend further theoretical study, covering an extended wavelength range and including a variety of cavity conditions, wavelength dependence, and filter bandwidth dependence. Such detailed theoretical investigations and simulations concerning these SFRL dynamics are now underway in our laboratories.

### Acknowledgements

We thank Prof. Michel Piché of Laval University for useful discussions.

\* Present address:  INO, 2740 rue Einstein, Quebec City, Quebec, G1P 4S4, CANADA

### References

1 D. Zhou, P.R. Prucnal, I. Glesk, *IEEE Photon. Technol. Lett.*, vol. 10 (1998), 781-783.

2 Z. Hu, L. Zheng, O. Tang, *Opt. Lett.*, vol. 25 (2000), 469-471.

3 R. Baribault, H. Chen, G. He, D. Gariépy, F. Babin, G. W. Schinn, *OFMC 2005*, 21-23 Sept. 2005, Teddington, UK.

4 W.Y. Oh, *et al*, *IEEE Photon. Technol. Lett.*, vol. 17 (2005), 678- 680.

5 L. Xu, I. Glesk, D. Rand, V. Baby, P.R. Prucnal, *Opt. Lett.*, vol. 28 (2003), 780-782.

6 H.L. Liu, *et al*, *IEEE Photon. Technol. Lett.*, vol. 18 (2006), 706-708.

7 Q. Xu, M. Yao, *IEEE J. of Quantum Electron.*, vol. 39 (2003), 1260-1265.

8. H. Chen, *Phys. Lett. A*, vol. 320 (2003), page 333.

9. C. H. Henry, *IEEE J. Quantum Electron.* vol. 18 (1982), 259-264.

10. A. Champagne, J. Camel, R. Maciejko, K.J. Kasunic, D.M. Adams, B. Tromborg, *IEEE J. of Quantum Electron.*, vol.38 (2002), 1493-1502.

11. G. Contestabile, M. Presi, E. Ciaramella, *IEEE Photon. Technol. Lett.*, vol. 16 (2004), 1775-1777.

**Prof. Ray T. Chen**
**80-micron Interaction Length Silicon Nano-Photonic Crystal Waveguide Modulator**
Yongqiang Jiang1, Wei Jiang1,2, Lanlan Gu1, Xiaonan Chen1, Ray T. Chen 1
1 Microelectronic Research Center, Department of Electrical and Computer Engineering,
The University of Texas at Austin, Austin, TX 78758, USA
2 Omega Optics, Austin, TX 78758, USA

An ultra-compact silicon electro-optic modulator is experimentally demonstrated based on silicon photonic crystal (PhC) waveguides for the first time to our knowledge. Modulation operation was demonstrated by carrier injection into an 80 µm-long silicon PhC waveguide of a Mach-Zehnder interferometer (MZI) structure. The modulation depth operating at 1567 nm is 92%. The _ phase shift driving current, I_, across the active region is as low as 0.15 mA, which is equivalent to a V_ of 7.5 mV when a 50_impedance-mate structure is applied.
Keywords: Photonic crystal (PhC), optical modulator, Mach-Zehnder interferometer (MZI), group velocity, carrier injection, phase modulation, plasma dispersion effect.

# Analysis of the Influences of Polarization-dependent Nonlinear Gain and Nonlinear Polarization Rotation on Optical Sampling in Semiconductor Optical Amplifier

Maotong Liu, Aiying Yang, Yu-nan Sun

Department of Photo-electronic Engineering, School of Information Science and Technology
Beijing Institute of Technology, Beijing, 100081, China, alicelau@bit.edu.cn

**Abstract** A new theoretical model is presented to describe polarization-dependent nonlinear gain in semiconductor optical amplifiers (SOAs) on subpicosecond timescales. Based on this model, the influences of nonlinear polarization rotation (NPR) on pulsed four-wave mixing (FWM) in SOA used for optical sampling are discussed. The numerical results are in good agreement with reported experimental measurements.

## Introduction

During the last few years, a variety of investigations have been performed to study the nonlinear polarization rotation (NPR) in semiconductor optical amplifiers (SOA). Considerable attention has focused on applications which may emerge from the NPR. This effect has been employed to realize optical time domain demultiplexing [1], all-optical wavelength conversion [2]-[4], all-optical logic [5]-[7], all-optical flip-flop memory [8], and all-optical label processing [9], etc., and has been experimentally demonstrated to have a significant potential in future optical networks. Several papers [10]-[13] have already been devoted to the theoretical analysis of NPR in SOA. Compared with the other methods, rate-equation method applied in [10] is simplest, yet powerful. Just as pointed out by [14], the main approximation in using the rate-equation approach as compared to the full semiconductor Bloch equations lies in the use of the adiabatic approximation for the interband polarization dynamics. Therefore, it is necessary to introduce reasonable assumptions into discussion and make some modifications to guarantee the applicability of this theory. The rate equations presented in [15] are based on two-level systems, but simple two-band model does not, by itself, give any polarization dependence. So the valance band (VB) may be sub-divided into a heavy-hole band (HHB) and a light-hole band (LHB), where the TE transitions predominantly occur with the heavy holes and the TM transitions occur predominantly with the light holes. The band structure is shown in Fig.1. Though there is only one conduction band (CB), we divide the local carrier density $n_c$ of conduction band into $n_c^{11}$ and $n_c^{12}$ which

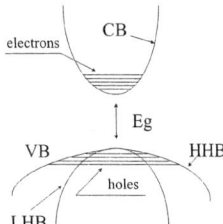

*Fig.1. Band structure for unstrained semiconductor material*

are more likely to interact with $n_v^{11}$ and $n_v^{12}$ respectively. Besides this, we also assume that: i) the optical axes of the device are those corresponding to the TE (the component parallel to the layers in the waveguide) and the TM (the perpendicular component) as expected in a bulk SOA. It has been proved by [16] that for a strong power injection, the TE and TM axes are no longer the optical axes of the structure (at least in some part of the amplifier). Here we consider pump beam power is not strong enough to induce this effect. ii) Assume that the polarized optical field can be decomposed into TE and TM components, and they interact with each other indirectly via the gain saturation [10].

The optical anisotropy of SOA is mainly attributed to three reasons [17]-[21]: i) To reduce the polarization dependence, the band structure is designed by introducing tensile strain in MQW or bulk SOA to enhance TM material gain, and it may result in an optical anisotropy; ii) Structure-induced birefringence which is represented by the difference in the confinement factor of TE and TM mode; iii) anisotropy in the waveguide nonlinearity which is caused by the third-order susceptibility tensor; These effects may cause gain discrepancy between the optical axes of waveguide and different values of refractive indices for the modes, and then induce NPR in SOA. In this paper we carry out numerical experiments on 1.55 μm lattice-matched bulk InGaAsP/InP SOA whose basic parameters can be obtained from [22], and do not take the strain-induced birefringence into account. Differ from [10] who use an empirical imbalance factor $f$ to distinguish the calculation of unstrained bulk material ($f = 1$) and the case of tensile strained ($f < 1$), we attribute the differences to the different basic parameters of bulk and strained SOA such as carrier energies, effective masses, energy gap, etc. All the parameters used here are obtained by using theoretical equations. To obtain the best fit with the experiment results, a fitting procedure is necessary which adjusts some parameters within reasonable limits and it is beyond the scope of this paper.

## Model

Based on the assumptions we have introduced above, we can obtain rate equations

$$\frac{\partial n_\beta^{ij}}{\partial t} = -\frac{n_\beta^{ij} - \bar{n}_\beta^{ij}}{\tau_{1,\beta}} - v_g^{ij} g_{ij} S_{ij} - n_\beta^{ij} \sigma_\beta v_g^{ij} S_{ij} \quad (1)$$

$$\frac{\partial N_{ij}}{\partial t} = \frac{I}{qV} - \frac{N_{ij}}{\tau_s} - v_g^{ij} g_{ij} S_{ij} + \frac{\Gamma_2}{\Gamma^{11}} \beta_2^{11} v_g^{11} S_{11}^2 + \quad (2)$$

$$\frac{\Gamma_2 \sqrt{v_g^{11} v_g^{12}}}{\sqrt{\Gamma^{11} \Gamma^{12}}} 2\beta_{2\perp} S_{11} S_{12} + \frac{\Gamma_2}{\Gamma^{12}} \beta_2^{12} v_g^{12} S_{12}^2$$

$$\frac{\partial T_\beta^{ij}}{\partial t} = \left(\frac{\partial U_\beta}{\partial T_\beta}\right)_N^{-1} \left\{ \sigma_\beta \hbar \omega_0 \left( N_{11} v_g^{11} S_{11} + N_{12} v_g^{12} S_{12}\right) + \left[\left(\frac{\partial U_\beta}{\partial N}\right)_{T_\beta} - E_\beta^{ij}\right] \right.$$

$$\left(v_g^{11} g_{11} S_{11} + v_g^{12} g_{12} S_{12}\right) + \left[ E_{2,\beta}^{ij} - \left(\frac{\partial U_\beta}{\partial N}\right)_{T_\beta}\right]\left[\frac{\Gamma_2}{\Gamma^{11}} \beta_2^{11} v_g^{11} S_{11}^2 + \right.$$

$$\left. + \frac{\Gamma_2 \sqrt{v_g^{11} v_g^{12}}}{\sqrt{\Gamma^{11} \Gamma^{12}}} 2\beta_{2\perp} S_{11} S_{12} + \frac{\Gamma_2}{\Gamma^{12}} \beta_2^{12} v_g^{12} S_{12}^2 \right]\right\} - \frac{T_\beta^{ij} - T_L}{\tau_{h,\beta}}$$

$$(3)$$

$$i = 1; \, j = 1, 2$$

where $\beta = c, v$ correspond to conduction and valence band. '11' represents TE mode, and '12' represents TM mode. The definitions of parameters here are similar to [23]. To obtain the expressions of the slowly varying envelope of the optical pulse $A_{ij}(z,t)$, model gain $G_{ij}(z,t)$, and material gain $g_{ij}(z,t)$, by the procedure used by [14], [23] and take $\sigma_v = 0$, we obtain

$$\frac{\partial A_{ij}}{\partial z} + \frac{1}{v_g^{ij}} \frac{\partial A_{ij}}{\partial t} - \frac{i}{2} \beta'_{2ij} \frac{\partial^2 A_{ij}}{\partial t^2} = \frac{A_{ij}}{2} \frac{G_{ij} - \Gamma^{ij} a_N^{ij} \tau_{1,c} n_c^{ij} \sigma_c S_{ij} + \Gamma^{ij} \Delta g_{hij}}{1 + \varepsilon_{SHB} S_{ij}} \quad (4)$$

$$-\frac{i}{2}\left(\alpha_N G_{ij} - \alpha_{T_c} \frac{\varepsilon_{SHB} P_{ij} G_{ij} + \Gamma^{ij} a_N^{ij} \tau_{1,c} n_c^{ij} \sigma_c S_{ij} - \Gamma^{ij} \Delta g_{hij}}{1 + \varepsilon_{SHB} S_{ij}}\right) A_{ij}$$

$$-\left[\Gamma_2 \left(\beta_2^{11} P_{11} + \beta_2^{12} P_{12}\right) + i\Gamma'_2 \frac{\omega_0}{c}\left(n_2^{11} P_{11} + n_2^{12} P_{12}\right)\right] \frac{1}{\sigma} A_{ij} - \frac{1}{2} \alpha_{int}^{ij} A_{ij}$$

$$\frac{\partial G_{ij}}{\partial t} = \frac{G_{0ij} - G_{ij}}{\tau_s} - a_{ij} v_g^{ij} S_{ij} \frac{G_{ij} - \Gamma^{ij} a_N^{ij} \tau_{1,c} n_c^{ij} \sigma_c S_{ij} + \Gamma^{ij} \Delta g_{hij}}{1 + \varepsilon_{SHB} S_{ij}} \quad (5)$$

$$+ a_{ij} \Gamma^{ij} \Gamma_2 \left(\frac{1}{\Gamma^{11}} \beta_2^{11} v_g^{11} S_{11}^2 + \frac{\sqrt{v_g^{11} v_g^{12}}}{\sqrt{\Gamma^{11} \Gamma^{12}}} 2\beta_{2\perp} S_{11} S_{12} + \frac{1}{\Gamma^{12}} \beta_2^{12} v_g^{12} S_{12}^2\right)$$

$$g_{ij} = \frac{g_{lij}\left(N_{ij}\right) - a_N^{ij} \tau_{1,c} n_c^{ij} \sigma_c S_{ij} + \Delta g_{hij}}{1 + \varepsilon_{SHB} S_{ij}} \quad (6)$$

$$\Delta g_{hij} = \sum_\beta \Delta g_{\beta,hij} \quad (7)$$

$$\frac{\partial \Delta g_{\beta hij}}{\partial t} = -\frac{\Delta g_{\beta hij}}{\tau_{h,\beta}} - \frac{\varepsilon_{sc,\beta}^{ij}}{\tau_{h,\beta}}\left(v_g^{11} g_{11} S_{11} + v_g^{12} g_{12} S_{12}\right) - \frac{\varepsilon_{fca,\beta}^{ij}}{\tau_{h,\beta}} \hbar \omega_0 \quad (8)$$

$$\left(v_g^{11} N_{11} S_{11} + v_g^{12} N_{12} S_{12}\right) - \frac{1}{\tau_{h,\beta}}\left(\varepsilon_{tpa,\beta}^{11} v_g^{11} S_{11}^2 + \varepsilon_{tpa,\beta\perp} \sqrt{v_g^{11} v_g^{12}} S_{11} S_{12}\right.$$

$$\left. + \varepsilon_{tpa,\beta}^{12} v_g^{12} S_{12}^2\right)$$

$$N_{ij} = \frac{g_{lij}}{a_{ij}} + N_{tr} \quad (9)$$

$$G_{ij} = \Gamma^{ij} g_{lij}\left(N_{ij}\right) \quad (10)$$

$$S_{ij} = \frac{P_{ij}}{\hbar \omega_0 v_g^{ij} \sigma} = \frac{|A_{ij}|^2}{\hbar \omega_0 v_g^{ij} \sigma} \quad (11)$$

$$P = \sqrt{P_{11}^2 + P_{12}^2} \quad (12)$$

Where

$$\varepsilon_{SHB}^{ij} = \sum_\beta a_N^{ij} \tau_{1,\beta} \quad (13)$$

$$\varepsilon_{SE,\beta}^{ij} = -\tau_{h,\beta} \frac{\partial g}{\partial T_\beta}\left(\frac{\partial U_\beta}{\partial T_\beta}\right)_N^{-1}\left[\left(\frac{\partial U_\beta}{\partial N}\right)_{T_\beta} - E_\beta^{ij}\right] \quad (14)$$

$$\varepsilon_{FCA,\beta}^{ij} = -\tau_{h,\beta} \sigma_\beta \frac{\partial g}{\partial T_\beta}\left(\frac{\partial U_\beta}{\partial T_\beta}\right)_N^{-1} \quad (15)$$

$$\varepsilon_{TPA,\beta}^{ij} = -\tau_{h,\beta} \frac{\partial g}{\partial T_\beta} \frac{\Gamma_2}{\Gamma^{ij}} \beta_2^{ij}\left(\frac{\partial U_\beta}{\partial T_\beta}\right)_N^{-1}\left[E_{2,\beta}^{ij} - \left(\frac{\partial U_\beta}{\partial N}\right)_{T_\beta}\right] \quad (16)$$

$$\varepsilon_{TPA,\beta\perp}^{ij} = -\tau_{h,\beta} \frac{\partial g}{\partial T_\beta} \frac{\Gamma_2}{\sqrt{\Gamma^{11}\Gamma^{12}}} 2\beta_{2\perp}\left(\frac{\partial U_\beta}{\partial T_\beta}\right)_N^{-1}\left[E_{2,\beta}^{ij} - \left(\frac{\partial U_\beta}{\partial N}\right)_{T_\beta}\right] \quad (17)$$

We consider that pulsewidths can be shorter than temperature relaxation times $\tau_{h,\beta}$ but are much longer than the intraband scattering times $\tau_{1,\beta}$ (typically 50—100fs). Therefore, calculations of $n_\beta$ can be simplified

$$n_\beta^{ij} = \frac{\bar{n}_\beta^{ij} - \tau_{1,\beta} v_g^{ij} g_{ij} S_{ij}}{1 + \tau_{1,\beta} \sigma_\beta v_g^{ij} S_{ij}} \quad (18)$$

To obtain the values of $\bar{n}_\beta^{ij}$, we use relation [14]

$$\Delta g_{\beta,hij} = a_N^{ij}\left(\bar{n}_\beta^{ij} - \bar{n}_{\beta,L}^{ij}\right)/v_g^{ij} \quad (19)$$

and an approximate method presented in [24]. We can use

$$E_c^{ij} + E_v^{ij} + E_g = \hbar \omega_0$$
$$E_{2c}^{ij} + E_{2v}^{ij} + E_g = 2\hbar \omega_0 \quad (20)$$

to calculate carrier energies. Using the relation of G and g, we can calculate the values of $N_0$ in each step of the split-step Fourier method which are taken as a constant in [10]. In the simpler approach, we take the same values of $v_g$, $a_N$, $\alpha_{int}$, $\beta'_2$, $a$. The values and expressions of the other parameters can be found in [15][22][25]-[27]. According to [19]-[21], the values of $\beta_2$ and $n_2$ for TE polarized light predicted to be 50% larger than for TM polarized light, and $\beta_{2\perp}$ is obtained by $\left(\beta_2^{11} + \beta_2^{12}\right)/6$.

Based on the polarization-dependent SOA model, we attempt to analyze polarization-dependent FWM performance. Generally, there are two types FWM in SOA[28], continuous-wave FWM and pulsed FWM. The latter one can be used to realize optical sampling. There have been several theoretical descriptions[29]-[30] on this FWM process, but they are all inadequate to describe polarization effects. The discussion here is based on theory presented in [30]. If both pump and probe fields are linearly polarized, the FWM signal is also linearly polarized[31]. For the simplest situation, the dynamic gratings can only be formed through beating of the same components of the pump and probe waves, namely both TE or both TM, and the TE (TM) component of the pump can be scattered only into TE (TM) polarized FWM[32]. Based on this rule, we obtain

$$\frac{\partial A_{0ij}}{\partial z} + \frac{1}{v_g}\frac{\partial A_{0ij}}{\partial t} - \frac{i}{2}\beta_2'\frac{\partial^2 A_{0ij}}{\partial t^2} = \frac{A_{0ij}}{2}\frac{G_{0ij} - \Gamma^{ij}a_{N0}\tau_{1c}n_{c0}^{ij}\sigma_c S_{0ij} + \Gamma^{ij}\Delta g_{h0ij}}{1+\varepsilon_{SHB}S_{0ij}}$$

$$-\frac{i}{2}\left[\alpha_N G_{0ij} - \alpha_{T_c}\frac{\varepsilon_{SHB}S_{0ij}G_{0ij} + \Gamma^{ij}a_{N0}\tau_{1c}n_{c0}^{ij}\sigma_c S_{0ij} - \Gamma^{ij}\Delta g_{h0ij}}{1+\varepsilon_{SHB}S_{0ij}}\right]A_{0ij}$$

$$-\left[\Gamma_2\left(\beta_2^{11}P_{011} + \beta_2^{12}P_{012}\right) + i\Gamma_2'\frac{\omega_0}{c}\left(n_2^{11}P_{011} + n_2^{12}P_{012}\right)\right]\frac{1}{\sigma}A_{0ij} - \frac{1}{2}\alpha_{int}A_{0ij}$$

(21)

$$\frac{\partial A_{sij}}{\partial z} + \frac{1}{v_g}\frac{\partial A_{sij}}{\partial t} - \frac{i}{2}\beta_2'\frac{\partial^2 A_{sij}}{\partial t^2} = \frac{A_{sij}}{2}\frac{G_{sij} - \Gamma^{ij}a_{N0}\tau_{1c}n_{c0}^{ij}\sigma_c S_{0ij} + \Gamma^{ij}\Delta g_{h0ij}}{1+\varepsilon_{SHB}S_{0ij}}$$

$$-\frac{i}{2}\left[\alpha_N G_{sij} - \alpha_{T_c}\frac{\varepsilon_{SHB}S_{0ij}G_{sij} + \Gamma^{ij}a_{N0}\tau_{1c}n_{c0}^{ij}\sigma_c S_{0ij} - \Gamma^{ij}\Delta g_{h0ij}}{1+\varepsilon_{SHB}S_{0ij}}\right]A_{sij}$$

$$-\frac{1}{2}\alpha_{int}A_{sij} - \left[\Gamma_2\left(\beta_2^{11}P_{011} + \beta_2^{12}P_{012}\right) + i\Gamma_2'\frac{\omega_j}{c}\left(n_2^{11}P_{011} + n_2^{12}P_{012}\right)\right]\frac{1}{\sigma}2A_j$$

$$-\left(\Gamma_2\beta_2 + i\Gamma_2'\frac{\omega_j}{c}n_2\right)A_{0ij}^2 A_{3-s\cdot ij}^* - \frac{1}{2}\eta_{s0}^{ij}P_{0ij}A_{sij} - \frac{1}{2}\eta_{0,3-s}^{ij}A_{0ij}^2 A_{3-s\cdot ij}^*$$

s=1,2  (22)

$$\frac{\partial G_{sij}}{\partial t} = \frac{G_{0sij} - G_{sij}}{\tau_s} - a_{sij}v_g^{ij}S_{sij}\frac{G_{0ij} - \Gamma^{ij}a_{N0}\tau_{1c}n_{c0}^{ij}\sigma_c S_{0ij} + \Gamma^{ij}\Delta g_{h0ij}}{1+\varepsilon_{SHB}S_{0ij}}$$

(23)

$$+a_s\Gamma^{ij}\Gamma_2\left(\frac{1}{\Gamma^{11}}\beta_2^{11}v_g^{11}S_{011}^2 + \frac{\sqrt{v_g^{11}v_g^{12}}}{\sqrt{\Gamma^{11}\Gamma^{12}}}2\beta_{2\perp}S_{011}^2 S_{012}^2 + \frac{1}{\Gamma^{12}}\beta_2^{12}v_g^{12}S_{012}^2\right)$$

s=0,1,2

The expressions for the coupling coefficient are

$$\eta_{ss'}^{ij} = \eta_{ss'}^{CDij} + \eta_{ss'}^{SEij} + \eta_{ss'}^{FCAij} + \eta_{ss'}^{SHBij}$$  (24)

$$\eta_{ss'}^{CDij} = \frac{a_s\tau_s}{\hbar\omega_s\sigma}\frac{G_{sij} - \Gamma^{ij}a_{Ns}\tau_{1c}n_{cs}^{ij}\sigma_c S_{sij} + \Gamma^{ij}\Delta g_{hsij}}{1+\varepsilon_{SHB}S_{sij}}(1-i\alpha_N)$$  (25)

$$\frac{1}{\left[-i(\omega_s - \omega_{s'})\tau_s + 1\right]\left[-i(\omega_s - \omega_{s'})\tau_1 + 1\right]}$$

$$\eta_{ss'}^{SEij} = \frac{\varepsilon_{SE}^{ij}}{\hbar\omega_s\sigma}\frac{G_{sij} - \Gamma^{ij}a_{Ns}\tau_{1c}n_{cs}^{ij}\sigma_c S_{sij} + \Gamma^{ij}\Delta g_{hsij}}{1+\varepsilon_{SHB}S_{sij}}(1-i\alpha_T)$$  (26)

$$\frac{1}{\left[-i(\omega_s - \omega_{s'})\tau_h + 1\right]\left[-i(\omega_s - \omega_{s'})\tau_1 + 1\right]}$$

$$\eta_{ss'}^{FCAij} = \frac{\varepsilon_{FCA}^{ij}N_{sij}\Gamma^{ij}}{\sigma}(1-i\alpha_T)$$  (27)

$$\frac{1}{\left[-i(\omega_s - \omega_{s'})\tau_h + 1\right]\left[-i(\omega_s - \omega_{s'})\tau_1 + 1\right]}$$

$$\eta_{ss'}^{SHBij} = \frac{\varepsilon_{SHB}^{ij}}{\hbar\omega_s v_g^{ij}\sigma}\frac{G_{sij} - \Gamma^{ij}a_{Ns}\tau_{1c}n_{cs}^{ij}\sigma_c S_{sij} + \Gamma^{ij}\Delta g_{hsij}}{1+\varepsilon_{SHB}S_{sij}}(1-i\alpha_{SHB})$$  (28)

$$\frac{1}{\left[-i(\omega_s - \omega_{s'})\tau_1 + 1\right]}$$

Where i=1, j=1, 2 describe the TE mode, '11', and TM mode '12'. s=0, 1, 2 correspond to the pump, the probe and the conjugate pulses. The parameters in (21-28) are derived in the same way as [14] [23] [30], and can be calculated using the values in [15][22][25]-[27]. The influence of probe and conjugate pulses on the evolution of pump pulses and the values of quasi-equilibrium density are neglected because their small energies in comparison with pump pulse. The FWM conversion efficiency $\eta_{FWM}$ is defined as the ratio between the output energy of the conjugate pulse and the input energy of the probe pulse

$$\eta_{FWM} = \frac{\int_{-\infty}^{\infty}P_2(L,t)dt}{\int_{-\infty}^{\infty}P_1(0,t)dt}$$  (29)

where L is the length of the SOA. Many papers [33]-[35] have focused on using pulsed FWM theory in SOA for optical sampling. To obtain the influences of polarization dependence on pulsed FWM process, based on the optical sampling principle described in [35], we can set the concrete simulation situations. To solve the equations (21)-(23), we use the same method as which is used to solve (4)-(8), the split-step Fourier method.

**Simulation results and discussion**

During the calculation, because we did not use the assumption used by [14] [23], our model, on theory, is believed to be valid for analyzing the propagation of sub-picosecond polarized optical pulses, but for the pulse duration which is as short as intraband scattering times, our model will not be applicable. From (8), we can see that the item $\Delta g_{\beta hij}$ includes the interaction between TE and TM mode. Because this item describes the changes of gain due to electron-phonon collisions with a characteristic time $\tau_{h,\beta}$, if the number of subsections is big enough ($\gg 30$), to simplify the calculation, in each step, we can use the $g_{ij}$ of previous section to calculate the values of $\Delta g_{\beta hij}$.

In our simulations, the SOA length is 500 μm, and the mode cross section is 1.5 μm$^2$. We will consider optical pulses with Gaussian shape [500fs, full-width at half-maximum (FWHM)] as input. The SOA pump current is 100mA. The confinement factor $\Gamma^{12}$ is chosen to be 30% less than $\Gamma^{11}$[10]. In Fig.2, the SOA gain is presented as a function of the pulse energy.

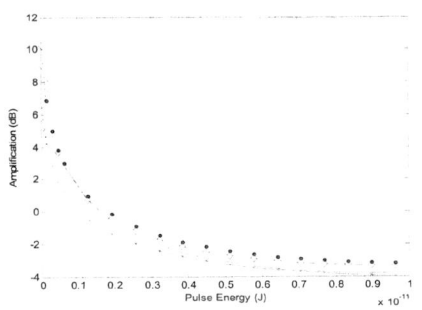

Fig.2. Computed polarization dependent amplification as a function of the inputted pulse energy.
-◇- TE mode
-x- TM mode
• TM mode, with $\Gamma^{11} = \Gamma^{12}$
— 45° with respect to the TE and TM polarization axes with $\Gamma^{11} = \Gamma^{12}$

The results are in agreement with the results presented by [10] [36] [37]. It has been proved by [10] & [36] that for high energetic optical pulses the FCA and TPA terms will dominate. Because the FCA

162

parameters in this discussion are same for TE and TM mode, for large pulse energies, the difference between TE and TM mode mainly attributes to TPA and different confinement factors. For bulk SOA, the way to realize polarization insensitivity is to use a submicrometer square active layer [38]. In this structure, there is no difference in the confinement factor between TE and TM polarization, and anisotropy in the waveguide nonlinearity will be the main reason for NPR. Many experiments have been carried out to investigate this effect [39] [40] [41]. From Fig.2, we can see that by taking $\Gamma^{11} = \Gamma^{12}$, the amplification of TM mode is enhanced effectively and the difference between the net amplifications of TE and TM mode nearly vanishes. But for the $45^o$ polarized input with $\Gamma^{11} = \Gamma^{12}$, the difference on net amplifications becomes more obvious with increased pulse energies. As the discussion above, it is mainly caused by anisotropy of TPA. It has been pointed out by [16] that a very small index difference is sufficiently high to induce a significant birefringence, and a total absence of birefringence is virtually impossible to obtain, though a reasonable polarization independent gain is technologically achievable.

The results which are presented so far are used to analyze the second cause and the third cause of optical anisotropy of SOA successfully and also can be used to explain the self-induced polarization rotation which has been investigated in [42]. In the following numerical experiments we investigate the polarization dependent performance of pulsed FWM in an unstrained bulk SOA which is polarization insensitive. It is well known that to obtain maximum conversion efficiency, the two input waves should have the same state of polarization while they interact in the SOA. We simulate copolarized input waves without time delay here. Pulsewidth for pump is 500fs, and for probe is 2.5ps. Input energy is 1pJ for pump signal, and 10fJ for probe signal. In Fig.3, the FWM conversion efficiency is presented as a function of the frequency detuning (defined as $\omega_{probe} - \omega_{pump}$).

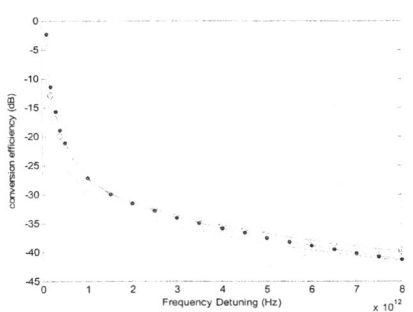

Fig.3. simulation results of $\eta_{FWM}$ versus frequency detuning for different input polarizations of copolarized pump and probe waves.
-◇- pump and probe are TE polarized
-•- pump and probe are TM polarized with $\Gamma^{11} = \Gamma^{12}$

— pump and probe are linearly polarized under an angle of $45^o$ with the TE and TM polarization axes with $\Gamma^{11} = \Gamma^{12}$

The results shown here are for frequency up-conversion (with negative frequency detuning) and coincide well with the experimental results presented in [41]. We can see that the FWM efficiency of the $45^o$ off-axis injection is even lower than TM polarized situation. The reason for this phenomenon is that the states of polarization of pump and probe signals change along the length of the amplifier, though the difference in gains of pump, probe and conjugate signals is very small. The main reason for the changed polarization states of pump wave is self-polarization modulation (SPoIM), and for probe and conjugate waves, it is attributed to cross-polarization modulation (XPoIM). For large pulse energies, the difference in TE and TM copolarized conversion efficiencies increases. It indicates the presence of birefringence in polarization insensitive unstrained bulk SOA. High conversion efficiency is very important for optical sampling. Maximum FWM efficiency occurs when pump and probe are polarized to be parallel to one of the principal axes.

**Summary and conclusions**

Based on a new assumption, we derived a new set of polarization dependent SOA model. Using this model, except the strain-induced birefringence, we analyze the other two causes of anisotropy of SOA successfully. Our model also can be used to analyze strained SOA with the basic material parameters. To further the understanding of optical sampling performance using pulsed FWM theory in SOA, we modify the pulsed FWM theoretical model, taking both SPoIM and XPoIM into account. From the simulations, we can see that the FWM conversion efficiency is polarization dependent, and this effect is more noticeable with increased frequency detuning. Owing to the very efficient extinction that can be obtained using a polarizer, many new applications of SOA polarization dependent effect have emerged. The models presented in this paper also can be used to analyze the issues relative to these applications.

**References**

1 D. M. Patrick et al., Electron. Lett., Vol. 30(1994), pp. 341-342.

2 Y. Liu et al., IEEE Photon. Technol. Lett., Vol. 15(2003), pp. 90-92.

3 Xiang Teng et al., ECOC Proc., Vol. 3(2005), pp. 661-662.

4 Chia Chien Wei et al., IEEE Photon. Technol. Lett., Vol. 8(2005), pp. 1683-1685.

5 H. Soto et al., IEEE Photon. Technol. Lett., Vol. 13(2001), pp. 335-337.

6 H. J. S. Dorren et al., IEEE J. Select. Topics

Quantum Electron., Vol. 10(2004), pp. 1079-1186.

7 H. Soto et al., IEEE Photon. Technol. Lett., Vol. 14(2002), pp. 498-500.

8 H. J. S. Dorren et al., IEEE J. Quantum Electron., Vol. 39(2003), pp. 141-148.

9 N. Calabretta et al., IEEE Photon. Technol. Lett., Vol. 18(2006), pp. 436-438.

10 X. Yang et al., Opt. Commun., Vol. 223(2003), pp. 169-179.

11 T. D. Visser et al., IEEE J. Quantum Electron., Vol. 35(1999), pp. 240-249.

12 Yutaka Takahashi et al., IEEE J. Quantum Electron., Vol. 34(1998), pp. 1660-1673.

13 Takaaki Kakitsuka et al., IEEE J. Quantum Electron., Vol. 38(2002), pp. 85-92.

14 Antonio Mecozzi et al., IEEE J. Select. Topics Quantum Electron., Vol. 3(1997), pp. 1190-1207.

15 J. Mørk et al., SPIE Vol. 2399(1995), pp. 146-159.

16 H. Soto et al., IEEE Photon. Technol. Lett., Vol. 11(1999), pp. 970-972.

17 Li-Qiang Guo et al., J. Lightw. Technol., Vol. 23(2005), pp. 4037-4045.

18 K. L. Hall et al., Opt. Commun., Vol. 111(1994), pp. 589-612.

19 D. C. Hutchings et al., Phys. Rev. B, Vol. 49(1994), pp. 2418-2426.

20 D. C. Hutchings et al., J. Opt. Soc. Am. B, Vol. 9(1992), pp. 2065-2074.

21 D. C. Hutchings, IEEE J. Select. Topics Quantum Electron., Vol. 10(2004), pp. 1124-1132.

22 G. P. Agrawal and N. K. Dutta, Long-Wavelength Semiconductor Lasers (Van Nostrand Reinhold, New York, 1986).

23 J. M. Tang et al., IEEE J. Quantum Electron., Vol. 34(1998), pp. 1263-1269.

24 W.B.Joyce et al., Appl. Phys. Lett., Vol. 31(1977), pp. 354-356.

25 Shun Lien Chuang, Physics of optoelectronic devices (Wiley-Interscience publication, 1995).

26 A. Uskov et al., IEEE Photon. Technol. Lett., Vol. 4(1992), pp. 443-445.

27 A. Mecozzi et al., J. Opt. Soc. Am. B, Vol. 13(1996), pp. 2437-2451.

28 Chongjin Xie et al., Opt. Commun., Vol. 164(1999), pp. 211-217.

29 J. Mørk et al., IEEE J. Quantum Electron., Vol. 33(1997), pp. 545-555.

30 J. M. Tang et al., IEEE J. Quantum Electron., Vol. 35(1999), pp. 1032-1040.

31 Y. Z. Hu et al., Phys. Rev. B, Vol. 49(1994), pp. 382-386.

32 Roberto Paiella et al., IEEE J. Select. Topics Quantum Electron., Vol. 3(1997), pp. 529-540.

33 M. Jinno et al., Electron. Lett., Vol. 30(1994), pp. 1489-1491.

34 Leaf A. Jiang et al., IEEE J. Quantum Electron., Vol. 37(2001), pp. 118-126.

35 Hitoshi Kawaguchi et al., SPIE Vol. 3283(1998), pp. 477-484.

36 A. K. Mishra et al., IEEE J. Select. Topics Quantum Electron., Vol. 10(2004), pp. 1180-1092.

37 F. Romstad et al., IEEE Photon. Technol. Lett., Vol. 12(2000), pp. 1674-1676.

38 Masayuki Itoh et al., IEEE Photon. Technol. Lett., Vol. 14(2002), pp. 765-767.

39 B. F. Kennedy et al., IEE Proc. Optoelectron., Vol. 151(2004), pp. 114-118.

40 R. J. Manning et al., Electron. Lett., Vol. 37(2001), pp. 229-231.

41 S. Diez et al., IEEE Photon. Technol. Lett., Vol. 10(1998), pp. 212-214.

42 N. Calabretta et al., J. Lightw. Technol., Vol. 22(2004), pp. 372-381.

# Gain leveling in a two-level system for EDFA using Quantum-interference effects

X.M. Su (1), X.W. Niu (1), Jung Bog Kim (2), Z.C. Zhuo (1)

*1: Physics College, Jilin University, Changchun 130023, PR China*
*2: Department of Physics, Korea National University of Education, Korean*

**Abstract**: Much work has been focused on gain leveling for EDFA for its importance in long distant transmission in dense wavelegth division multiplexing (WDM) system. For 1480nm pumping, energy from the pump field excites the erbium ions in the low lying levels of the $^4I_{15/2}$ manifold to the high lying ones in the $^4I_{13/2}$ manifold, and light amplification occurs from 1520nm to 1560nm because of the stokes shift existing between the emission and the absorption bands of this transition. However, the gain profile of EDFA is uneven and therefore leads to error performance in different WDM channels in 1530-1560 nm. Here we propose a scheme of solving gain leveling for EDFA by one of quantum interference effects. We apply an additional strong coherent field to drive the transitions from $^4I_{15/2}$ manifold to $^4I_{13/2}$ manifold. The susceptibility $\chi$ induced by the strong field is reverse with the original gain spectrum as a filter does. This scheme can reach a flat gain operating around 1.53 $\mu m$.

## Introduction

Gain equalization for EDFA [1] causes more and more attention and therefore a lot of solutions are proposed by various available technology and new methods [2-9]. Of these solutions, from technology view, one of effective and applicable equalization methods is to use various optical filters as equalizers for gain leveling in that their transmission spectrum matches with the inverted erbium gain spectrum in 1530-1560 nm band and counteract the peak profiles of EDFA. Effective equalizers, for instance, include the use of Mach-Zehnder filter [3], blazed fiber Bragg gratings[4], long period fiber gratings [5] and Electromagnetically Induced Transparency (EIT) technique [9].

It is well known that EIT is a powerful technique to render a system transparent to a resonant transition as a result of quantum interference between various pathways [10]. It leads to the modulation of the optical properties of the medium when a strong coherent field coupled coherently with the upper or lower level of a resonant transition. The basic EIT effect on a two-level system interacting with a strong field was discussed [11]. In this paper, we use EIT technique for gain leveling of EDFA in a two-level system with 1480nm pumping. An additional strong coherent field is employed to drive the transitions from $^4I_{15/2}$ manifolds to $^4I_{13/2}$ manifold. Due to the presence of a strong coherent field, the two levels related with laser radiation split into doublets. The susceptibilty at the frequency of signal field induced by the strong field is reverse with the original gain spectrum as a filter does. This paper is different from Ref. [9] where a third level $^4I_{11/2}$ has to be considered for 980nm pumping. Here corresponds to $\lambda_p$=1480nm pumping scheme, $Er^{3+}$ ions are excited directly within the $^4I_{13/2}$ - $^4I_{15/2}$ laser transitions. The energy from the pump laser excites the erbium ions in the low lying levels of the $^4I_{15/2}$ manifold to the high lying ones in the $^4I_{13/2}$ manifold. Then, it is possible to divide the manifolds of metastable level $^4I_{13/2}$ into two individual states, and the higher and lower lying ones are corresponding to the pumping and signal level [12], respectively.

## Theory and discussions

The energy schematic of two-level model of EDFA for gain leveling is shown as in Fig. 1, when it is pumped at 1480nm. Level |1> is the ground level $^4I_{15/2}$ and level |2> is metastable level $^4I_{13/2}$, level |2a> and |2b> are the pumping and signal level, respectively. The transitions between manifolds of level |2> to |1> are driven by the weak signal field with frequency $\omega_s$ and amplitude $\varepsilon_s$ and the strong coherent field with frequency $\omega_c$ and Rabi frequency $\Omega$. Incoherent pump is corresponding to the transition |2b>-|1> and its pumping rate denotes as $\Lambda$. Without the strong field, Fig. 1 shows a typical two-level EDFA system with population inversion between manifolds of level |2a> and |1> due to the fast unradiative theomalization process within |2a> and |2b>.

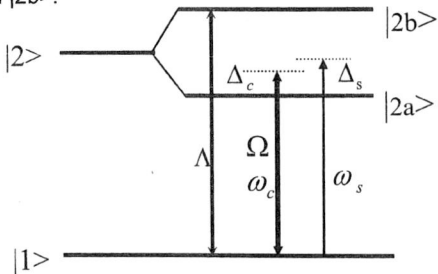

Fig. 1 Two level system of gain leveling for EDFA

Considering two-level system shown as fig. 1, in the interaction picture, the equations of the densty matrix elements are [12]:

$$\dot{\sigma}_{a1} = (i\Delta_{ac} - \gamma_{a1})\sigma_{a1} - i[g\varepsilon_s e^{-i\delta t} + \Omega](\sigma_{aa} - \sigma_{11})$$

$$\dot{\sigma}_{b1} = (i\Delta_{bc} - \gamma_{b1})\sigma_{b1} - i\Omega(\sigma_{bb} - \sigma_{11})$$

$$\dot{\sigma}_{11} = \Lambda(\sigma_{bb} - \sigma_{11}) + \Gamma_{a1}\sigma_{aa} + \Gamma_{b1}\sigma_{bb}$$
$$i[(g\varepsilon_s^* e^{i\delta t} + \Omega^*)\sigma_{a1} + \Omega^*\sigma_{b1} - c.c.]$$

$$\dot{\sigma}_{aa} = \gamma_{ba}\sigma_{bb} - \Gamma_{a1}\sigma_{aa} -$$
$$i[(g\varepsilon_s^* e^{i\delta t} + \Omega^*)\sigma_{a1} - c.c.]$$

$$\sigma_{11} + \sigma_{aa} + \sigma_{bb} = 1$$

$$(1)$$

Where $\Gamma_{a1}$, $\Gamma_{b1}$ are the radiative rates corresponding to the respective transitions; $\Gamma_{ba}$ is the nonradiative rate for the theomalization process; $\Delta_{ac} = \omega_c - \omega_{a1}$ , $\Delta_{bc} = \omega_c - \omega_{b1}$ , $\delta = \Delta_p - \Delta_c$, $\Delta_s = \omega_s - \omega_{a1}$;

We apply perturbation method to solve equation (1) which are valid to all order of coherent field and to first order to probe field in order to find the linear susceptibilities to the transition |2 a>- |1> in that the signal field is weak comparing with the coherent field. The density matrix elements are expressed as:

$$\sigma_{jj} = \sigma_{jj}^{(0)} + \sigma_{jj}^{(1)}e^{-i\delta t} + \sigma_{jj}^{(-1)}e^{i\delta t}, (j = a, 1)$$

$$\sigma_{a1} = \sigma_{a1}^{(0)} + \sigma_{a1}^{(1)}e^{-i\delta t} + \sigma_{a1}^{(-1)}e^{i\delta t}$$

$$(2)$$

Where $\sigma^{(0)}$, $\sigma^{(1)}$ and $\sigma^{(-1)}$ are corresponding to zero order, one order and one order conjugated of the signal field. Inserting (2) into (1), and being noted that $\Gamma_{b1} = \Gamma_{a1} = \Gamma$ ; $\gamma_{b1} = \gamma_{a1} = \gamma = \Gamma/2$ , we have:

$$\sigma_{a1}^{(1)} = g\varepsilon_s(\sigma_{22}^{(0)} - \sigma_{11}^{(0)})\frac{1 - D_1/D_2}{\Delta_s + i\gamma}$$

$$D_1 = i|\Omega|^2(-2i\delta + 2\Gamma + 2\gamma_{ba} + 3\Lambda - 3\Gamma_{b\Omega})(\frac{1}{\Delta_c - i\gamma} - \frac{1}{\Delta_s - i\gamma})$$

$$D_2 = (-i\delta + 2\Lambda - 2\Gamma_{b\Omega} + \Gamma - \Gamma_s)(-i\delta + \gamma_{ba} + \Lambda - \Gamma_{b\Omega}) - (2i\delta + 2\gamma_{ba} + 3\Lambda - \Gamma_{b\Omega})(\Lambda - \Gamma_{b\Omega} + \Gamma_{ba} - \Gamma_a + \Gamma_s)$$

$$\sigma_{11}^{(0)} = \frac{(\Lambda - \Gamma_{b\Omega} + \Gamma)(\Gamma + \gamma_{ba} - \Gamma_{a\Omega}) - \gamma_{ba}(\Lambda - \Gamma_{b\Omega} + \Gamma_{\Omega})}{(2\Lambda - 2\Gamma_{b\Omega} + \Gamma - \Gamma_{a\Omega})(\Gamma + \gamma_{ba} - \Gamma_{a\Omega}) - (\gamma_{ba} + \Gamma_{a\Omega})(\Lambda - \Gamma_{b\Omega} + \Gamma_{\Omega})}$$

$$\sigma_{22}^{(0)} = \frac{\gamma_{ba} - (\gamma_{ba} + \Gamma_{a\Omega})\sigma_{11}^{(0)}}{\Gamma + \gamma_{ba} - \Gamma_{a\Omega}}$$

$$(3)$$

Where $\Gamma_{\Omega b} = \frac{2\gamma|\Omega|^2}{\Delta_{bc}^2 + \gamma^2}$ , $\Gamma_{\Omega a} = \frac{2\gamma|\Omega|^2}{\Delta_{ac}^2 + \gamma^2}$

$$\Gamma_s = \frac{2\gamma|\Omega|^2}{\Delta_s^2 + \gamma^2}$$

The macroscopic polarization of the medium is $P = N < \mu >$ where $N$ is the $Er^{3+}$ density and $< \mu > = tr(\sigma\mu) = N(\mu_{21}\sigma_{12} + \mu_{12}\sigma_{21})$ is average dipole moment. By the definition $P = \chi(\omega_s)E(\omega_s)$, the susceptibility of the singal field at frequency $\omega_s$ as $\chi(\omega_s) = \frac{4\hbar gN}{\varepsilon_0\varepsilon_s}\sigma_{a1}^{(1)}$ , we have

$$\chi(\omega_s) = \frac{4\hbar g^2 N}{\varepsilon_0}\sigma_{a1} \qquad (4)$$

It can be obtained, from equaiton (4), for the susceptibility $\chi_{jk}(\omega_s)$ corresponding to the respective transition between jth sublevel of |2a> and kth sublevel of $^4I_{15/2}$. The total susceptibility results from the sum $\chi_{jk}(\omega_s)$. The dipole moment matrix element for each transiton $\mu_{jk} = 2\hbar g$ can be replaced by the oscillation strength $f_{jk} = \frac{2m\omega_{jk}\,|\,\mu_{jk}\,|^2}{3\hbar e^2}$ for the respective atomic transition |2j> -|1k> and the each parameter of $f_{jk}$ is specific for $Er^{3+}$. It is assumed each single transiton has the same radiative decay rate, so $\Gamma_{2j\to1k} = \Delta\omega_{hom}^{-1}/7$, where $\Delta\omega_{hom}$ is the homogeneous broading.

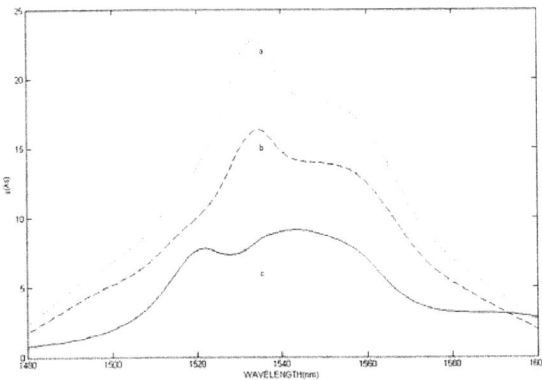

Fig. 2. $\text{Im}(\chi(\lambda_s))$ vs the wavelength of the singal field: the parameters used are (a) $\Omega = 0$, $\Delta_c = 0$; (b) $\Omega = 16GHz$, $\delta = 0$; (c) $\Omega = 24GHz$, $\delta = 0$. Other Parameteter used are $\Gamma = 1.0$, $\Gamma_{ba} = 10.0$, $\Lambda = 10.0$.

We draw the imaginary part of the total susceptibility $\chi$ vs. the wavelength of the signal field in figure 2. It shows that the position of the gain peak emergies near to 1.53 $\mu m$ with the coherent field $\Omega = 0$ (curve (a)). The peak gain excursion can be reduced by using a coherent field with a proper the intensity and frequency (curve (b) and (c)). The

parameters for figure 2 are chosen as: Γ=1.0, Γ$_{ba}$=10, Λ=10; Parameters of strong coherent field ($\Omega$ :Rabi frequency; $\delta$ : angle frequency difference between coherent field and probe field) for curve (2a)-(2c) are : (a) $\Omega = 0$, $\Delta_c = 0$; (b) $\Omega = 16GHz$, $\delta = 0$; (c) $\Omega = 24GHz$, $\delta = 0$. Figure 2 shows that a large laser intensity ( which is proportional to $\Omega$ ) with a proper positive frequency detuning $\delta$, benefits to the filter using EIT scheme we proposed here. The Rabi frequency of coherent field for figure (2c) is in the order of $10^{10}Hz$, which is corresponding to several hundreds of megawatts of the coherent field to be used [9]. It demonstrates that using EIT technique to flate the gain profile of EDFA with 1480nm pumping is feasible.

**Conclusions**

In this paper, we prove that Electromagnetically Induced Transparency can be used as a filter to flat gain spectrum for two level erbium-doped fiber amplifier (EDFA) system with 1480nm pumping, by applying an addition strong coupling field. With a proper the frequency and amplitude of the coherent field, this scheme can be realized a flat gain operating around 1.53 $\mu m$ considering the Stark-splitting sublevels of laser levels for $Er^{3+}$.

**References**

1 R. J. Mcars et al, Electron. Lett, **23** (1987), 1026

2 J. R. Arnitage, IEEE. J. Quantum Electron., **26** (1990), 423

3 K. Inoue et al, IEEE Photon Tech Lett., **3**(1991), 718

4 R. Kashyap et al, Electron. Lett. **29** (1993), 1025

5 A. M Vengsarkar et al, Opt. Lett. **21** (1996), 336

6 M. Yamada et al, IEEE, Photon Tech Lett., **8** (1996), 882

7 J. X. Cai et al, IEEE Photon Tech Lett., **9** (1997), 916

8 H. F. Zhang et al, Phys. Rev. A, **65** (2002), 043812

9 Z. C. Zhuo et al, Phys. Lett., 336 (2005) 25

10 S. E. Harris, Phys. Today, **50** (1997), 36

11 Ryan S. Bennink et al, Phys. Rev. A, **63**, 033804 (2001)

# Sidelobes Suppression of Periodically Placed Resonators Side-Coupled to Photonic Crystal Waveguide

H. H. Li, K. Xu, J. Wu, J. T. Lin

Key Laboratory of OCLT, Ministry of Education, Beijing University of Posts and Telecommunications, Beijing, 100876, P. R. China, email: hanhui.li@gmail.com

**Abstract** *Apodization technique is applied to suppress sidelobes in the reflection spectrum of periodically placed resonators side-coupled to photonic crystal waveguide. Good agreement is obtained between theoretical analysis by transfer-matrix method derived from coupled mode theory in time and simulation experiments.*

## Introduction

Some types of dielectric structures, whose dielectric constant vary periodically in space, own a photonic band gap in which electromagnetic wave can not propagate. These types of structures are referred to photonic band gap materials or photonic crystals (PhC) [1-3]. Great interests have been attracted in research on PhC based optical devices, such as PhC waveguides [4] and filters [5-6], due to their compactness and the potential application in photonic integrated circuits and all-optical communication network.

Optical filters are key components, to access the signal on one particular channel or multiple channels, in wavelength division multiplexed optical communication system. In PhC, optical filters are normally composed of microcavities (also known as resonators) and waveguides. Reflection filters based on resonators side-coupled to PhC waveguide are analyzed by Xu et al [5] and L. Lin et al [6]. In this kind of filters composed of periodically cascaded resonators side-coupled to a single mode PhC waveguide, sidelobes exist due to abruptly beginning and ending of the uniform structure. That is not desirable in application like wavelength division multiplexing.

In this paper, reflection filter based on multiple single mode resonators side-coupled to two-dimensional (2D) PhC waveguide is analyzed. In our proposed filter structure, the resonators are periodically placed on both sides of the 2D PhC waveguide. The configuration of our optical system is similar to that of optical filters composed of ring resonators [7-8] or distributed feedback resonators (DFB) [9]. The direct coupling between two neighboring resonators along the waveguide is prohibited by placing them on two sides of the PhC waveguide respectively. The indirect coupling between resonators through guided mode of the PhC waveguide is analyzed by transfer-matrix method based on coupled mode theory (CMT) in time [9-10]. Apodization technique will be used to suppress sidelobes in order to optimize the reflection spectrum. We first give our design and theoretical analysis.

Furthermore, first-principle finite-difference time-domain (FDTD) [11] experiment is carried to verify the theoretical analysis.

## Theoretical analysis

Here shows a simple resonant reflection filter composed of three identical single mode resonators side-coupled to a PhC waveguide. The dashed lines (also depicted in the model of the optical system) going through the resonators are reference planes. $L$ is the distance between two neighbouring resonators along PhC waveguide direction. $a_i$ is the mode amplitude of corresponding resonator. $S_{+i} / S_{-i}$ is the amplitude of incoming/outgoing wave corresponding to specific resonators.

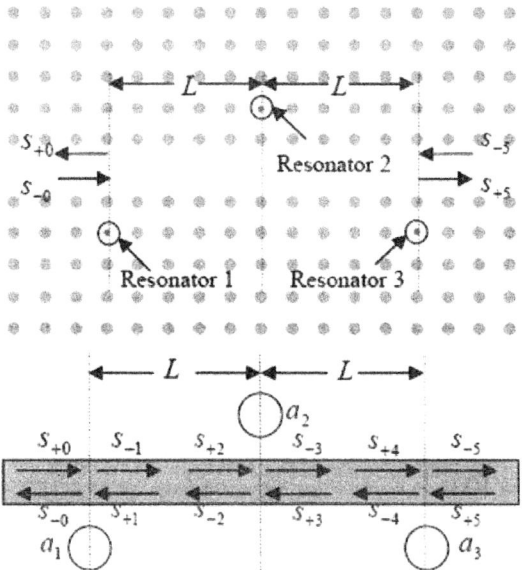

*The three-resonator reflection filter in a PhC square lattice, and the corresponding theoretical model of the structure.*

Furthermore, the direct coupling through evanescent waves tunneling between two neighbouring resonators is prohibited by placing them on the two sides of the PhC waveguide respectively. So the resonators can only interact through the guided mode of the PhC waveguide. Based on these conditions, the proposed reflection filter can be modeled.

To analyze one general reflection filter composed of $N$ single mode resonators, a transfer-matrix method derived by CMT in time is applied to study characteristics of the optical system because it is tedious to obtain reflection coefficient through CMT in time directly.

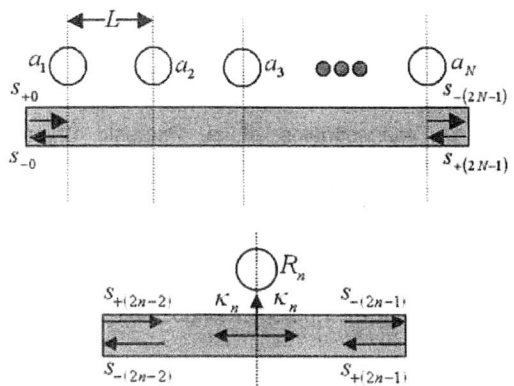

Reflection filter composed of periodically placed resonators side-coupled to line waveguide, and an arbitrary resonator in the periodically placed resonators.

Assuming no direct coupling between two neighboring resonators through evanescent wave tunnelling in the optical system composed of multiple resonators side-coupled to waveguide, the relationship between incoming and outgoing waves can be described by following equation,

$$\begin{pmatrix} s_{+0} \\ s_{-0} \end{pmatrix} = \left( \prod_{n=1}^{N} T_n D_n \right) \begin{bmatrix} s_{-(2N-1)} \\ s_{+(2N-1)} \end{bmatrix} \quad (1)$$

where $T_n$ is the transfer matrix related to $n$-st side-coupled resonator and $D_n$ represents phase retardation between $n$-st resonator and next one. $T_n$ and $D_n$ are expressed by

$$T_n = \begin{pmatrix} \alpha_n & \eta_n \\ -\eta_n & \gamma_n \end{pmatrix} \quad (2)$$

and

$$D_n = \begin{bmatrix} \exp(j\phi_n) & 0 \\ 0 & \exp(-j\phi_n) \end{bmatrix} \quad (3)$$

respectively in which

$$\alpha_n = 1 + [2/\tau_w]/[j(\omega - \omega_0) + 1/\tau_0],$$
$$\eta_n = [2/\tau_w]/[j(\omega - \omega_0) + 1/\tau_0],$$

$$\gamma_n = 1 - [2/\tau_w]/[j(\omega - \omega_0) + 1/\tau_0]$$ and $\phi_n$ is phase retardation between $n$-st resonator and next one. In the above discussion, Equation (1) and (2) are directly derived from CMT [9-10] in time. $\omega_0$ is the resonance frequencies of the resonators. $1/\tau_0$ is the decay rate due to internal loss and $1/\tau = 2/\tau_w + 1/\tau_0$ in which $1/\tau_w$ is decay rate of an arbitrary resonator into one port of the waveguide under the condition that symmetry along waveguide is assumed. In the optical system, $\kappa_n$ is the coupling coefficient between the $n$-st resonator and one port of the waveguide. It is related to $\tau_w$ by $\kappa_n = \sqrt{2/\tau_w}$.

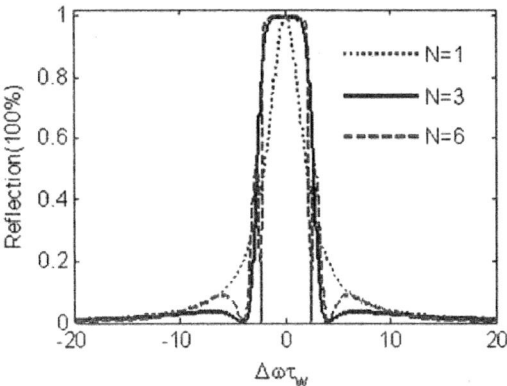

Reflection as a function of deviation from resonance of reflection filter composed of $N$ resonators periodically placed on both sides the waveguide.

As for one optical system composed of multiple resonators side-coupled to PhC waveguide, sidelobes appear due to abruptly beginning and ending of the uniform structure. Apodization technique can be used to suppress sidelobes by breaking the uniform of the optical system. For simplicity, we only use one common Gaussian taper function to see the effects. In the following section, the numerical simulation is conducted to verify the theoretical analysis.

**Design and simulation results**
As an illustration, a simple Gaussian function of coupling coefficient between each resonator and waveguide is obtained to analyze the performance of optical system. In the following example, the coupling coefficient $\kappa_n$ between $n$-st resonator and waveguide is expressed by

$$\kappa_n = \kappa_0 \exp\{-G[n - (N+1)/2]^2\} \quad (4)$$

where $\kappa_0$ can be adjusted to give the desired bandwidth of reflection spectrum, $N$ is the number of resonators in the analyzed optical system and $G$

169

is one real value to set full-width at half-maximum of the Gaussian function.

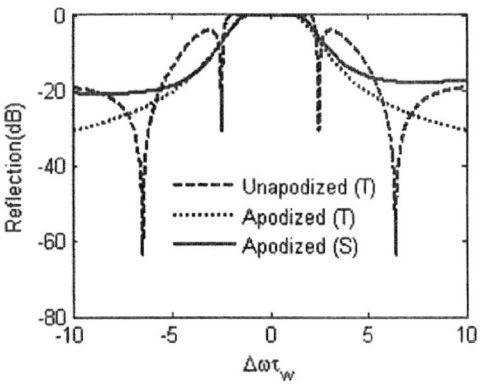

*Reflection spectra for a five-resonator reflection filter with apodization and without apodization, the results are obtained through theoretical (T) analysis and simulation (S).*

In the above figure show the reflection spectra of the apodized five-resonator reflection filter in comparison with un-apodized one. It can be seen that sidelobes have been removed. Furthermore, numerical simulation by FDTD [11] is done to verify the theoretical analysis. Good agreement is obtained between the theoretical analysis (dotted curve) and numerical results (solid curve). The higher reflection from numerical simulation compared with theoretical analysis at the side band may be due to the intrinsic reflection in the FDTD simulation [12].

**Conclusions**

Sidelobes exist in reflection spectrum of the optical system composed of periodically placed resonators side-coupled to PhC waveguide. These are not desirable characteristics in application like WDM system. Based on one simple transfer-matrix method derived from CMT in time, performance of the optical system is analyzed. To suppress sidelobes, apodization technique is applied to optimize the performance of the optical reflection filter by breaking the intrinsic periodicity of the system. Good agreement between theoretical analysis and FDTD simulation is obtained.

**References**
1  E. Yablonovitch, Phys. Rev. Lett., 58 (1987), 2059.
2  S. John. Phys. Rev. Lett., 58 (1987), 2486.
3  J. D. Joannopoulos et al, Photonic Crystals, (Princeton U. Press, Princeton, N.J., 1995).
4  A. Mekis et al, Phys. Rev. Lett., 77 (1996), 3787.
5  Y. Xu et al, Phys. Rev. E, 62 (2000), 7389.
6  L. Lin et al, Phys. Rev. B, 72 (2005), 165330
7  B. E. Little et al, Opt. Lett., 25 (2000), 344,.
8  G. Griffel, IEEE Photon. Technol. Lett., 12 (2000), 810.
9  H. A. Haus et al, IEEE J. Quantum Electron., 28 (1992), 205.
10  H. A. Haus, Waves and Fields in Optoelectronics (Englewood Cliffs, NJ: Prentice-Hall, 1984).
11  A. Taflove et al, Computational Electrodynamics: The Finite-Difference Time-Domain Method, 2nd ed. (Artech, Boston, Mass., 2000).
12  A. Mekis et al, IEEE Microwave Guided Wave Lett., 9 (1999), 502.

# Photon Tunnelling Effect in One Dimensional Photonic Crystal Containing Partially Negative Permittivity Uniaxial Materials

Guoan Zheng, Ke Chen

Electronic and Optical Engineering, Zhejiang University, HangZhou, China, 310027

Email: guoanzheng@msn.com

**Abstract** *The band structure for a one-dimensional periodic structure with partially negative permittivity uniaxial materials is studied. It is shown that the photon tunnelling propagation modes can exist in the present structure. Since the material with partially negative permittivity tensor can be fabricated with low loss, potential applications in optical devices are expected to be drawn from our result.*

## Introduction

When light comes from an optically denser material to another material with an incidence angle larger than the critical angle, the phenomenon of total reflection occurs. No energy can transfer to the second media and the electromagnetic field decay exponentially. When a third media with a high refraction index is placed behind a thin slab of the second media, the photons can tunnel through the second media into the third one. The second media is similar to a band gap and the phenomenon that the photo goes through this band gap is called tunnelling effect [1-3] which has been known for a long time. Recently, Left-handed materials (LHM) with negative permittivity and negative permeability have attracted many attentions [4-6] and photon tunnelling effects in multilayered structure containing LHM [7-9] are then observed. However, the LHM in the optical or infrared part of the spectrum is usually accompanied by big loss since it is difficult to have negative magnetic resonance in high frequencies [10]. Some scholars propose to use the nonmagnetic anisotropic material whose permittivity tensor is partially negative to achieve the property of the LHM [11].

In this paper we study the band structure of the one-dimensional (1D) photonic crystal containing the anisotropic material whose permittivity tensor is partially negative. It is shown that the photon tunnelling modes exist in the present structure and thus form the numeral results; we will show that different orientation of the optical axis of the anisotropic media can lead to different band structure. Since the non-magnetic anisotropic material can be fabricated with low loss, our result may have potential applications in optical device such as attenuated total reflectance spectroscopy [12] and scanning photon tunnelling microscopy [13].

## Dispersion relation for the 1D photonic crystal containing uniaxial materials

Consider the dielectric constant of a uniaxial media as follow (the media we consider here is non-magnetic e.g. μ=1)

$$\overset{=}{\varepsilon} = \begin{pmatrix} \varepsilon_\perp & 0 & 0 \\ 0 & \varepsilon_\perp & 0 \\ 0 & 0 & \varepsilon_\parallel \end{pmatrix} \tag{1}$$

The tensor is given in principle coordinate (x-y-z system). The optical axis is along the z direction with dielectric constant $\varepsilon_\parallel$. For a conventional uniaxial media, $\varepsilon_\perp$ and $\varepsilon_\parallel$ are both positive. However, $\varepsilon_\parallel$ can be negative [11] in certain cases. To just give one example [11-14], for the composite of 10% of $S_iC$ nanospheroids with an aspect ratio of 1/2, aligned with their shorter axis along the x axis and embedded in quartz, for a wavelength of $CO_2$ laser of 12 $\mu m$, we obtain $\varepsilon_\parallel = -2.7 + 6 \times 10^{-4} i$, $\varepsilon_\perp = 1.6 + 1 \times 10^{-5} i$.

We consider a 1D periodic structure with alternate layers of air and anisotropic media, as shown in Fig.1, $d_1$ and $d_2$ are the widths of the two inclusion layers respectively and a=d$_1$+d$_2$ is the period. The angle between the optical axis of the anisotropic media and the z axis is θ (note that the optical axis is in plane x-z).

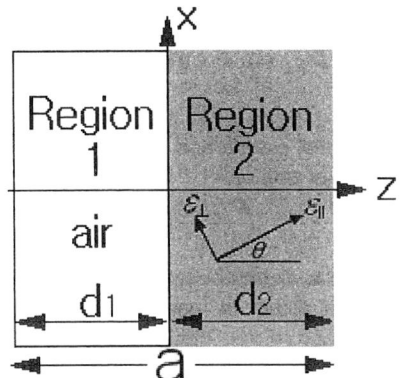

Fig.1 A 1D periodic structure consisting of alternate layers of air and anisotropic media with $\varepsilon_\parallel$ <0

We consider an oblique propagation of monochromatic electromagnetic field (with time dependence $e^{-i\omega t}$) in a periodic structure with oblique wave vector k$_x$ along the x axis. For the H-polarization case, more details are covered in reference 7.

$$H_{1y} = [e^{ik_1 z} + A e^{-ik_1 z}]e^{ik_x x} \tag{2a}$$

171

$$E_{1x} = \frac{k_1}{\omega \varepsilon_1}(e^{ik_{1z}z} - A e^{-ik_{1z}z})e^{ik_x x} \tag{2b}$$

$$E_{1z} = -\frac{k_x}{\omega \varepsilon_1}(e^{ik_{1z}z} + A e^{-ik_{1z}z})e^{ik_x x} \tag{2c}$$

$$H_{2y} = [B e^{ik_{2z}^i z} + C e^{ik_{2z}^r z}]e^{ik_x x} \tag{3a}$$

$$D_{2x} = \frac{1}{\omega}(B k_2^i e^{ik_{2z}^i z} + C k_2^r e^{ik_{2z}^r z})e^{ik_x x} \tag{3b}$$

$$D_{2z} = -\frac{k_x}{\omega}(B e^{ik_{2z}^i z} + C e^{ik_{2z}^r z})e^{ik_x x} \tag{3c}$$

The dispersion equation for the air and the anisotropic media can be expressed as

$$k_x^2 + k_{1z}^2 = (\omega/c)^2 \tag{4a}$$

$$k_x^2(\varepsilon_\perp \cos^2\theta + \varepsilon_\parallel \sin^2\theta) + k_{2z}^2(\varepsilon_\parallel \cos^2\theta + \varepsilon_\perp \sin^2\theta)$$
$$+ 2\sin\theta\cos\theta k_x k_{2z}(\varepsilon_\parallel - \varepsilon_\perp) - \varepsilon_\parallel \varepsilon_\perp (\omega/c)^2 = 0 \tag{4b}$$

where c is the speed of light in vacuum. For a given $k_x$, there are two solutions for $k_{1z}$, which are $k_1$, $-k_1$, and two solutions for $k_{2z}$ which are $k_1^i$, $k_2^r$.

The permittivity tensor of the anisotropic media can be expressed as

$$\overline{\overline{\varepsilon}} = \begin{pmatrix} \varepsilon_\perp \cos^2\theta + \varepsilon_\parallel \sin^2\theta & 0 & -\sin\theta\cos\theta(\varepsilon_\perp - \varepsilon_\parallel) \\ 0 & \varepsilon_\perp & 0 \\ -\sin\theta\cos\theta(\varepsilon_\perp - \varepsilon_\parallel) & 0 & \varepsilon_\parallel \sin^2\theta + \varepsilon_\perp \cos^2\theta \end{pmatrix} \tag{5}$$

Applying $\overline{D} = \overline{\overline{\varepsilon}}\,\overline{E}$, we can get the x component of electric field ($E_{2x}$) as bellow

$$E_{2x} = D_{2x}\left(\frac{\cos^2\theta}{\varepsilon_\perp} + \frac{\sin^2\theta}{\varepsilon_\parallel}\right) + D_{2z}\sin\theta\cos\theta\left(\frac{1}{\varepsilon_\parallel} - \frac{1}{\varepsilon_\perp}\right) \tag{6}$$

The tangential electric and magnetic fields should be continuous at z=0, i.e.

$$H_{1y} = H_{2y}, \quad E_{1x} = E_{2x}\,(z=0) \tag{7}$$

To obtain the dispersion relation for this 1D photonic crystal, we need to use the following periodic conditions according to the Bloch theorem

$$H_{2y}(z=d_2) = H_{1y}(z=-d_1)e^{iqa}$$
$$E_{2x}(z=d_2) = E_{1x}(z=-d_1)e^{iqa} \tag{8}$$

where q is in the first Brillouin zone $-\pi/a \le q \le \pi/a$
From Eq. (7,8), we have

$$A + 1 = B + C \tag{9a}$$

$$\frac{k_1}{\varepsilon_1}(1-A) = (Bk_2^i + Ck_2^r)\left(\frac{\cos^2\theta}{\varepsilon_\perp} + \frac{\sin^2\theta}{\varepsilon_\parallel}\right)$$
$$- k_x(B+C)\left(\frac{\sin\theta\cos\theta}{\varepsilon_\perp} - \frac{\sin\theta\cos\theta}{\varepsilon_\parallel}\right) \tag{9b}$$

$$B \cdot e^{ik_{2z}^i d_2} + C \cdot e^{ik_{2z}^r d_2} = (e^{-ik_1 d_1} + A \cdot e^{ik_1 d_1})e^{iqa} \tag{9c}$$

$$\frac{k_1}{\varepsilon_1}(e^{-ik_1 d_1} - Ae^{ik_1 d_1})e^{iqa} = (Bk_2^i e^{ik_{2z}^i d_2} + Ck_2^r e^{ik_{2z}^r d_2})\left(\frac{\cos^2\theta}{\varepsilon_\perp} + \frac{\sin^2\theta}{\varepsilon_\parallel}\right)$$
$$- k_x(Be^{ik_{2z}^i d_2} + Ce^{ik_{2z}^r d_2})\left(\frac{\sin\theta\cos\theta}{\varepsilon_\perp} - \frac{\sin\theta\cos\theta}{\varepsilon_\parallel}\right) \tag{9d}$$

We can solve for A, B, C using Eq. (9a-9c). Using Eq. (9d), we can get the band structure of the present 1D photonic crystal.

## The band structure of the present 1D dimensional photonic crystal

From analytical Eq. (9d) we can find out how k1 depends on q: if q is a real number, the modes can propagate through this photonic crystal; if q has an imaginary part, it will stand for a band gap case. In Fig. 2(a-d), we study the band structure of the present 1D photonic crystal with different orientation of the optical axis, i.e. with different θ. We only consider the real solutions for q because the wave can only propagate through the present structure in this case.

In Fig.2 (a), the band structure shows that there is no tunnelling mode since $k_1$ is a real number (the black line). However, in Fig.2 (b-d), the tunnelling modes exist in the present structure (with imaginary $k_1$ and real q, the red lines) and it means the field which decreases in the air layer is amplified in a anisotropic layer with $\varepsilon_\parallel < 0$. It is well-known that the photon tunnelling effect exists in two cases. One is the long-range surface plasma waves supported by thin metallic films [15-16] and the other is the optical modes of long wavelength in supperlattices [17-19]. In both of the two cases, the optical field passes through the structure indirectly by the coupling of surface plasma modes or by the excitation of photons. However, in the present case, the photon tunnelling effect is based on a very different mechanism—the amplification of the evanescent wave in the anisotropic layer. Since the anisotropic layer with $\varepsilon_\parallel < 0$ can be fabricated with low loss [11], fundamental research results here are expected to be the design guidelines in other nano-photonic or sub-wavelength device applications.

172

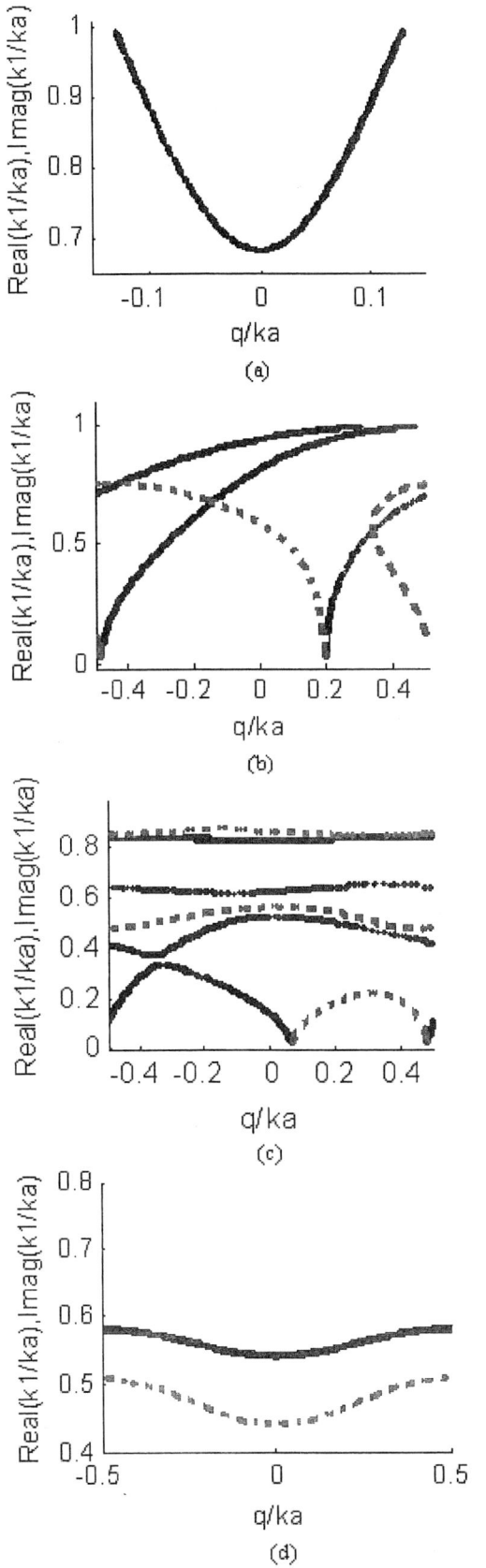

$q$ and $k_1$ are normalize by $ka = 2\pi / a$. The solid lines stand for the solution of a real number for $k_1$. The dashed lines stand for the solution of a imaginary number for $k_1$, i.e. photon tunnelling modes. In this calculation, $\varepsilon_{\parallel} = -2.7$, $\varepsilon_{\perp} = 1.6$, $d_1 = d_2 = a/2$ and $a = \lambda = 12\mu m$.

## Conclusions

We have found the photon tunnelling modes in the band structure by analyzing the explicit dispersion equation for a 1D periodic structure with alternating air and anisotropic media with $\varepsilon_{\parallel} < 0$. Unlike the left-handed material, the anisotropic media here is non-magnetic and it can be fabricated with low loss. Our result may have some applications in optical device.

## References

1 E.G.Cravalho et al, J. Heat Transfer 89, 351 (1967)
2 J.B.Pentry, J.Phys.:Condens.Matter 11,6621 (1999)
3 P.Yeh, Optical Waves in Layered Media( Wiley, New York, 1988)
4 J.B.Pendry, Phys.Rev.Lett. 85,3966 (2000)
5 Bae-Ian Wu et al, J. of Appl. Phys., 93, No.11, June 2003.
6 R.A.Shelby et al, Science 292,77 (2001)
7 Liang Wu et al, Phys. Rev. B,67,235103 (2003)
8 Michael W.Feise et al, Phys. Rev. B,66,035113 (2002)
9 Z.M. Zhang et al, Appl.Phys.Lett., 80,1097(2002)
10 J.B. Pentry et al, Phys. Today 57(6),37(2004)
11 Viktor.A. et al, Phys. Rev. B, 71, 201101 (2005)
12 P.R.Griffiths et al, Fourier Transform Infrared Spectroscopy(Wiley, New York, 1986)
13 R.C. Reddick et al, Rev. Sci. Instrum. 61,3669 (1990)
14 O.Levy et al, Phys. Rev. B, 56, 8035 (1997)
15 D. Sarid, Phys. Rev. Lett. 47, 1927 (1981)
16 F. Yang et al, Phys. Rev. B 44, 5855 (1991)
17 G. C. Cho et al, Phys. Rev. Lett. 65, 764 (1990)
18 A. V. Kuznetsov et al, Phys. Rev. B 51, 7555 (1995)
19 K. J. Yee et al, Phys. Rev. B 60, 8513 (1999)
20 Jense Li, Phys.Rev.Lett. 90, 083901 (2003)

Fig.2 The band structure with (a) $\theta = 0°$,(b) $\theta = 40°$, (c) $\theta = 60°$,(d) $\theta = 90°$. The Bloch propagation constant

# High Performance and Low-cost Fiber Grating Laser Module Package Employing a Hyperbolic Fiber Microlens

H.M. Yang[1], C.W. Lee[2], Z.G. Tsai[2], and W.H. Cheng[2]

[1]Department of Communication Engineering, I-SHOU University, Kaohsiung, Taiwan

[2]Institute of Electro-Optical Engineering, National Sun Yat-sen University, Taiwan

[1]No1, Section 1, Hsueh-Cheng Rd.,Ta-Hsu Hsiang, Kaohsiung, Taiwan, 84008, R.O.C

Tel: +86-7-6577711 ext 6767, Email: hmyang@isu.edu.tw

**Abstract** This study demonstrates that it is possible to fabricate a fiber grating external cavity laser (FGECL) module with a low cost while still maintaining a good performance by using a low-cost AR-coated ($5 \times 10^{-3}$) laser and a tapered hyperbolic-end fiber (THEF) microlens.

## Introduction

Using a temperature-insensitive and a low-chirp fiber Bragg grating reflector as the external cavity for the semiconductor gain element, fiber grating external cavity lasers (FGECLs) have been developed for 2.5Gbit/s WDM with a low bit error rate penalty [1,2]. However, the AR coating process on the fiber tips makes the fabrication process complicated, and raises the price of production. Therefore, a compact integration of the diode laser and the fiber grating external cavity, which allow the ease of optical coating and the reduction of back reflection from the fiber tip, is of great research interest for optimizing the FGECL performance. Fiber microlens which provides an efficient coupling mechanism as to match the spot size of the diode laser to the fiber is commonly used for optical alignment [3,4].

In this study, we demonstrate that it is possible to fabricate FGECL modules in low cost while still maintaining a good performance by employing a low-cost AR coated ($5 \times 10^{-3}$) laser and a tapered hyperbolic-end fiber (THEF) microlens. Previously, the high-performance FGECL modules have only been available by using complicated AR-coated ($1 \times 10^{-5}$) laser process that lead to a high packaging cost [9]. The THEF has demonstrated up to 86% coupling efficiency which the FP laser beam has a far-field divergence of 25°×38° (lateral × vertical)..Low-cost FGECL modules with a good performance were achieved primarily due to the THEF microlenses.when compared to the currently available hemispherical microlenses having low coupling efficiency (typically 50%) [5,6]. Therefore, the packaged FGECL module was suitable for use in low cost 2.5 Gbit/s lightwave transmission systems.

## Fiber grating external cavity laser (FGECL) structure

The fiber grating used in the FGECL is comprised of a single-mode fiber grating, and of an uncoated THEF microlens. They are fabricated on the same photosensitive single-mode fiber of mode-field diameter of 9.6 $\mu$m at $\lambda$ = 1.55 $\mu$m. Fig.1 shows an FGECL module configuration. The FGECL module consisted of an HR/AR-coated diode laser, a THEF microlens, a fiber grating, a photodiode, and a TE cooler, as shown in Fig. 1 (a). A laser welded butterfly-type FGECL module package is shown in Fig. 1 (b). The FP laser diode has a 90% HR coating at the back facet, and a low-cost AR-coated front facet of 0.5%, which was mounted on a TE cooler to change the substrate temperature.

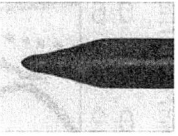

Fig. 1 An AR-coated FGECL module, (a) a laser welded butterfly-type FGECL module,and (b) a THEF microlens.

## Performances of the fiber grating external cavity laser (FGECL)

The AR-coated FGECL had a high SMSR at T from 20℃ to 40℃, as shown in Fig. 2. At 20℃ to 40℃, the SMSR varies with a value from 38.5 dB to 47.4 dB. The FGECL module showed the extremely superior single longitudinal mode characteristic. In the range of T from 20℃ to 40℃, little variation of the $\lambda_o$ was observed. The average variations of $\Delta\lambda/\Delta T \sim 0.0042$ nm/℃, for the temperature dependence of non-AR-coated FGECL.

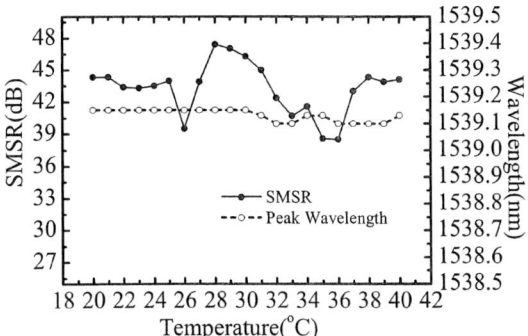

Fig. 2 The lasing peak wavelengths ($\lambda_o$'s) and SMSR of AR-coated FGECL module as a function of temperature at the injection current of 40 mA.

Fig. 3 shows the BER performances of the AR-coated and non-AR-coated FGECL modules with a transmission rate of 2.488 Gbit/s. The measured conditions were NRZ format, $2^{23}-1$ pseudo-random binary sequence (PRBS) pattern, with a modulation current of 20 mA, and an operation current of 40 mA at 25℃.

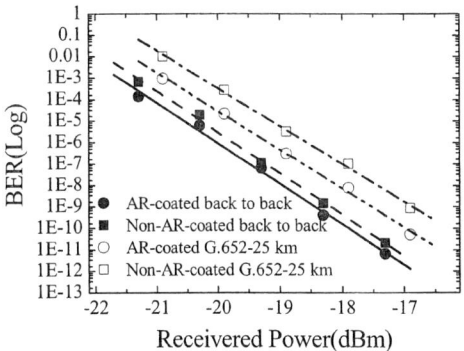

Fig. 3 The BER performances of the AR-coated and non-AR-coated FGECL modules with a transmission rate of 2.488 Gbit/s.

In general, a high-performance fiber grating external cavity laser (FGECL) requires a low AR coated ($1\times10^{-5}$) laser [1]. However, such packages were complicated processes which made their cost too high. This study demonstrates the possibility to fabricate FGECL modules in low cost while still maintaining a good performance by employing a low-cost AR-coated ($5\times10^{-3}$) laser and a tapered hyperbolic-end fiber (THEF) microlens. Compared with previous works [2-4], our low-cost FGECL modules exhibited better SMSR (>44dB), a good wavelength stability (almost fixed from

18~70mA), and an excellent temperature stability ($\Delta\lambda/\Delta T \sim 0.0042$nm/℃). In our previous work [7], the FGECL showed a higher SMSR (>55dB) and a better wavelength stability ($\Delta\lambda/\Delta I \sim 0.0062$nm/mA from 8~250mA), however, both the fiber tip and the diode laser's front facet were required a low AR coating ($1\times10^{-5}$). This makes the fabrication process complicated and causes a rise in the expense of the production.

## Conclusions

In summary, we have demonstrated that it is possible to fabricate FGECL modules at a low cost while still maintaining a good performance by employing a low-cost AR-coated ($5\times10^{-3}$) laser and a tapered hyperbolic-end fiber (THEF) microlens. Previously, the high-performance FGECL modules have only been available by using a complicated AR-coated ($1\times10^{-5}$) laser process that lead to a high packaging cost [1,7]. Low-cost FGECL modules with good performance were achieved primarily due to the THEF microlenses having a high coupling efficiency (typically 75%) to enhance the feedback power from the fiber grating external cavity to the HR/AR-coated laser when compared to the currently available hemispherical microlenses having a low coupling efficiency (typically 50%) [6,8]. This clearly indicates that the packaged FGECL modules are suitable for use in low cost 2.5 Gbit/s lightwave transmission systems, such as local area networks (LANs), Metropolitan area networks (MANs), and fiber to home (FTTH) applications.

## Acknowledgements

This work was partly supported by the National Science Council, Taiwan under Contracts NSC 95-2221-E-214 -067.

## References

1  F.N. Timofeev et al Fiber and Integrated Optics, 19 (2000), pp. 327.

2  H. Bissessur et al, IEEE Photon. Technol. Lett., Vol. 11 (1999), pp. 1304.

3  K. Shiraishi et al, IEEE J. Lightwave Technol., Vol. 13 (1995), pp. 1736.

4  H.M. Yang et al, J. Electron. Mater., Vol. 30 (2001), pp. 271.

5  Huei-Min Yang et al,, IEEE J. Lightwave Technol., Vol. 22, no. 5 (2004), pp. 1395.

6  C.A.Edwards et al,, IEEE J. Lightwave Technol., Vol. 11 (1993), pp. 252.

7  P.A. Morton et al, Appl. Phys. Lett., Vol. 64 (1994), pp. 2634.

8  H. Izadpanah et al, SPIE, Vol. 836 (1987), pp. 306.

### FTTP Optical Transceivers:
### When, how and if integration can make a real impact on performance-cost ratio?

**Wei-Ping Huang**
Department of Electrical and Computer Engineering
McMaster University
Hamilton, Ontario, Canada
Email: huang@ece.eng.mcmaster.ca

Optical transceivers are key components that drive the performance-cost ratio for FTTx networks. With the development of the advanced network architectures such as the GPON and WDM-PON, the requirements for the functionality and performance of the optical transceivers become increasingly more demanding. Further, the volume and unit cost of the optical transceivers for the optical network terminals (ONTs) is another challenging issue to be resolved. The existing PON transceiver technologies are based on discrete optoelectronic devices, stand-alone TO-cans, coaxial packaged BOSAs, and manually assembled modules. There have been intensive research and development for innovative and disruptive technologies, which promised to replace the current OSAs made of discrete chips and parts with a highly integrated solutions based on hybrid or monolithical integration. The new technologies may reduce the number of discrete components, improve the manufacturing yield, increase the production throughput, and reduce overall cost. In this presentation, I am going to take a close look at the current status of the transceiver technologies with a systematic discussion for both the conventional and the emerging technologies in the context of FTTP PON applications. In particular, I will like to offer my views about when, how and if the photonic integration can deliver the promise on performance-cost ratio of the transceivers.

**About the Author:**

Dr. Wei-Ping Huang received his bachelor's degree from Shandong University, China in 1982, a master's degree in 1984 from the University of Science and Technology of China, and Ph.D degree in 1989 from the MIT, USA. He has held a variety of faculty positions at University of Waterloo and McMaster University, Canada as well as numerous visiting, adjunct and consulting positions with several academic and industrial institutions in Canada, U.S., Japan and China. He has also founded several venture companies in the area of photonics and served in various executive, advisory, and consulting positions throughout the different stages of the business. Dr. Huang is internationally known for his contributions and expertise in photonic devices and integrated circuits. He has authored and co-authored over one hundred twenty journal papers and seventy conference papers and holds seven US patents.

**River Huang Ph.D**
**CEO**
**Shenzhen First Mile Communications Ltd.**

**Optional Solutions for FTTH Market in China**

1.Features of FTTH market in China
Introduce the features of FTTH market through analyzing the feature of housing district in China.

2.Comparison of Optional Solutions for FTTH Market in China
Clarify the advantages and weakness of EPON, GPON, EAON based on the features of FTTH market in China.

3.Customer demand low cost FTTH solution
High equipment cost embarrasses the deployment widely of FTTH, here one low cost FTTH solution (EAON) suggested which matches the features of FTTH market in China.

# Design and Implement of practical EPON Network Management System

Sun jie (1), Zou junni (2), Ye jiajun (3), Zhai xuping (4)

1: School of Communication and Information Engineering, Shanghai University, Shanghai
200072,beautifulhome1982@yahoo.com.cn
2: School of Communication and Information Engineering, Shanghai University, zoujn@shu.edu.cn
3: School of Communication and Information Engineering, Shanghai University
4: School of Communication and Information Engineering, Shanghai University, zhaixp@staff.shu.edu.cn

**Abstract** *EPON is considered as the strategy in solving the "last mile" of Access Wideband Network. This paper introduces the basic framework of EPON Network Management System and concentrates on the process explanation of the key function. Tornado is also used to simulate and test the program effectively and successfully.*

## Introduction

Ether Passive Optical Network (EPON) is a novel technology of fiber access network. It combines many advantages of PON and ether network technologies, including low cost, high bandwidth, compatibility with now existing ether network, easy-maintenance. This technology has become the most challenging solution strategy of client access network.

Although EPON is widely welcomed in communication field, it really needs more time to get practicality and population. The main cause leading this is the question to EPON from telecommunication businesses. As we all known, ether access technology is applied to the access network. The bottle neck of this technology is the lack of network management. To realize the efficient management of EPON equipment, to make sure the steady, efficient and safe of network, this paper shows a solution of EPON network management system, which is orient to telecommunication businesses.

## Section I: Design of EPON network management system

The design scheme of EPON network management system is shown in Figure 1[1]. It has the ability to deal with OLT and ONU, including setting parameters, watching performances, warning malfunction, managing safety, etc.

The management of EPON by NMS is realized based on SNMP protocol[2] through Web[3]. It mainly consists of two modules: the module of NMS controlling and the module of NMS agent. The former can realize the manager-end function of SNMP protocol and Web Server function. The later can realize agent-end function of SNMP protocol. Meanwhile it also deals with the communication between EPON equipment.

According to the EPON protocol, IEEE 802.3ah, the management to ONU by OLT is realized by communication of OAM, namely, OLT receives OAM frame from ONU and gets the state information of ONU. The NMS agent module may process these messages further. Finally NMS control module completes the whole massage-interaction of NMS system by the transmission from NMS to administrator through HTTP protocol[4].

Nowadays, the manufactories of OLT and ONU kernel control chip in the world are only Teknovus from America and Passave from Israel, from which Teknous has the overwhelming advantages and leading technologies in the world. Regard as this, we choose the chip TK3721 (OLT kernel chip) and TK3713 (ONU kernel chip) from Teknovus to enhance the universal purpose of this design. The OAM message interaction between OLT and ONU can be achieved mainly by these two chips. The following paper mainly discusses the design and realization of NMS control module and NMS agent module.

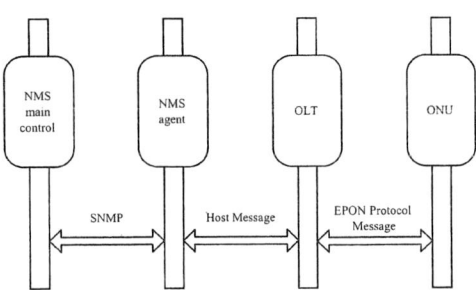

Figure 1. The Scheme of EPON Network System

## Section II: The consist of system function modules

### A. The main control module

The main control module is realized in the central control panel of telecommunicated-level machine case. Its functions include the echoes of controlling instructions from remote controlling platform, running the functions of SNMP manager, watching and warning the physical states of machine case, OLT main board and remote ONU equipment. It mainly consists of 4 sub-modules, including Management Application (MA) module, SNMP manager module, Alarm module and Web Server module. The relationship among these is shown in Figure 2.

Web Server provides administrator with friendly interactive interface and support the remote control of manager. It also communicates with MA module and sends the administrator's control requires to MA.

MA is the central module in the main control module and it completes the watch of work and the management of up-level protocol. It echoes the control instructions from remote control platform, SNMP manager module and Alarm module and sends to proper modules.

Alarm module watches the working state of machine case. When something abnormal occurs, MA module can display warning message in the client operation platform.

The main function of SNMP manager module is to

178

support MA to make real-time communication with MPU unit on OLT board through SNMP protocol, to realize the function of watching and managing OLT. Meanwhile OLT board can also send warning messages to MA through SNMP.

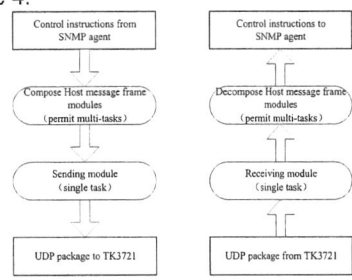

Figure 2. The Function Module of EPON Network Management System

## B. Agent module

The agent module, developed on the OLT board which contains MPU and embedded system, has many functions, including sampling and maintenance OLT and ONU equipment physical state parameters, running SNMP agent functions, communication with TK3721 chip. Agent module consists of SNMP agent module, Application and Proceeding module(APP), MIB, EPON Proxy module, shown in Figure 2.

SNMP agent module is the interface module between MA and OLT board.

APP module is responsible of dealing with messages from SNMP and sending the instructions which is needing to be operated to EPON Proxy through interface programs.

MIB saves information of OLT board physical state, of business configure, of business performance, of malfunction. This information may be inquired by up-level module.

EPON Proxy module is responsible of supplying corresponding APIs to APP and TK3721 chip.

## Section III: The realization of network management

In this system, when SNMP agent gets the SNMP Set messages sent by manager, it first sends the needing operations to EPON Proxy through interface programs. EPON Proxy then reads information package and sends configure require to TK3721. After receiving the return information of successful operation from TK3721, EPON Proxy will modify MIB, save information into SNMP PDU and finally return it to manager.

Because EPON Proxy can send Get to require TK3721 by turns at regular time and refresh MIB, SNMP agent directly acquires information from MIB when it receives SNMP Get message from manager. SNMP agent finally puts what it gets into SNMP PDU and returns it to manager.

According to the product manual of TK3721, UDP communication is applied between EPON Proxy and TK3721. The data format of Host Interface Message is shown in Figure 3[1]. Every massage consists of head and data. The head, sent from EPON Proxy to TK3721, contains 4 parts: Type、Tag、Length、ID3721. Each of first 3 parts occupies 2 Bytes and ID3721 1 Byte. ID3721 is used to recognise the different OLT chips on the same board (Regard as the present technology, one single board can contain 4 TK3721 chips at most and can support 4 EPON networks). On the other hand, the message, which is the response from TK3721 to EPON Proxy, has Return Code of 2 Bytes without ID3721. Host Interface Message has 203 different types totally. The data field consists of label part and parameter table depending on the actual situation.

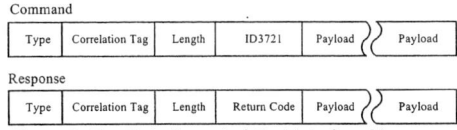

Figure 3. The Data Format of Host Interface Message

## Section IV: The example of programs

EPON Proxy module is the direct communication way between network manager and managed objects and is also important to the successful management of EPON equipment. After EPON Proxy sends Get/Set Message to TK3721, TK3721 can also make corresponding Get/Set Reply as response to EPON Proxy[1]. So EPON Proxy must compose/de-compose every sent/received message. The process is shown in Figure 4.

Figure 4. The process of the message sending/receiving

## Section V: Debugger results

We test the connection between EPON Proxy and TK3721 as an example. Within this communication, EPON Proxy will send the instruction (Set OLT MDIO), which can control the standard MII register, to set specific port of TK3721. After finishing programming and compiling, we also use VxSim in Tornado developed by Vxworks to simulate the actual situation.

According to the product manual of TK3721 Host Interface from Teknovus[1], the format of Set OLT MDIO (the address of Set OLT MDIO in Host Interface is 0x4c) is listed in Table 1.

| Offset | Size | Description |
|--------|------|-------------|
| 0000 | 8 | Port label |
| 0008 | 2 | Register address |
| 0010 | 2 | Value |

Table 1. The information format of Set OLT MDIO

From this table, Port Label is separated into 2 parts, including OLT MAC address of 6 Bytes which is set as 544b372101a0 and port number of 2 Bytes which is set as 0000. The Register Address is set 0000 and Value is the information, which is set from EPON Proxy to TK3721. The whole data field is 12 Bytes. Because the head of Host Interface Message has 7 Bytes, the total Message size is 0x13. The information of Set OLT MDIO can be shown through the simulation of VxSim in Tornado.

Figure 5 shows the data package (displayed in Hex) which is the result of simulation of VxSim in Tornado.

Figure 5. The simulated data package

From this figure, the first two Bytes are 004c, which is the Type of this Message. The following 0001 is Correlation Tag, which is used to identify the order of received packages. The next two Bytes, which are 0013, is the whole size of the Message. 01 is the address of sending target TK3721. The next 544b372101a0 is OLT MAC address and 0000 is the number of port. The following 0000 is the Register Address. The last 1040 is the information of setted OLT MDIO.

It can be verified that TK3721 can make the correct responce of request from EPON Proxy and can also do the corresponding operations.

### Section VI: Conclusions
With the fast development of EPON, it is meaningful to design a steady and efficient network management system. This article designs an EPON network management system based on embedded Vxworks platform and explains its whole framework and realization scheme particularly. The process of communication between key module-EPON Proxy and TK3721 is mainly discussed. Finally, we use Tornado to simulate and verify the function.

### References
[1]Teknovus products manual, www.teknovus.com
[2]William Stallings, *SNMP Network Management*, Beijing: CEPP, 2001
[3]Yan xinghui, Tong xiaonian, *The research of SNMP Network Management Performance Based on Web/Java*, Modern computer, 2003(9)
[4]Tang erchang, Shao ming, Pan zhihao, *Design and implement of EPON Network Management System based on Embedded System*, VIDEO ENGINEERING, 2005(8)

# Design of a Hybrid Optical CDMA/WDMA System with the Position Code

Po-Hao Chang, Jun-Ren Chen, Yen-Hao Huang

Department of Electrical Engineering, National Dong Hwa University, 1,Sec.2, Da-Hsueh Rd., Shou-Feng, Hualien, Taiwan, R.O.C.

Phone: +886-3-8634076, Email: po@mail.ndhu.edu.tw

**Abstract** *In this paper, we proposed an Optical Code Division Multiple Access (OCDMA) system with the position code to re-utilize the time slots wasted. We find that the performance of the new system will be better than that of original system. For example, it can accommodate at least 50 more users in the worst performance when BER is 10-9, and suppress the narrowband interference (NBI) caused by Wavelength Division Multiple Access (WDMA) signals in the hybrid OCDMA/WDMA system.*

## I. Introduction

The Optical Code Division Multiple Access (OCDMA) with spectral phase coding in the frequency domain has been developed in [1]. The disadvantages of this OCDMA are the cost and the low bit rates caused by the low repetition rates of ultrashort pulse laser source. A hybrid OCDMA/WDMA system [2] and a Multi-slot OCDMA system [3] based on Optical Time Division Multiple Access (OTDMA) technology [4] are proposed for improvement. Now, we design a new system called a hybrid OCDMA/WDMA system with the position code by using the time slots wasted. In this way, our system will utilize a two-dimensional code representing both phase and position.

According [1], the transmitter of the OCDMA system consists of a band-limited signal source, a data source, and a spectral-phase encoder. The output of the band-limited source is multiplied (modulated) by a data source that takes on two values, namely "0" or "1," for on-off keying. The spectral-phase encoder adds a determinate phase shift to each spectral component of electric field of the ultrashort light pulse.

We will integrate our OCDMA system into the WCDMA system used nowadays. The OCDMA signals and WDMA signals will be combined in the star coupler. For the receiver of the OCDMA signals, the WDMA signals will be viewed as the narrow band interference (NBI). On the other hand, for the receiver of the WDMA signals, the OCDMA signals will be viewed as the wide band interference (WBI) as shown in Fig. 1. We will discuss how to suppress the narrow band interference caused by the WDMA signals in next section.

The remainder of the paper is organized as follows. In Section II, we describe our system model. The performance analysis of the hybrid OCDMA/WDMA system with the position code will be shown in Section III. Finally, we provide a conclusion in Section IV.

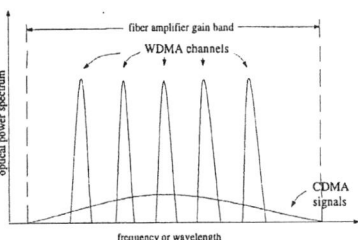

Fig. 1 : Hybrid OCDMA/WDMA scheme in which OCDMA and WDMA channels are overlaid in the same spectral region .

## II. Hybrid OCDMA/WDMA systems with position code

As mentioned before, the OCDMA system is mainly used in the transmission of the LANs. Its data rate is limited by the low repetition rate of the ultrashort light pulse source, and its capacity depends on the length of the phase code. We also know that many timeslots will be wasted. So, we would like to design a new structure to utilize other timeslots efficiently as shown in Fig. 2. We utilize some optical devices such as the fiber Bragg grating (FBG, [5][6]) which can separate one broadband light pulse into some sub-pulses with different bandwidths. Hence, we may reuse other timeslots for encoding so that the capacity and the data rate may be enhanced. We call this process as the position coding.

First, the ultrashort light pulse goes through the position encoder and is divided into some sub-pulses. Then, each sub-pulse is encoded by a pulse shaping apparatus which generates the phase code. After we finish two encoding processes, the optical signal goes into the network. It combines with the signals from other users via the star coupler on the fiber. Thus, the signal will be jammed by the MAI in the fiber network. There will be a phase decoder and position decoder in the receiver.

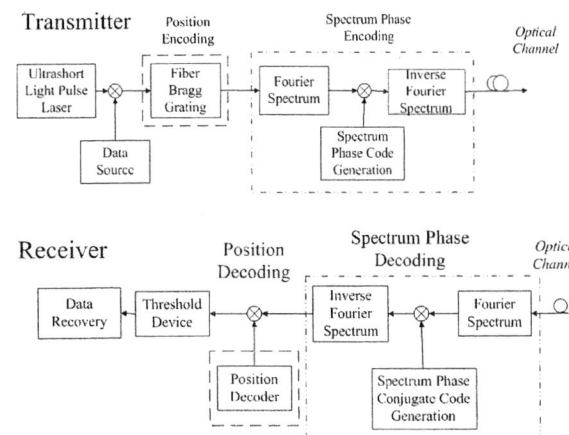

**Fig.2:** *The OCDMA system with the position code.*

Because each sub-pulse is located on different bands, we view them as independent to each other. First, we take only one sub-pulse into consideration and obtain its bit error rate. Then, we combine other sub-pulses with the position code. In this way, we make sure that each sub-pulse has a specific timeslot for its position. Hence, the detection device must determine in which timeslot each sub-pulse is so that the best performance will be achieved.

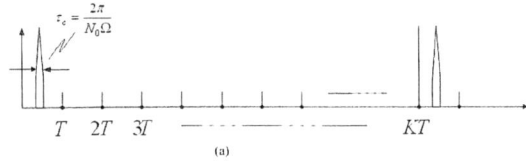

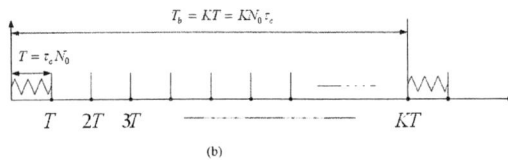

**Fig.3:** *(a) The schematic timing diagram of the utrashort light pulse; (b) The schematic timing diagram of the utrashort light pulse encoded by the phase code.*

First, we use a new device called Fiber Bragg Grating (FBG) to separate an ultrushort light pulse into some sub-pulses with different bandwidth, and place them on different timeslots. Then, we called it "one" when there is a sub-pulse in a timeslot, and "zero" when there is nothing in a timeslot as shown in Fig. 4. We will call this process as the position coding. After the position encoder, each sub-pulse will be encoded by the phase encoder, and each sub-pulse will be distributed in a section of the entire grating of the phase modulation as shown in Fig. 5.

In the receiver, we will also use the FBG as the decoder. For example, if the position code is

[01101000] in the transmitter, the corresponding code is also [01101000] in the receiver. In this way, we will get a maximum unit light power and decide the signals transmitted.

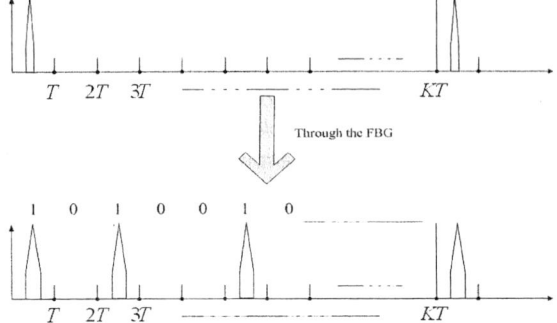

**Fig. 4:** *The schematic timing diagram of the position code.*

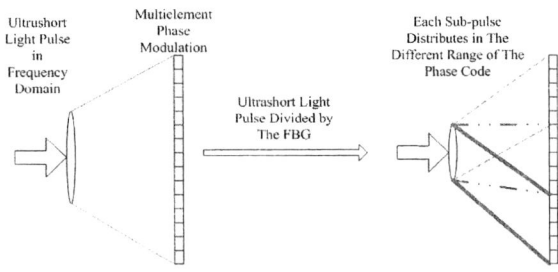

**Fig. 5:** *A schematic diagram of each sub-pulse in the phase modulation.*

## III. Simulation results

In our system, we would like to receive the OCDMA signals and regard the WDMA signals as the interference. We assume that all M OCDMA with the position code users have identical bit rates and signal formats, all L WDMA users have identical bit rates and signal formats and they are sufficiently separated so that any adjacent-channel interference (caused by overlap of the adjacent WDMA channel) can be neglected. Without loss of generality, we assume that the OCDMA user 1 is the desired signal and all other OCDMA users will produce multi-access interference (MAI). The WDMA users can be viewed, from OCDMA user 1's standpoint, as narrow-band interference (NBI).

We assume that we will separate the ultrashort light pulse into $\lambda$ sub-pulses by the FBG, and the bandwidth becomes $W/\lambda$ in ideal case, where $W$ is the bandwidth of the ultrashort light pulse. Because each sub-pulse is in the different bands, we may view them as independent to each other.

Fig. 6 shows the BER versus threshold for the original code-length N0=512 (the code-length of each sub-

182

pulse will be equal to 512/3), time slot K=25, and the number of the OCDMA users M=150. We find that the BER of our new system would be near by 10-11.

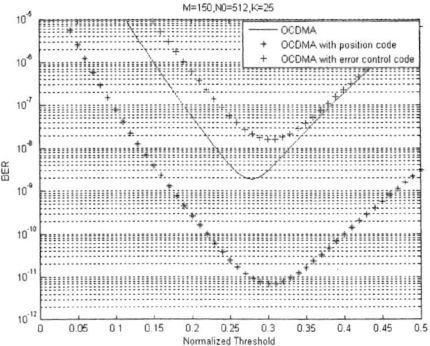

*Fig. 6   BER versus normalized threshold (Ith/P0) for code-length N0=512 (the code-length of each sub-pulse will be equal to 512/3), K=25, M=150.*

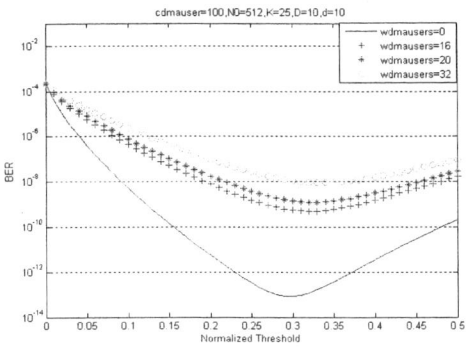

*Fig. 7   BER versus normalized threshold (Ith/P0) for code-length N0=512 (the code-length of each sub-pulse will be equal to 512/3), K=25, M=100, D=10, d=10.*

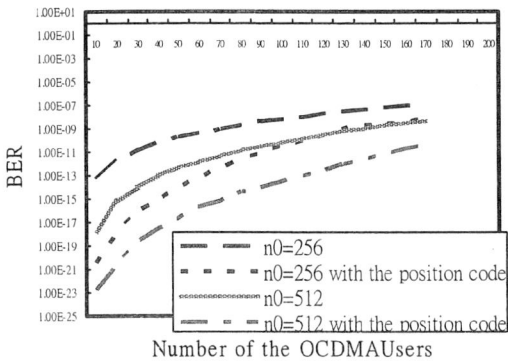

Number of the OCDMAUsers

Fig.8   The BER versus the number of the OCDMA users.

Fig. 7 shows the BER versus threshold for the original code-length N0=512 (the code-length of each sub-pulse will be equal to 512/3), time slot K=25, the number of the WDMA users L=0 (pure OCDMA), L=16, L=20, L=32, and the number of the OCDMA

users M=100. We find that the BER of our new system would be near by 10-13, 10-9, 10-9,10-8 when the ratio of the power of OCDMA system and WDMA system d=10 and the ratio of the bit rate of OCDMA system and WDMA system D=10.

Fig. 8 shows the BER versus the number of the OCDMA system users. We find that the BER of our new system is better than the original system.

## IV. Conclusions

In our proposed new system, we find that the disadvantage of wasted time slots in the original OCDMA system can be improved by the position code, and we may achieve better performance than before. It can accommodate at least 50 more users in the worst performance when BER is 10-9. Note that our new system must be processed in the synchronous mode because of the position code.

In the hybrid OCDMA/WDMA system, we find that the position code can also suppress the NBI, but this depends on the ratios of the power and data rate between the OCDMA signals and the WDMA signals, to which we must pay attention.

## References

[1] J. A. Salehi, A. M. Weiner, and J. P. Heritage, "Coherent Ultrashot Light Pulse Code-Division Multiple Access Communication System", *IEEE/OSA J. Lightw. Techcol.*, pp.478-491, March 1990.

[2] Po-Hao Chang and E. J Coyle, "Performance Analysis of Hybrid Code-Division Multiple-Access/Wavelength-Division Multiple-Access System Based on Spectral Encoding," submitted to *IEEE/OSA J. Lightw. Technol.*.

[3] Hung-Shiang Chen, "Design and Performance Analysis of a Hybrid Multi-slot Optical CDMA/WDMA System." Master Thesis, Institute of Electrical Engineering National Dong-Hwa University, Taiwan, R.O.C., July 23, 2005.

[4] W. Huang, M. H. M. Nizam, I. Andonovic, and M. Tur, "Coherent Optical CDMA System Used for High-Capacity Optical Fiber Networks System Description, OTDMA Comparison, and OCDMA/WDMA Networking." *IEEE/OSA J. Lightw. Technol.*, vol. 18, no. 6, pp. 765-778, June 2000.

[5] L. R. Chen, "Technologies for Hybrid Wavelength/Time Optical CDMA Transmission," *2001 Canada Conference on Electrical and Computer Engineering*, vol. 1, 13-16 May 2001, pp. 435-440.

[6] H. Fathallah, L. A. Rusch and S. LaRochelle, "Passive Optical Fast Frequency-Hop CDMA Communication System," *IEEE/OSA J. Lightw. Techonol.*, vol. 17, no. 3, pp. 397-205, 1999.

# A Novel Implementation of VLAN-Based Multicast Carried on LLID in EPON

Min Zhu, Junni Zou, and Rujian Lin

School of Communication and Information Engineering, Shanghai University

Shanghai 200072, China, E-mail: nuptzhuminxuan@126.com

**Abstract** *As one of the best solutions for FTTH, EPON provides 1 Gbit/ s upstream/ downstream bandwidth. Meanwhile, all kinds of high-bandwidth services such as VOD, teleconferencing and distance learning have been running over an access network and over LANs. In this paper, we present a novel implementation of VLAN-based multicast in EPON system. Using LLID (Logical Links Identifier) to carry multicast VLAN, which is mapped with IP multicast address, RS sub-layers can exactly filter and distribute multicast data frames. The basic ideas, detailed process of implementation are given in this paper, and the verification and test procedure is also provided.*

## I Introduction

With the rapid development of broadband video applications and the explosive growth of broadband access users, IP multicast technology has become one of the most important Network technology over Internet, backbone networks and access networks. IP multicast technology achieves efficient data transmission across the point-to-multipoint network, saving network bandwidth and effectively reducing network load. However, in the broadband access network (EPON) and LANs, multicast data has been broadcast to ONTs (Optical Network Terminals) or users. This will greatly increase the burden of the ONT. In the presence of large amounts of multicast service, OLT will broadcast all service data, and these data may lead to the entire EPON system collapse.

For this reason, many research literatures presented kinds of implementation method of multicast in the EPON system. Literature [1] analyzes in detail three prerequisites of multicast realization in EPON system, that is ONU multicast filtering mechanism, ONU joining/leaving multicast group mechanism and OLT creating/maintaining multicast group mechanism, and proposed an implementation of multicast in EPON. In this implementation, OLT creates/ maintains Multicast Forward Table (MFT) using IGMP Snooping technology and re-definition of OAM frame, controls ONU to join/leave multicast group. OLT establishes MFT by the new definition of multicast LLID, and ONU maintains a corresponding table of multicast LLID and host MAC address.

In the study of various implementations, it was found that there are many drawbacks. For example, OLT indirectly manages multicast group that end users participate in by sending OAM frame; ONU broadcasts multicast data to end users (ONU only play a filtering role); and OLT forward IGMP Report messages to the core network causing bandwidth loss.

To overcome the drawbacks above, we propose a novel implementation of multicast in EPON, in which functional and structural improvements are made in ONU. With IGMP Snooping, ONU automatically creates and maintains a VLAN table corresponding to IP multicast addresses and sends multicast VLAN ID carried on LLID to OLT. OLT also creates and maintains the same MFT in the same way. The rest of this paper is structured as follows. In Section 2, we present the basic idea and detailed process of implementation. In Section 3, we design an experiment and give an analysis of experiment procedures. In the last Section, a summary of our work is given.

## II Key Technology of VLAN-Based Multicast

EPON is a point-to-multipoint access network. In the downstream, service data will be broadcast to ONUs. ONU decide whether to forward data to the upper layer according to a unicast/ multicast/ broadcast filtering rule. This means that the network-sharing characteristic makes implementation of multicast easier.

Two key issues must be solved in the implementation of multicast in EPON. First is to create and maintain MFT. Second is to filter and forward multicast data. For the former, MFT is created and maintained generally by monitoring multicast Join or Leave message from end users, tracking changes in the multicast group and members of the group. For the latter, multicast data is filtered and forwarded according to the MFT and multicast identifier (IP address, MAC address, LLID, VLAN) of data frame. Referring to two issues above, this section introduces implementation of multicast of EPON in three aspects.

### A IGMP Snooping and Proxy Mechanisms Analysis

#### A.1 IGMP Snooping

IGMP Snooping is that Layer 2 network equipment (switch, OLT/ONU) create the mapping table of VLAN and IP multicast address by intercepting three IGMP messages: Join message (Report), Leave message and Query

184

message, and maintain IGMP protocol state entities between routers and hosts consistency. So Layer 2 network equipment can forward multicast data to corresponding user ports according to the multicast group topology structure, achieving the purpose of restraining diffusion of multicast data.

OLT and ONU all have IGMP Snooping capability in the proposed implementation. When Leave or Join messages passes through EPON system, OLT and ONU monitors multicast IP addresses and forms a corresponding table of multicast IP addresses and multicast VLAN, which is generated by ONU and carried on LLID and sent to OLT.

### A.2    OLT IGMP Proxy

The diagrams below illustrate the difference between a pure snooping implementation at the OLT and a proxy implementation. The proxy acts on behalf of all terminals on the EPON span, aggregating multiple IGMP Reports from EPON terminals to a single Report in response to Queries from the video server (if a group is needed by any terminal in the network). The OLT IGMP Proxy also terminates queries from upstream servers, and generates Reports based on its current multicast forwarding table. To maintain this table, the OLT Proxy generates its own Queries to downstream Hosts. Similarly, the OLT proxy terminates all Leave messages from downstream Hosts. When the OLT proxy decides to stop forwarding a group, because no downstream Host needs the group any longer, the proxy will generate a Leave message upstream to the video server.

As can be seen from the two graphics, IGMP proxy produces advantages of quicker response time to the report, as well as reduced traffic into the core network and the video server, increasing the number of terminals a single server can support.

Figure1 illustrates a pure snooping implementation, in which all Reports pass through the Snooping OLT to the video server. Figure 2 below shows the effects of IGMP proxy, in which the video server sees only a single Report.

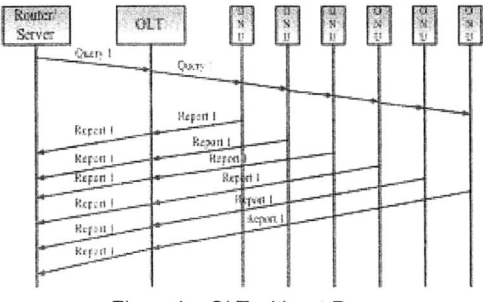

Figure1    OLT without Proxy

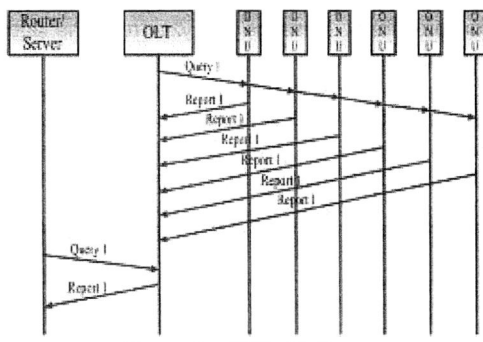

Figure 2    OLT with Proxy

### B    Internal Structure of Switched-ONU

Given typical functional drawbacks of ONU module, this paper presents a new type of ONU modules having VLAN-based switch chip. The structure of ONU module is as follows in Figure 3.

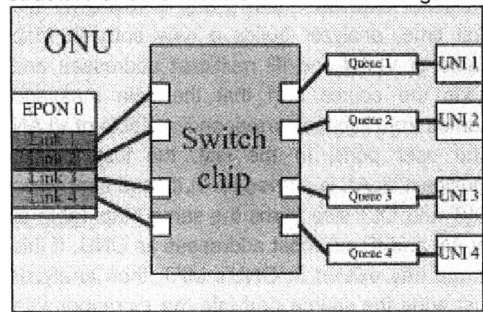

Figure 3    Internal Structure of Switched-ONU

### B.1    multi-LLID one-EPON port

Presented ONU module contains multiple logical links identifiers (LLIDs) within the same device. Multiple logical links allow different services to have independent control from the OLT's Dynamic Bandwidth Allocation (DBA) engine. For example, a link can be configured for low latency and low data rate to support voice service, while a second link can be configured for high efficiency and high data rate to support standard data traffic. These multiple LLIDs are completely independent. There is no performance difference between this configuration and multiple separate ONUs on the same EPON. So we see multiple LLIDs as multiple traditional ONUs (each has one LLID only), then the switch chip switches multiple ONUs equivalently.

### B.2    Switch chip with VLAN capability

The switch chip is composed of four elements: switch engine (ASIC, Layer 3 aware); message analyzer; multicast forward table (CAM table, containing multicast VLAN and UNI ports); lots of ports (numbered 0, 1, 2, ……, n). It is assumed that switch engine is connected with the analyzer by internal port0 and with user ports and links by N external ports. Switch engine receives all messages from external ports and redirects IGMP message to the internal port, otherwise, forwards other messages according to MFT and the forward policy. Message analyzer receives and analyzes IGMP message redirected from the internal port. After analyzing, message analyzer creates and maintains two tables (applied to all LLIDs in the same ONU), that is, Map Table of VLAN and IP multicast addresses (Figure 4); Table of VLAN and user ports (Figure 5).

| IP Multicast | VLAN ID | Tgroup |
| --- | --- | --- |

Figure 4    Map Table of VLAN and IP multicast address

| VLAN ID | User Port | Tport |
| --- | --- | --- |

Figure 5    Table of VLAN and User Ports

Specific process is as follows: message analyzer monitors Join message, which contains IP multicast address. If this group is found for the first time, analyzer builds a new entry in Map Table of VLAN and IP multicast addresses and adds the source port that the Join message comes from into the corresponding Table of VLAN and user ports. In the last, the just formed multicast VLAN is carried on LLID and sent up to OLT and OLT also forms the same Map Table of VLAN and IP multicast addresses as ONU. If this group has existed in ONU's MFT, then analyzer just adds the source port into the corresponding Table of VLAN and user ports.

Message analyzer monitors Leave message. First, analyzer removes the user port from the corresponding Table of VLAN and user ports and terminates the Leave message. Second, if the set of user ports is empty, analyzer also removes the corresponding entry from VLAN's table. So ONU no longer accepts the group message. In the last, removed multicast VLAN is carried on LLID and sent up to OLT and OLT also removes the corresponding entry.

For unicast message, the switch chip can automatically forwards the unicast data using MAC address self-learning capability.

### B.3    multi-UNI

Proposed ONU module has multiple UNI ports, and each one has the corresponding FIFO queue. Although each port can accommodate 1 Gbps, some of the logical links are not designed for such high data rates and sending upstream data in the burst mode. So it is necessary that each UNI is configured with a FIFO queue used to buffer upstream data. According to the varied rate of UNI port, each FIFO queue can be configured with different size.

### C    Mapping of multicast VLAN and LLID

ONU module intercepts IGMP messages that come from user port, and automatically creates and maintains Map Table of VLAN and IP

multicast addresses, in which the just formed multicast VLAN is carried on LLID and sent up to OLT. Original LLID contains 16 bits, in which the most significance bit is mode bit ("1"denotes broadcast, "0" unicast). To add multicast LLID, in consistence with the standards, we use the 2 most significance bits as mode bit, that is, "11"debonts broadcast LLID, "00" unicast LLID and "10" multicast LLID. Therefore, we map 12 bits VLAN into the 12 less significance bits of LLID, and the 4 most significance bits are provisioned with "1000". The mapping relation of multicast LLID and multicast VLAN are as shown in Figure 6.

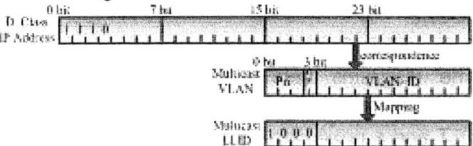

Figure 6    Mapping Relation of Multicast LLID and Multicast VLAN

Multicast VLAN is carried on LLID and sent up to OLT and OLT also forms the same Map Table of VLAN and IP multicast addresses as ONU by IGMP Snooping capability. So when downstream multicast data pass through OLT, OLT will attach a multicast LLID that carries the multicast VLAN onto the downstream multicast message. EPON port containing multicast VLAN Table (applied to all LLIDs) can accurately filter data based on the multicast LLID in RS sublayer. The switch chip forwards the multicast data to UNIs according to the attached multicast VLAN carried on LLID.

### III    Experimental Result

To test the reliability and validity of the implementation of multicast in EPON, we designed and implemented the proposed EPON multicast in an experimental access network platform as shown in Figure 7. For convenience, we only consider a video server that provides multicast service in the experiment. The video server provides multi-channel programs which correspond to IP multicast address. Users apply to join the group before viewing the program guide. It is assumed that the video server has three program channels (C1, C2, C3), which respectively correspond to multicast address 224.5.6.110, 224.5.6.111 and 224.5.6.112. ONU module has IGMP Snooping protocol, monitoring all received IGMP messages and tracks each data-exchange and state-change process. The following is a description and analysis of proposed implementation.

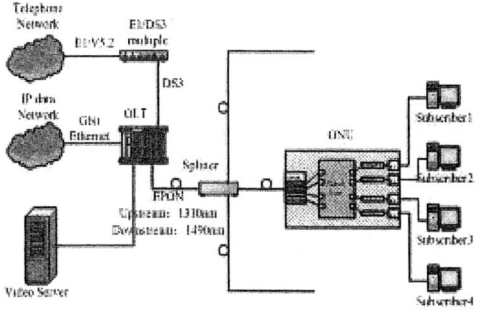

Figure 7: The Topology of Experimental Network

When initialized, user1 (UNI1) sends multicast Join message to join multicast group 22.5.6.110. ONU "hears" the message and assigns a multicast VLAN to that group, adds user port to the multicast VLAN table. In EPON port, ONU maps the multicast VLAN into LLID and sends it up to OLT. If monitoring the 4 most significant of LLID as "1000", OLT recognizes the message as multicast message, and forms the same map table of VLAN and IP multicast addresses as ONU by IGMP Snooping. Then OLT continues to send the message up to video server. It is same to other users.

When downstream-multicast data is sent to host users, OLT will attach corresponding multicast LLID to the message. ONU filters and forwards multicast data to UNI according to the multicast VLAN carried on LLID.

When hearing the multicast Leave message which indicates that subscriber 1 is leaving group 224.5.6.110 from UNI 1, ONU removes the user port from MFT and no longer forwards that group data. Meanwhile if no other user ports exist in MFT, ONU removes multicast VLAN and no longer accepts that group data.

## IV  Conclusion

On the basis of improvement in existing Layer 2 multicast, this paper analyzes key technical issues of implementing multicast in EPON system and proposes a novel implementation of VLAN-based multicast. The main features of this implementation are as follows: ①OLT and ONU all have IGMP Snooping and Proxy capabilities which are key technologies of Layer 2 multicast, and significantly reduce the bandwidth requirements of multicast control messages. ② ONU has been greatly improved in functional and structural aspects to accommodate multiple LLIDs and also partition multiple VLAN's virtual network for multicast forward. ③By mapping multicast VLAN into LLID, RS sublayer can accurately filter multicast data based on the multicast LLID.

In the last, an experiment validates the accuracy and reliability of the proposed implementation and realizes efficient multicast services in EPON system.

## References

1 Zhang Wei, "The implementation of LLID multicast in EPON", STUDY ON OPTICAL COMMUNICATIONS, 2005, Sum. No. 129, Page(s): 28 –30

2 Jun Wang, "IGMP Snooping: A VLAN-Based Multicast Protocol", High Speed Networks and Multimedia Communications 5th IEEE International Conference, 3-5 July 2002 Page(s):335 – 340

3 Tan Minqiang, "Implementation of IP Multicast in Switching Ethernet with IGMP Snooping and Proxy", Journal of Beijing University of Posts and Telecommunications, Jun. 2003, Vol. 26 No. 2, Page(s): 28 –32

4 Du xu, "The implementation of IGMP Snooping Protocol", Computer Applications, Vol.24 June, 2004, Page(s): 14 –16

# Simple analysis of the distribution of birefringence in fiber optics links

Krzysztof Perlicki (1)

1 : Warsaw University of Technology, Institute of Telecommunications, Nowowiejska 15/19, 00-665 Warsaw, Poland, perlicki@tele.pw.edu.pl

**Abstract** *Simple method to calculation of the spatial distribution of birefringence is presented. The birefringence distribution may be deduced from the calculation of the round-trip Mueller matrices derived from only two power evolution traces.*

### Introduction

Polarization optical time domain reflectometry (POTDR) was proposed several years ago as a suitable method for measuring birefringence in optical fibers by means of a analysis of the spatial evolution of a polarized signal through the fiber being tested [1]. The major problem of the POTDR technique is to simple determine, from the analysis of the polarization properties of the backscattering optical signal, the distribution of the local birefringence of single-mode optical fiber [2]. Presented method allowing to quick quantify the birefringence of an optical fiber link by analysing the optical power of the backscattering light. The analytical results are compared with numerical results. Comparisons with numerical results give excellent agreement. We demonstrate that the spatial birefringence distribution may be deduced from the calculation of the round-trip Mueller matrices derived from only two different angles of the polarizer.

### Analytical model

In general, an optical fiber exhibits axially-varing birefringence and can be represented by a series of short and homogeneous elements. Each element can be described as randomly rotated waveplate. The Mueller matrix of the i-th element may be written as:

$$M_{OR} = \begin{bmatrix} 1 & 0 & 0 & 0 \\ 0 & A^2 + B^2 C & AB(1-C) & BD \\ 0 & AB(1-C) & B^2 + A^2 C & -AD \\ 0 & -BD & AD & C \end{bmatrix} \quad (1)$$

where: $A = \cos(2\Theta i)$, $B = \sin(2\Theta i)$, $C = \cos(\phi i)$, $D = \sin(\phi i)$, $\Theta i$ is the angle between the so called „fast axis" and x-axis of the reference frame and $\phi i$ is the total phase shift between the two polarization mode.
The angle $\Theta i$ is uniformly distributed in $(0, 2\pi)$. The phase shift between the two polarization mode is equal to:

$$\phi i = bi \cdot Lel \quad (2)$$

where bi is the local birefringence and Lel is the element length.

To describe the backscattering process we consider the optical fiber as a cascade of backscattering elements so to obtain a cascade of distributed reflectors [3]. For light propagating forward to the end of the N-th element the round-trip Mueller matrix has the following form [4]:

$$M_{RT,N} = M_{OR,1}{}^T \cdot .... \cdot M_{OR,N}{}^T \cdot M_R \cdot M_{OR,N} \cdot ... \cdot M_{OR,1}, \quad (3)$$

where $M_{OR,N}{}^T$ is the transpose of $M_{OR,N}$ and $M_R$ is the Mueller matrix of a reflection; $M_R = \text{diag}(1,1,1,-1)$ [3].
Let us consider the polarizer/analyzer imposing 0° linear input state of polarization with respect to an arbitrary x-axis. The first component Stokes vector which representing the optical power is given by (fiber losses are neglected):

$$P_0 = \cos^2(\phi) + \cos^2(2\Theta)(1-\cos^2(\phi)) \quad (4)$$

In the same way, then the polarizer/analyzer imposes input linear state of 45°, the detected power is:

$$P_{45} = \cos^2(2\Theta)(1-\cos^2(\phi)) \quad (5)$$

From eqns. (4-5) the angle $\phi$ may be expressed via:

$$\phi = \arccos(\sqrt{P_0 - P_{45}}) \quad (6)$$

and the angle $\Theta$ can be described by:

$$\Theta = \frac{1}{2}\arccos\left(\frac{\sqrt{P_{45}}}{\sin(\phi)}\right) \quad (7)$$

After measuring optical power $P_{N-1}$ at point $z_{N-1}$ ($z_{N-1}$ is the length of the optical fiber segment, which includes elements from 1 to N-1) and $P_N$ at point $z_N$ (for the polarizer angle 0° and 45°) we can calculate $\phi_{1:N-1}$ and $\Theta_{1:N-1}$ of the segment, which includes elements from 1 to N-1 and $\phi_{1:N}$ and $\Theta_{1:N}$ of the segment, which includes elements from 1 to N, respectively.
The Mueller matrix of the optical fiber segment, which includes elements from 1 to N ($M_{RO,1:N}$) may be written by:

188

$$M_{RO,1:N} = M_{RO,N} \cdot M_{RO,1:N-1} \qquad (8)$$

where $M_{RO,N}$ is the Mueller matrix of the N-th element. From eqns. (6-8) we can evaluate of $\phi_N$ and $\Theta_N$ of the N-th element. Additional, from eqn. (2) value of local birefringence may be calculated.

## Results

In order to verify described method for evaluation of the distributed optical fiber birefringence we performed numerical test. We modeled optical fiber as a cascade of randomly rotated waveplates.

We assume the local birefringence to be a strictly stationary stochastic process [2]. We suppose the optical fiber to be affected only by linear birefringence [5]. The local birefringence satisfy the following differential equation [6]:

$$\frac{dbi(z)}{dz} = -\rho bi(z) + \sigma \eta i(z) \qquad (9)$$

where: i=1, 2, $\eta i$ is the white-noise, Gaussian-distributed, statistically independent processes with zero mean value and unitary standard deviation.
The parameters $\rho$ and $\sigma$ describe statistical properties of bi(z). The integration step was set at 1 m.
All numerical investigations were performed on 20 km fiber link.
Figure 1 and 2 show evolution of rms value of the birefringence (<b>) along optical fiber length for two cases: medium and small values of <b>.

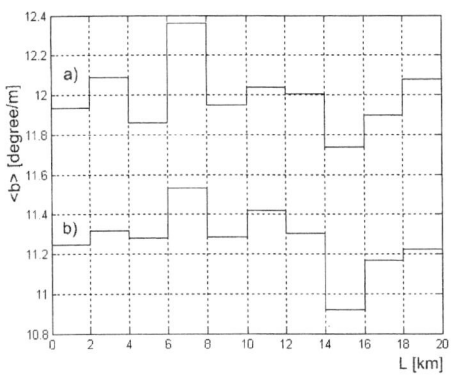

Fig 1 Evolution of <b> versus optical fiber length for a medium birefringence, a) numerical results,
b) values calculated from eqns. (6-8) and eqn. (2)

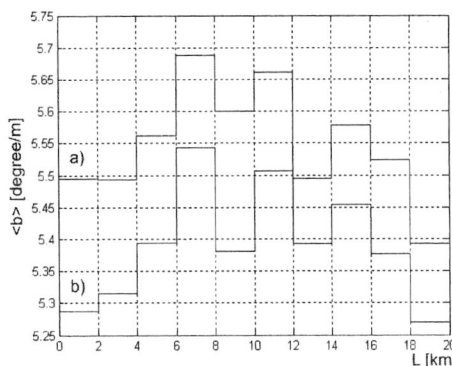

Fig 2 Evolution of <b> versus optical fiber length for a for a small birefringence, a) numerical results,
b) values calculated from eqns. (6-8) and eqn. (2)

Remarkable agreement is obtain in the rms value of the birefringence obtained by numerical simulation and from our method, particulary in the range of a small birefringence. Difference in value of <b> is 8% for Fig. 1 and less than 4% for Fig. 2.

## Conclusions

In conclusion, a simplified analytical method to evaluate an optical fiber birefringence was presented. It allows a significant simplification of the measurement of the spatial distribution of birefringence in optical fiber links; particularly for the small value of the birefringence. The spatial birefringence distribution may be deduced from the measurement of the round-trip Mueller matrices derived from only two detected power traces corresponding to the angle of the polarizer equals to 0° and 45°.

## References

1 B. Huttner et al, JLT, 17 (1999), 1843-1848.
2 A. Galtarossa et al, OFT, 9 (2003), 119-142.
3 M. van Deventer, JLT, 11 (1993), 1895-1899.
4 F. Corsi et al, JLT, 16 (1998), 1832-1843.
5 M. Wuilpart et al, PTL, 13 (2001), 836-838.
6 F. Corsi et al, JLT, 17 (1999), 1172-1178.

# Polarization tracking based on Stokes curve analysis

Krzysztof Perlicki

Institute of Telecommunications, Warsaw University of Technology

Nowowiejska 15/19, oo-665 Warsaw, Poland, perlicki@tele.pw.edu.pl

**Abstract** *A method to recognize polarization sets is proposed. Presented method is based on analysis of torsion and curvature of three-dimensional curve. The space curve is defined by states of polarization on the Poincare sphere.*

## Introduction

The optical transmission techniques in which states of polarization (SOP) of light is used to carry information include Polarization Division Multiplexing and Polarization Shift Keying. Polarization multiplexing allows simultaneous communication of multiple signals in the same channel by transmitting each with a different SOP. In systems employing Polarization Shift Keying the SOP is directly modulated to carry information [1,2]. An inherent problem with polarization transmission systems is that the conventional optical fiber communications channel causes a random, time varying change of the SOP. The polarization control systems based on either optical or electronic polarization tracking are usually required to track the SOP of the output signal. But, the cost of polarization control prevents widespread applications these polarization techniques in practical communications systems [3]. Presented method is based on analysis of torsion and curvature of three-dimensional curve segments. The space curve is defined by SOPs on the Poincare sphere.

## Calculation of torsion and curvature

At each point of a three-dimensional curve can be defined three mutually perpendicular unit length vectors: the tangent $\vec{T}$, the normal $\vec{N}$, and the binormal $\vec{B}$. The tangent $\vec{T}$ shows the direction the curve is moving in, the normal $\vec{N}$ lies along the direction which the curve is currently bending in, and the binormal B is a vector perpendicular to $\vec{T}$ and $\vec{N}$ ($\vec{T},\vec{N},\vec{B}$ is known as the Frenet frame). A three-dimensional curve can be described by its two parameters: torsion and curvature [4]. The torsion is the rate of tuning of binormal B along the length of the space curve. In turn, the curvature is the rate of tuning of tangent $\vec{T}$ along the length of the space curve. To estimate torsion and curvature asymptotic analysis of space curves described in [4] is used.

Let a segment of three-dimensional curve is interpolated by four points $P_{n-1}$, $P_n$, $P_{n+1}$ and $P_{n+2}$ with the corresponding edges $\overrightarrow{P_{n-1}P_n}$, $\overrightarrow{P_nP_{n+1}}$, $\overrightarrow{P_{n+1}P_{n+2}}$ and their lengths denoted by C, D, E (Fig.1). The tangent $\vec{T}_n$ is estimated at point $P_n$ by:

$$\vec{T}_n = \frac{P_{n+1} - P_{n-1}}{|P_{n+1} - P_{n-1}|} \qquad (1)$$

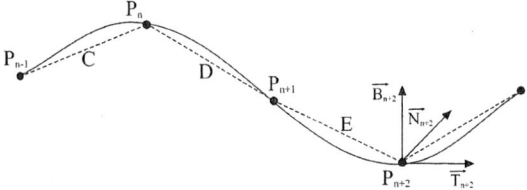

Fig.1. A space curve interpolated by points $P_{n-1}$, $P_n$, $P_{n+1}$ and $P_{n+2}$. $\overrightarrow{T_{n+2}}$, $\overrightarrow{N_{n+2}}$, $\overrightarrow{B_{n+2}}$ – Frenet frame at point $P_{n+2}$

The normal $\vec{N}_n$ at point $P_n$ can be estimated by:

$$\vec{N}_n = \frac{\overrightarrow{P_nP_{n+1}} - \overrightarrow{P_{n-1}P_n}}{|\overrightarrow{P_nP_{n+1}} - \overrightarrow{P_{n-1}P_n}|} \qquad (2)$$

The binormal $\vec{B}_n$ is given by:

$$\vec{B}_n = \vec{T}_n \times \vec{N}_n . \qquad (3)$$

The torsion ($\tau$) located at the edge $\overrightarrow{P_nP_{n+1}}$ depends on the angle between $\overrightarrow{B_n}$ and $\overrightarrow{B_{n+1}}$ and can be approximated as [4]:

$$\tau_{n,n+1} = \frac{3\arcsin\left(|\overrightarrow{B_{n+1}} \times \overrightarrow{B_n}|\right)}{C + D + E} . \qquad (4)$$

The curve curvature ($\kappa$) at point $P_n$ depends on the angle between $\overrightarrow{T_{n-1}}$ and $\overrightarrow{T_n}$, in other words: curvature, at point $P_n$, depends on the angle between the edges $\overrightarrow{P_{n-1}P_n}$ and $\overrightarrow{P_nP_{n+1}}$. The curvature can be approximated as [4]:

$$\kappa_n = \frac{2\arccos\left(\frac{\overrightarrow{P_{n-1}P_n} \cdot \overrightarrow{P_nP_{n+1}}}{C \cdot D}\right)}{C + D} . \qquad (5)$$

The space curve segment, which is interpolated by the four points $P_{n-1}$, $P_n$, $P_{n+1}$ and $P_{n+2}$, can be described using torsion estimated at the edge $\overrightarrow{P_nP_{n+1}}$ and curvature estimated at point $P_n$.

## Simulation results

In our theoretical analysis the three-dimensional curve is defined by series of sixteen points where points represent a SOPs. The SOPs are uniformly distributed on the Poincare sphere, according to algorithm described in [5]. Four polarization sets are considered. Each set represents a single polarization channel consisting of four SOPs (in other words: a single channel transmits four-level polarization modulated signal). The first channel consists of polarization: $P_1$, $P_2$, $P_3$ and $P_4$, the second channel: $P_5$, $P_6$, $P_7$, $P_8$, the third channel: $P_9$, $P_{10}$, $P_{11}$, $P_{12}$ and the fourth channel: $P_{13}$, $P_{14}$, $P_{15}$, $P_{16}$. Each channel is described by the pair of parameters: torsion calculated at the edge $\overrightarrow{P_n P_{n+1}}$ (i.e.: $\overrightarrow{P_2 P_3}$, $\overrightarrow{P_6 P_7}$, $\overrightarrow{P_{10} P_{11}}$ and $\overrightarrow{P_{14} P_{15}}$) and curvature calculated at point $P_n$ (i.e.: $P_2$, $P_6$, $P_{10}$, $P_{14}$). The values of torsion, curvature pair $(\tau, \kappa)$ are equal to: (0.6447, 3.0996) for the first channel, (0.1834, 1.1042) for the second channel, (0.1777, 1.0599) for the third channel, (0.5840, 1.6729) for the fourth channel, respectively. The input SOPs constellation is shown in Fig.2a. Next, the input signal is transmitted through a birefringent optical fiber. The optical fiber is emulated by means of 20 randomly birefringent plates. At the fiber output the locations of polarization channels on the Poincare sphere are changed (Fig.2b). The pure fiber birefringence transformation only causes a rigid rotation of the polarization constellation [2]. The output SOPs change but the values of torsion and curvature are preserved. The correct configurations of the SOPs are found by appropriate values of torsion and curvature. The torsion and curvature values of the each input channel are compared with the values of torsion and curvature calculated for each set consisting of four SOPs, which are selected from sixteen output SOPs. Number of considered output sets is equal to 43680. For the each input channel the value of its curvature repeats 26 times, the value of its torsion repeats 2 times and the value of its $(\tau, \kappa)$ pair repeats once in the output sets. The described method was also tested for more input channels. The similar results were achieved. For example, for 6 channels (24 SOPs) the number of the channel curvature, torsion and $(\tau, \kappa)$ pair repetition is equal to: 42, 2 and 1, respectively. The most important is that the values of the $(\tau, \kappa)$ pairs are unique. The analysis of the $(\tau, \kappa)$ pairs permits us to recognize the polarization configuration (i.e. polarization channel) consisting of the four SOPs. In real telecommunication systems the SOPs drift is quite often a time-varying, random process. Because of this, impact of random fluctuation of the SOPs locations on torsion and curvature detection was also checked.

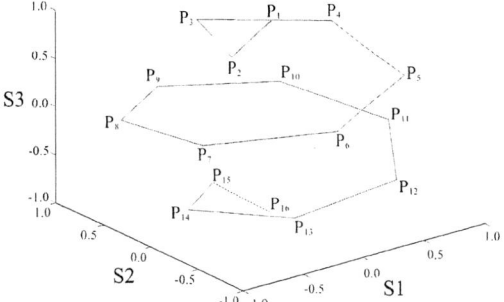

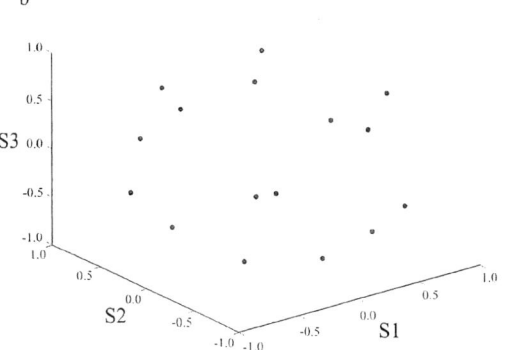

Fig.2. Input (a) and output (b) states of polarization constellations

There was found that described method is correct when the SOPs drift on the Poincarè sphere is less than 0.01°.

## Conclusions

The novel method to recognize polarization configurations is proposed. This method is based on analysis of the two space curve parameters: torsion and curvature. The obtained results indicate that the analysis of $(\tau, \kappa)$ pair value permits recognizing the polarization configuration. To estimate torsion of the space curve we need at least four points (in turn, to estimation curvature only three points). For this reason described method is correct only for polarization configuration consisting of minimum four SOPs.

## References

1. G. D. VanWiggeren, Phys Rev Lett, 88 (2002), 097903-1-097903-4.
2. R. J. Blaikie, IEEE T Commun, 45 (1997), 95-102.
3. Y. Han, OFC/NFOEC 2005, JWA39.
4. H. C. Crenshaw, J Exp Biol, 203 (2000), 961-982.
5. E. B. Saff, Math Intell, 19 (1997), 5-11.

# PMD Distribution Measurement by an OTDR with Polarimetry Considering Depolarization of Rayleigh Backcattered Waves

Takeshi Ozeki, *Member, IEEE*, Satoshi Seki and Kimiaki Iwasaki

*Abstract*— The p-OTDR waveforms were fitted well by modified Mueller matrixes, under the assumptions of scattering by cloud of spherical particles and a reduction in the DOP due to the interference of the backscattered waves with the phase fluctuation. We demonstrated a PMD distribution measurement based on the p-OTDR Jones Matrix Eigenvalue (JME) method with assumption of slowly-varying PMD along fiber length, and were success in regeneration of PMD distribution of test fibers fabricated by splicing constituent fibers with known PMDs. Surprisingly, the regenerated PMD distribution along the fiber length shows a linear accumulation of constituent fiber PMD in the short range of 6 km.

*Index Terms*— Polarization Mode Dispersion, Rayleigh back- scattering, interference, degree of polarization, optical fiber,

## I. INTRODUCTION

IN up-grading operating large scale WDM networks, mapping of PMD distribution by p-OTDR is desirable for confirming the transmission line qualities of installed fibers. Mapping of PMD along fibers makes it possible to replace fiber cable segments with larger PMD for economical realization of system upgrading. And also, it is useful to develop new functional fibers such as photonic crystal fibers with desirable polarization mode properties.

The PMD measurement by the p-OTDR was reported for evaluating PMD reduction effect of spinning in cable fabrication process[1]. A trial of estimating PMD distribution along a fiber was reported by combining p-OTDR measurement and optical circuit synthesis[2]. Recently, a lot of studies of PMD measurement using p-OTDR were reported[3]. The accuracy and dependability seem to be difficult issues for practical applications.

We reported, by assuming non-spherical scattering particles, p-OTDR waveforms were fitted well, with the fitting error of less than $10^{-20}$, by the modified Mueller scattering matrixes, which means that the degree of fitting freedom is sufficient and that the description of the system is suitable[4]. As results, Jones matrixes for light propagation through a fiber and modified Mueller matrixes for scattering were determined simultaneously, with the accuracy limited by the received signal to noise ratio of about 9dB. In another word, we were able to determine an equivalent circuit of the birefringent fiber to calculate PMD based on JME (Jones Matrix Eigenvalue) method using a single wavelength p-OTDR[4,5] We confirmed the reliability of our p-OTDR JME PMD measurement method by repeated comparisons with the standard JME measurement using transmission-type polarimetry.

However, we cannot explain the pulse width dependence of DOP by the non-spherical backscattering assumption. We confirm that the DOP reduction due to the interference of the backscattered waves in Stokes measurements also makes it possible to fit the measured OTDR waveforms with sufficient accuracy even under the assumption of spherical scattering particles.

It is a very surprise that the PMD regenerated by the assumption of slowly-varying PMD along fiber length shows the linear accumulation of constituent fiber PMD in the short range of 6km. The reverse direction measurements show reasonable regeneration of supposed linear accumulation along fiber length, which strongly suggests the PMD distribution along fiber shows linear accumulation instead of the root-length accumulation in relatively short fiber.

## II. FORMULATION

The formulation was described in reference[4] so that briefly the modification is described here. An optical fiber is described by Jones matrix as follows, excluding polarization independent phase shift corresponding to chromatic dispersion:

$$J_m = \begin{bmatrix} \exp(-j\phi_m/2) & 0 \\ 0 & \exp(j\phi_m/2) \end{bmatrix} \cdot \begin{bmatrix} \cos\theta_m & -\sin\theta_m \\ \sin\theta_m & \cos\theta_m \end{bmatrix} \cdot \begin{bmatrix} \exp(-j\psi_m/2) & 0 \\ 0 & \exp(j\psi_m/2) \end{bmatrix} \quad (1)$$

where we divide the fiber into M segments, numbered by m from the input port to the end in order, as shown in Fig.1. $J_m$ denotes the Jones matrix from the input to the m-th segment. $\phi_m, \theta_m, \psi_m$ are the Euler's generalized phase shifts.

In case of spherical backscattering[7,8] and interference reduction of DOP, we can use $a_1 = a_2 = C_m$, and then we can use $\delta_m$, the effective DOP reduction factor, as the fourth variable in the Marquardt algorithm. Now, referring Fig.1, we can describe the received Stokes vector by converting Jones matrix $J_m$ to the modified Mueller matrix $M_m$ as shown in Eq.2, where $(Ix, Iy, U, V)^T$ is the transmitted modified Stokes vector. $M'_m$ denotes the converted Mueller matrix corresponding to $J_m^T$.

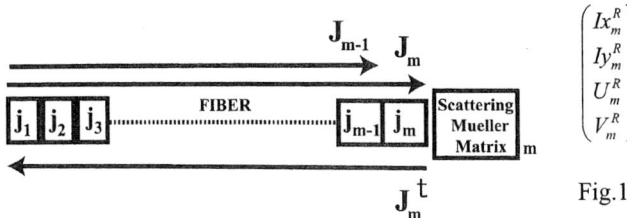

$$\begin{pmatrix} Ix_m^R \\ Iy_m^R \\ U_m^R \\ V_m^R \end{pmatrix} = M'_m \cdot a_1 \cdot \begin{bmatrix} 1 & 0 & 0 & 0 \\ 0 & 1 & 0 & 0 \\ 0 & 0 & -1 & 0 \\ 0 & 0 & 0 & -1 \end{bmatrix} \cdot M_m \cdot \begin{pmatrix} Ix \\ Iy \\ U \\ V \end{pmatrix} \quad (2)$$

Fig.1    Model of fiber backscattering

The received modified Stokes vectors are calculated by the effective reduction factor $\delta_m$ as follows:

$$\left( Ix_m' \; Iy_m' \; U_m' \; V_m' \right)^T = M_{scattering}^{spherical} \cdot M_m \cdot (Ix Iy U V)^T \quad (3)$$

$$Ix_m^R = |A2|^2 \, Ix_m' + |A4|^2 \, Iy_m' + \{ \mathrm{Re}(A2 \cdot A4^*) U_m' - \mathrm{Im}(A2 \cdot A4^*) V_m' \} \delta_m$$

$$Iy_m^R = |A3|^2 \, Ix_m' + |A1|^2 \, Iy_m' + \{ \mathrm{Re}(A3 \cdot A1^*) U_m' - \mathrm{Im}(A3 \cdot A1^*) V_m' \} \delta_m$$

$$U_m^R = \{ 2\mathrm{Re}(A2A3^*) I'x_m + 2\mathrm{Re}(A4A1^*) I'y_m + \{ \mathrm{Re}(A2A1^* + A4A3^*) U_m' - \mathrm{Im}(A2A1^* + A4A3^*) V_m' \} \delta_m \quad (4)$$

$$V_m^R = \{ 2\mathrm{Im}(A2A3^*) I'x_m + 2\mathrm{Im}(A4A1^*) Iy_m' + \{ \mathrm{Im}(A2A1^* + A4A3^*) U_m' + \mathrm{Re}(A2A1^* - A4A3^*) V_m' \} \delta_m$$

where the modified Stokes vector just after backscattered is denoted by $\left( Ix_m' \; Iy_m' \; U_m' \; V_m' \right)^T$. $A1$, $A2$, $A3$ and $A4$ are the Jones matrix

element[6] of $J_m$, and the subscript "m" is omitted for simplicity: $J_m = \begin{bmatrix} A2_m & A3_m \\ A4_m & A1_m \end{bmatrix}$ (5)

Here, Eq.4 is approximated in such form by the freedom limit in Marquardt algorithm considering $C_m = Ix_m^R + Iy_m^R = a_{1m}$, where only cross-polarized components are interfered to reduce their intensity. By substituting Eq.1 into Eq.4, we can confirm $a_{1m} = Ix_m^R + Iy_m^R = Ix_m' + Iy_m'$. This assumption is supported only by the reasonable regeneration of PMD distributions, as shown in the following. In case of spherical scattering particles and interference reduction of DOP, we can apply the Levenberg-Marquartd method to determine $\phi_m, \theta_m, \psi_m, \delta_{mm}$ using $(Ix_m^R, Iy_m^R, U_m^R, V_m^R)^T$ based on Eq.4. The segment Jones matrix is calculated by

$$j_m(\omega) = J_{m+1}(\omega) \cdot J_m^{-1}(\omega) \quad (6)$$

which is converted to an equivalent expression using the angular frequency $\omega$ of a tunable laser source and equivalent group delay differences of $t_{1,m}$ and $t_{2,m}$ as follows:

$$j_m(\omega) = \begin{bmatrix} \exp(-j\omega \cdot t_{1,m}/2) & 0 \\ 0 & \exp(j\omega \cdot t_{1,m}/2) \end{bmatrix} \cdot \begin{bmatrix} \cos\theta'_m & -\sin\theta'_m \\ \sin\theta'_m & \cos\theta'_m \end{bmatrix} \cdot \begin{bmatrix} \exp(-j\omega \cdot t_{2,m}/2) & 0 \\ 0 & \exp(j\omega \cdot t_{2,m}/2) \end{bmatrix} \quad (7)$$

Using this $j_m(\omega)$, we synthesize the Jones matrix $T_m(\omega)$ for the frequency $\omega$, $T_m(\omega) = \prod_{k=1}^{m} j_k(\omega)$. The PMD along the fiber distance based on JME is calculated as difference of the eigen values of the following PMD operator[5], $D_m = dT_m(\omega)/d\omega \cdot T_m(\omega)^{-1}$ (8)

## III.  EXPERIMENTS: PMD DISTRIBUTION ALONG TEST FIBERS

Hereafter, we regenerate the PMD distribution along the test fiber fabricated by splicing constituent fibers with known PMD using the same p-OTDR in reference [4]. Because of the limited signal to ratio in the measurement and the ambiguities in determination of $t_{1m}, t_{2m}, \theta_m$ due to their periodicities involved in Eq.6 and Eq.7, we tried the following algorithm to select the solution sets from the following candidates categorized by the sign of $\theta_m$:

$$\tan^2 \theta_m = -j_{m1,1} j_{m2,2} / j_{m1,2} j_{m2,1} \quad (9) \qquad \varepsilon = |J_m - j_m J_{m-1}|^2 + g^2 |J_{m+1} - j_m J_m|^2 + g^2 |J_{m-1} - j_m J_{m-2}|^2 \quad (10)$$

The sign of $\theta_m$ is selected by minimizing $\varepsilon$ in Eq.10, where $j_m$ is said to be used as the forward and backward predictors.

Fig.2 denotes two samples of the regenerated PMD distributions of the test fiber E-D-ABC and E-D-CBA. The dotted lines denote the linear accumulation of constituent fiber PMD. The test fiber ABC is fabricated by splicing constituent fibers with known PMD (ps) / length (km): the fiber A has 0.14ps/0.5km, the fiber B has 0.1ps/0.5km and the fiber C has 0.02ps/1km. The fiber D is a 2Km SMF with PMD of 0.12ps. The fiber E is a 2km SMF with PMD of 0.16ps. It is said that the coincidence of the regenerated PMD plotted by a red solid line and the expected PMD distribution of the test fiber plotted by dotted blue line is best for $g^2 = 2 \times 10^{-20}$, which is determined by considering the convergence criteria of the Marquadt algorithm ($10^{-20}$). This value of $g^2$ is commonly used through the experiment. The expected PMD distribution of the test fiber is assumed to be linear accumulation of

PMD along fiber length instead of the assumption of squared-root accumulation, which is commonly used. It is said that the PMD distribution mapping using p-OTDR JME method assuming spherical scattering particles and DOP reduction due to interference of backscattered waves is fairly applicable up to 6km with our experimental setup.

(a)                                    (b)

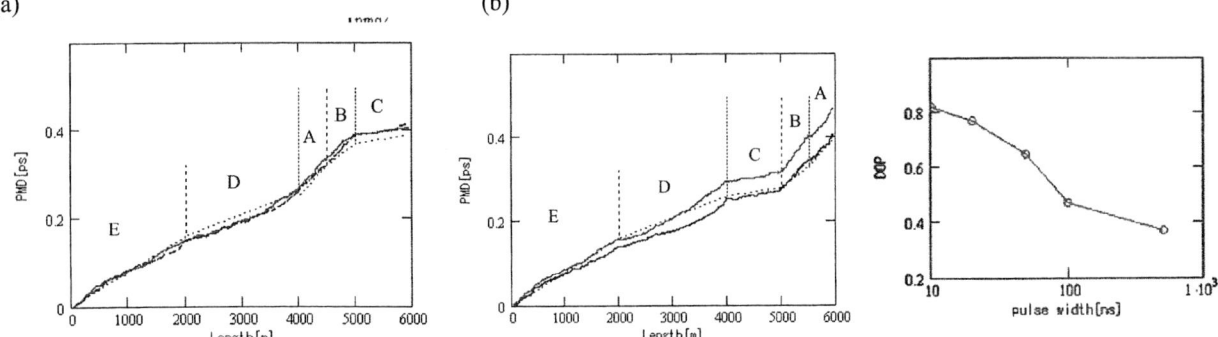

Fig.2 PMD Distribution mapping up to 6 km. (a) E-D-ABC (b) E-D-C-B-A          Fig.3 Pulse width dependence of DOP.

## IV. DISCUSSION

We measured the dependence of DOP on the optical pulse width as shown in Fig.3. The DOP is measured for a sample fiber with 0.03ps/rkm with the grating mirror feedback tunable laser diode. The spectrum width was measured by Anritsu Q7335 was 0.815MHz. The DOP decreases as increasing the optical pulse width. We set our pulse width to be 10ns from these measurements. There were a few reports discussed the degree of polarizations of scattered coherent light. However, there is the possible mechanism in OTDR cases as follows: The scattered lights from various points in the scatter cloud illuminated by the pulse have different phases and interfered to each other. But it is only superposition of coherent waves with various phases, and superposition of coherent waves with the same frequency does not reduce the degree of polarization. In our OTDR, the length of scattering cloud related to the superposition is 1m or 0.5m for time gating. The phases of backscattered lights distributed in this length are temporally fluctuated due to thermal and acoustic fluctuation cause the spectrum broadening of the Rayleigh scattering, which was discussed by H.Z.Cummins[9] who estimated the Rayleigh backscattering line width less than 100MHz and measured that the line width broadening of Rayleigh scattering of $CO_2$ was measured to be 6kHz. For CW Rayleigh backscattering we confirm the spectrum broadening to 2.21.MHz with 0.84 MHz deviation in 10 times measurements. Assuming that interference in the Stokes vector measurements is the main cause of DOP degradation, it is said that the measured DOP variation depending on pulse width may make sense. The further study is necessary to confirm the DOP degradation in OTDR.

## V. CONCLUSION

We demonstrated a PMD distribution measurement based on p-OTDR JME (Jones Matrix Eigenvalue) method with the assumption of spherical Rayleigh backscattering with interference DOP reduction, and success in regeneration of PMD distribution of test fibers fabricated by splicing constituent fibers with known PMDs. It is a very surprise that the PMD regenerated by the statistical correlation of slowly varying PMD along fiber length shows the linear accumulation of constituent fiber PMD in the short range of 6km. The reverse direction measurements show reasonable regeneration of supposed linear accumulation along fiber length, which strongly suggests the PMD distribution along fiber shows linear accumulation instead of root-length accumulation in relatively short fiber with smaller PMD.

### REFERENCES

[1]    A. Tardy, M. Jurczyszyn, F. Bruyere and M. Herz; OFC'95, paper ThD2 (1995)
[2]    T. Ozeki; A. Mitsuoka, M. Yoshimura, and T. Kudou; OFC'96, 295 (1996)
[3]    F. Corsi, A. Galtarossa, and L. PalmieriJ. Lightwave Technol.vol.16, 1832 (1998)B. Huttner, N. Gisin and J. P. von der Weid , Photonics Technol. Lett, vol.10, 1458 (1998)J. G. Ellison and A. S. Siddiqui;, J. Lightwave Technol.vol.18, 1226 (2000) H. Sunnerd, B. E. Olsson, M .K arisson, P. A. Andreksonand, J. Brentel; J. Lightwave Technol.vol.18, 897(2000)
[4]    T.Ozeki et.al , Optics Letters, vol.28, No.15,1293(2003)., T.Ozeki, K.Shinozuka,and K.Iwasaki, OFC2004,
[5]    T. Kudou, M. Iguchi, M. Musuda, and T. Ozeki;J. Lightwave Technol.vol.18, 614 (2000),    D.C.Poole: Electron. Letters, vol.22,1029(1986)
[6]    H.C.van de Hulst, "Light Scattering by Small Particles",pp.53-57, Dover Publish Co. (1981)
[7]    M.I.Mishchenko, J.W.Hovenier, and L.D.Travis;, pp.77-82, Academic Press, 2000. M .I. Mishchenko and J. W. Hovernier, Optics letters, 20, 1356-1358(1995)
[8]    H.C.van de Hust, "Light Scattering by small particles"pp. 44–45 and p.80, Dover Publication.Inc.New York, 1957.
[9]    H.Z. Cummins; "Laser Light Scattering Spectroscopy", p.247-294, Quantum Optics,   Academic Press,1969

**Geometry Modulation in Optic Communication**
**Dr. Mingwu Gao**
**Zhejiang University**

An optical modulation technique: Geometry Modulation (GM), taking advantage of different geometric field intensity shapes (GFISs) of different modes or their superposition on the same cross section of the Multimode Fiber (MMF), while WDM is still available. The bandwidth of MMF could be boosted dramatically.

**Professor Wei-I Lee**
Department of Electrophysics
National Chiao Tung University
Hsinchu, 30010 Taiwan

### Randomly non-lithographic masking and MOCVD regrowth of GaN micro-hillocks to improve light emitting diode efficiencies

In this work the multiple quantum well LED with micro-hillocks p-GaN on top surface was studied. We use a novel nonlithography masking processes and MOCVD to regrown the micro-hillocks structure. Because the micro-hillocks p-GaN on LED surface change the light path between semiconductor and epoxy materials, the external efficiency of LED can be dramatically increased almost 60%. Compare with use PEC wet etch technology we use regrowth to make micro-hillocks structure on LED also reduced the defect densities on surface, we get the excellent leakage current performance on micro-hillocks LED.

# High Efficiency Deep-Blue Organic Light-Emitting Devices

Yingfang Zhang [a] (1), Yi Zhao [a] (2), Feng He [b] (3), Yuanyuan Lin [a] (4), Chunyan Ruan [a] (5),

Yuguang Ma [b] (6), Shiyong Liu [a] (7)

[a]State Key Laboratory of Integrated Optoelectronics, Qianwei Campus, Jilin University, 2699#, Qianjin

Street, Changchun, 130012, People's Republic of China

[b]Key Lab for Supramolecular Structure and Materials of Ministry of Education, Jilin University,

Changchun, 130012, People's Republic of China

1:zyfanger@163.com, 2:zhaoyi2688@163.com, 3:hefeng@email.jlu.edu.cn, 4:lyy1122@tom.com,

5:ruanlili82@yahoo.com.cn, 6:ygma@jlu.edu.cn, 7:syliu@mail.jlu.edu.cn

**Abstract** *High efficiency, deep-blue organic light-emitting devices(OLEDs) with undoped and doped structures are fabricated. By using oligo(phenylenvinylene) derivatives as blue emitter, the maximum luminance efficiency of OLEDs attains 3.67 and 4.76 cd/A, corresponding to the undoped and doped device. The CIE coordinates of them are (0.15, 0.15) and (0.15, 0.12) at 1000 cd/m$^2$, respectively.*

### Introduction

In the past two decades, organic light-emitting devices (OLEDs) have attracted a great deal of interest.[1-3] For full-color flat-panel display applications, it is essential to deliver a set of primary RGB emitters with high efficiency, saturated color chromaticity, as well as sufficient lifetime. Up to date, especially after the discovery and application of phosphorescent materials in OLEDs, to achieve high efficiency and color saturation for the red and green emitters is no longer a major problem. Although blue phosphorescent device can also obtain higher efficiency comparing with fluorescent one, it is difficult for the former to obtain deep-blue emission. The reason for this lies in two factors: 1) there is no good host materials available that possesses sufficiently large triplet band gap energy to effectively prevent triplet exciton of the dopant from quenching by back energy transfer; 2)and the limitation of short-conjugated ligand selection for synthesizing transition metal-based metal-ligand charge-transfer complexes serves only to compound the already grave and aggravated situation.[4] So, in order to produce a saturated blue OLED with CIEy<0.15, fluorescent emitter should be considered.

Oligo(phenylenvinylene) (OPV) and its derivatives, as promising fluorescent materials in blue OLEDs, have attracted particular attention these years. We have reported sky-blue and white OLEDs based on OPV derivatives.[5-7] In this paper, we report high efficiency, deep-blue OLEDs based on OPV derivatives. At the same time, we show that a doping system comprising of host and guest with similar chemistry structure can efficiently improve the efficiency of the devices.

### Experimental details

Figure 1 shows the chemical structure of an OPV derivatives, 4'-N,N-diphenylaminostyryl-triphenyl (DPA-TP), as well as its PL spectrum in neat film. The PL spectrum has a peak at 451 nm with a full-width at half-maximum (FWHM) of 56 nm, indicating that DPA-TP can be used as a blue emitter in OLEDs. Detailed processes of fabrication and measurement have been described in our previous paper.[8]

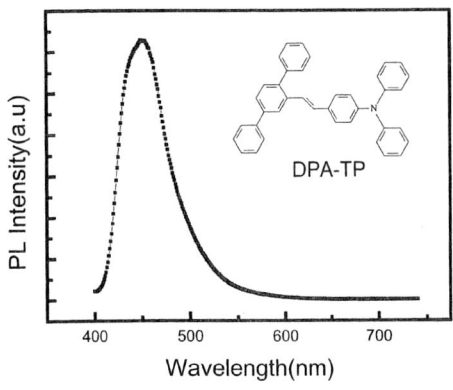

Figure 1. Chemical structure and PL spectrum in neat film of DPA-TP

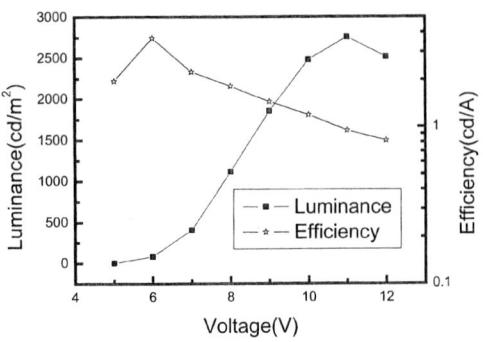

Figure 2. Luminance-efficiency-voltage characteristics of undoped blue OLED

## Results and discussion

### Undoped blue OLED

The device has a configuration of ITO/ PEDOT:PSS/ TCTA (40 nm)/ DPA-TP (20 nm)/ BAlq (30 nm)/ LiF (0.5 nm)/ Al. In this device, 20nm-thick DPA-TP is used as the blue emitter, PEDOT:PSS combining with 4,4',4''-tri(N-carbazolyl) triphenylamine (TCTA) as the hole injection/transporting layer (HIL/HTL), bis(2-methyl-8-quinolinolato)(p-phenylphenolate) aluminum(III) (BAlq) as exciton-blocking layer and electron-transporting layer (ETL), ITO and LiF/Al bilayer as anode and cathode, respectively.

Figure 2 shows the luminance (squar) and efficiency (star) versus driving voltage characteristics of the undoped device. The device attains a maximum luminance of 2756 cd/m$^2$ at 11V and a maximum luminous efficiency of 3.67 cd/A at 6V, as well as a CIE coordinates of (0.15, 0.15) at 1000 cd/m$^2$. It is in the region of deep-blue. As an undoped deep-blue OLED, this efficiency is fairly high. We attribute this to TCTA and BAlq, which both have a large energy gap and the former has a small electron affinity and the latter has a large ionization potential. Both of them can prevent carriers from diffusing into the opposite electrode, thereby confine carriers into the emitting layer and improve the efficiency.

### Doped blue OLED

As we known, doping is an effective method to improve the efficiency of OLEDs.[9] Using a host-guest doping system, through choosing a proper host material, it can not only depress the concentration quenching of the guest, enhance the efficiency, but also improve the color saturation of device. Hereinafter, we discuss a doped blue OLED.

Considering the advantage for the guest to disperse into the host easily, we choose another OPV derivative, 2,5,2',5'-tetrastyryl-biphenyl (TSB), as the host material in our experiment. The chemical structure of TSB is shown in Fig. 3. As both OPV derivatives and having similar chemical structure, the guest is expected to disperse into the host easily and uniformly.

TSB

Figure 3. Chemical structure of TSB

The structure of the doped device is ITO/ PEDOT: PSS/ TCTA (40 nm)/ TSB: 8wt.%DPA-TP (20 nm)/ BAlq (30 nm)/ LiF (0.5 nm)/ Al. In this device, DPA-TP doped TSB is used as blue emitter.

The current density versus voltage characteristics of the undoped and doped devices are depicted in Fig. 4. Because of the high conductivity of the host material, the current density of the doped device is higher than that of the undoped device at the same voltage. It is indicated that the injection and balance of carriers in the doped device is easily achieved. The higher current density also results in an enhanced luminance. The maximum luminance of the doped device attains 4000 cd/m$^2$ at 14 V, which is higher than that of the undoped one. At the same time, owing to the depression of concentration quenching by doping, the maximum luminance efficiency of the doped device also increases and attains 4.76 cd/A at 5 V.

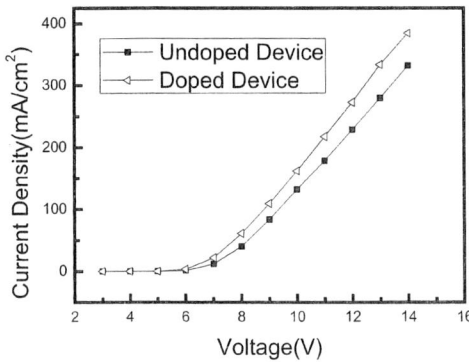

Figure 4. Current density versus voltage of the doped and undoped devices.

The other advantage of doping is improving the color saturation. Figure 5 shows the EL spectra at 100 mA/cm$^2$ of the two devices. Comparing with the PL spectrum in Fig. 1, we can see both devices denote the emission from DPA-TP, but the main peak of the doped device blue shifts from 460 nm to 448 nm, and the FWHM also decreases. As a result, the CIE coordinates of the doped device is (0.15, 0.12) at 1000cd/m$^2$, which is deeper than that of the undoped one. The improvement of the color saturation should owe to the dispersing of the DPA-TP in TSB, which decrease the probability of intermolecular aggregation.

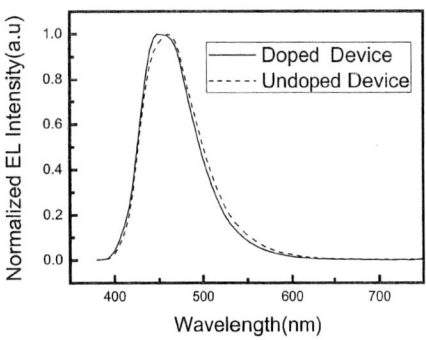

Figure 5. EL spectra at 100mA/cm$^2$ of the doped and undoped devices.

## Conclusions

High efficiency, deep-blue OLEDs have been fabricated based on OPV derivatives. Using large energy gap HTL and ETL to confine carriers into the emitting layer, undoped OLED with a multi-layer structure can obtain a maximum efficiency of 3.67 cd/A, with a CIE coordinates of (0.15, 0.15) at 1000 cd/m$^2$. Furthermore, by utilizing doping system comprising of host and guest with similar chemical structure, effectively decreasing the intermolecular aggregation of the guest and depressing the concentration quenching of it, improved efficiency and color saturation are achieved. The maximum efficiency of the doped device attains 4.76 cd/A with a CIE coordinates of (0.15, 0.12) at 1000cd/m$^2$. So, large energy gap carriers transporting materials and proper doping system are benefit for blue OLEDs.

## References

1 Chishio Hosokawa et al, Appl. Phys. Lett. **67**(1995) 3853

2 Yanqing Li et al,　Adv. Mater. **14**(2002),1317

3 Ying Kan et al, Appl. Phys. Lett. **84**(2004) 1513

4 Shi-Jay Yeh et al,　Adv. Mater. **17**(2005), 285

5 Gang Cheng et al, Semicond. Sci. Technol. **19**(2004) L78

6 Feng He et al, Chem. Commun. (2003) 2206

7 Yingfang Zhang et al, Appl. Phys. Lett. **88**(2006) 223508

8 Yingfang Zhang et al, Appl. Phys. Lett. **86**(2005) 011112

9 Jianmin Shi et al, Appl. Phys. Lett. **80**(2002) 3201

# 2×2 juxtaposed Type Optical Fiber Raman Temperature Sensor

zhang Lixun(1), Liu Yongzhi(1), Ou Zhonghua (3),Dai Zhiyong, Peng Zengshou

1.+862883204363, School of Opto-eletronic Information UESTC, Chengdu, Sichuan 610054,China,zlx@uestc.edu.cn

2.+862883204363,yzliu123@uestc.edu.cn, 3.+862883204363,ozh@uestc.edu.cn

**Abstract:** We report on a 2×2 juxtaposed type optical fiber Raman temperature Sensor. Two output ports of the 2×2 couple make coupling of the optical fiber respectively and two optical fiber made fascis will be a item of the sensor optical fiber .On the condition that the quasi-continuous wave inputs, the couple has the nonlinear switch theory, hence the signals received by the detector will take four times of what a single sensor optical fiber does. A new demodulated method for the simultaneous distributed temperature measurement in an optical fiber based on spatially resolving the Ant-Stokes signals of the spontaneous Raman back-scattering , the sensitivity of sensor is heightened. We make up the experimental system and achieve 5km of length,1m of distance resolution ,accuracy to $\pm 0.1\,°C$ .

**Keywords:** Raman backscattering ;Rayleigh scattering ; distributed optical fiber; temperature.

## 1. INTRODUCTION

Optical fiber temperature sensor is one of the distributed measurement apparatus that applied to the smart structure of citizen and military engineering. Most commercially available distributed temperature sensors are based on the principles of Raman Optical Time-Domain Reflectometry (OTDR) in which a short pulse light is launched into the sensing fiber. From the backscattered light spectrum, a very weak, but temperature-sensitive band (the so-called anti-Stokes Raman band) is selected. The anti-Stokes and Stokes scattered radiation is sensitive to the fiber temperature and be used to infer the temperature along the fiber length. For eliminating fiber characteristics which affect the signals, the ratio of the anti-Stokes to the Stokes signals has been reported by Dakin[1] and the ratio of the anti-Stokes to the Rayleigh signals has been reported by Hartog[2].their sensitivity are 1.065%、 0.862%[3], respectively. Rayleigh backscattered intensity in the 12/125 single mode fiber is estimated to −35dBm of the incident power while the anti-Stokes power is estimated to −65dBm of the incident power and Stokes power is estimated to −55dBm of the incident power[4-5]. The criteria[6] by which a distributed sensor is

measured include its spatial resolution, i.e. its ability to distinguish two closely spaced events along the fiber. Also of interest is the range, i.e. the length of fiber that can be addressed by a single instrument. The spatial resolution and the range determine the equivalent number of points that the sensor is capable of addressing. Most commercially available systems are capable of addressing at least 300 points and the highest performing systems, up to 30000 points. Clearly, the sampling rate of the digitizing electronics must be sufficient faithfully to replicate, in real time or by interleaving, the backscatter signals received. Further performance criteria include the resolution of the measure and which is usually limited by signal-to-noise ratio at the receiver although some suppliers have chosen to define this performance parameter in relation to the resolution of their display. This paper report that two output ports of the 2 × 2 couple make coupling of the optical fiber respectively and two optical fiber made fascis will be a item of the sensor optical fiber .On the condition that the quasi-continuous wave inputs, the couple has the nonlinear switch theory, hence the signals received by the detector will take four times of what a single sensor optical fiber does. A new demodulated method for the simultaneous distributed temperature measurement in an optical fiber based on spatially resolving the Ant-Stokes signals of the spontaneous Raman back-scattering , the sensitivity of sensor is heightened. We make up the experimental system and achieve 5km of length,1m of distance resolution ,accuracy to $\pm 0.1\,^{\circ}\!C$ .

## 2 .Principles of distributed temperature sensing

### 2.1 coupling nonlinear switch

To relate our calculations to the optical fiber temperature properties, we consider a pulse laser(1.55 μ m 10ns,quasi-continuous wave) coupled to the sensor optical fiber. Using the coupled mode theory[6], we relate the, electricfield $E_n$ (1,2,3,4,5)

$E_t , E_r$ (see Fig. 1) in the coupling region as

$$E_2 = i\sqrt{1-\rho}E_1 , E_3 = \sqrt{\rho}E_1 \quad (1)$$

$$E_4 = E_2 \exp(-\alpha_0 L/2 + in_1 kL + i\gamma \mid E_2 \mid^2 L) \quad (2)$$

$$E_5 = E_3 \exp(-\alpha_0 L/2 + in_1 kL + i\gamma \mid E_3 \mid^2 L) \quad (3)$$

$$E_r = (\sqrt{\rho}E_5 + i\sqrt{1-\rho}E_4)\exp(-\alpha_a L/2 + in_1 kL)\sqrt{K_a S v_a^4 R_a(T)} \quad (4)$$

$$E_t = (\sqrt{\rho}E_4 + i\sqrt{1-\rho}E_5)\exp(-\alpha_a L/2 + in_1 kL)\sqrt{K_a S v_a^4 R_a(T)} \quad (5)$$

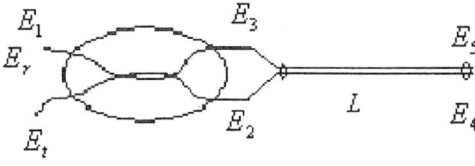

Fig. 1. Model for the $E_1$ coupled to the sensor optical fiber.

Where $\rho$ is the coupling coefficients, $n_1$ is

refraction index, $k$ is the azimuthal wavenumber, $\gamma$ is

the SPM coefficients. $S$ is back-scattering factor of

optical fiber, $L$ is the probe length of optical

fiber, $K_a$ is coefficient concern of Anti-Stokes

scattering cross section, $V_a$ is frequency of

Anti-Stokes scattering photon, $a_0 , a_a$ is loss

coefficient of optical fiber on the frequency of incident laser ,Anti-stokes scattering frequency, respectively;

$R_a(T)$ is population of lower and upper molecular

energy level, it is dependent on the local domain temperature of optical fiber:

$$R_a(T) = [\exp(h\Delta v/kT) - 1]^{-1} \quad (6)$$

Where $\Delta v$ is wavelength shift,

$\Delta v = 1.32 \times 10^{13} Hz$ , $h$ is the Plank's

constant, $k$ is Boltmann's constant, $T$ is absolute temperature. we obtain the ratio of the transmission

power $P_t$ to the incident power $P_1$ as

$$\frac{P_t}{P_1} = \left| \frac{E_t}{E_1} \right|^2 = 2\rho(1-\rho)(1+\cos(\gamma P_1 L(1-2\rho)))$$
$$\bullet \exp(-\alpha_0 L - \alpha_a L)K_a S v_a^4 R_a(T) \quad (7)$$

the ratio of the reflect power $P_r$ to the incident $P_1$

$$\frac{P_r}{P_1} = \left| \frac{E_r}{E_1} \right|^2 = (\rho^2 + (1-\rho)^2 - 2\rho(1-\rho)\cos(\gamma P_1 L(1-2\rho)))$$
$$\bullet \exp(-\alpha_0 L - \alpha_a L)K_a S v_a^4 R_a(T) \quad (8)$$

if $\rho = 0.5$, We can obtain these equations :

$$\frac{P_t}{P_1} = \exp(-\alpha_0 L - \alpha_a L)K_a S v_a^4 R_a(T) \quad (9)$$

$$\frac{P_r}{P_1} = 0 \quad (10)$$

hence the signals received by the detector will take four times of what a single sensor optical fiber does.

### 2.2 old demodulated method

Anti-Stokes Raman backscattering photon flux :

201

$$N_a(T,l) = N_e K_a S v_a^4 R_a(T)$$
$$\cdot \exp[-(a_0 + a_a)l] \qquad (11)$$

RayLeigh backscattering photon flux:

$$N_R = K_R \cdot S \cdot v_0^4 \cdot N_e$$
$$\cdot \exp[-(a_0 + a_0) \cdot l] \qquad (12)$$

Where $l$ is the length of optical fiber, $K_R$ is coefficient concern of Rayleigh scattering cross section, $v_0$ is frequency of Rayleigh.

Raman backscattering lights are separated from the Rayleigh line by the optical fiber coupler and the optical grating. The Raman light signals are converted to the electrical signal by the Silicon or InGaAs avalanche photodiode(APD).The electrical signals are averaged approximately $2^{12} \backsim 2^{15}$ times by a DSP to decrease the white noise which is mainly generated by the dark-current of APD. Anti-Stokes Raman back-scattering photon flux OTDR curve is demodulated by Rayleigh scattering photon flux OTDR curve

$$\frac{N_a(T,l)}{N_R} = \frac{K_a}{K_R}(\frac{v_a}{v_0})^4 R_a(T)\exp[-(a_a - a_0) \cdot l] \qquad (13)$$

$$r_a = \frac{N_a(T,l)/N_R}{N_a(T_0,l)/N_R} = \frac{R_a(T)}{R_a(T_0)} \qquad (14)$$

$T_0$ is know, The local domain temperature $T$ on the optical fiber is measurement from formula (14).

The electrical signal by digital average is

$$r = \frac{\eta \chi_{SNR} N_a(T,l)/N_R + e(T)}{\eta \chi_{SNR} N_a(T_0,l)/N_R + e(T_0)} \qquad (15)$$

Where $\chi_{SNR}$ is gene of signal-to-noise by digital average, $\eta$ is coefficient of optoelectronic converted quantity, $e(T)$, $e(T_0)$ is system noise under $T$, $T_0$ respectively.

According to formula (14) and (15),Error formula as following[5]:

$$\left| r - \frac{R_a(T)}{R_a(T_0)} \right| = \left| \frac{e(T) - re(T_0)}{\eta \chi_{SNR} N_a(T_0,l)/N_R} \right| \qquad (16)$$

If $T$ is more bigger than $T_0$, the $r$ is more great, so the demodulated error of the formula (14) is more large.

### 2. 3 New demodulated method

In the 0 °C $\backsim$ 120 °C range, $r \in [1,2.3]$ ,a demodulate method is that the temperature range would be divided in several orthogonal regions, a value of temperature would be obtained by formula (14) and the one must belong to a certain region; the up and down boundary temperature of the certain region take as the demodulated conditions, the $r$ comes being litter.

Set the measured temperature range $[T_0, T_N]$

$$[T_0, T_N] = \bigcup_{i=0}^{n-1} A_i, A_i = [T_i, T_{i+1}] \qquad (17)$$

where

$$T_{i+1} - T_i = \frac{T_N - T_0}{n}, i = 0,1,\cdots,n-1, T \in A_i$$

new demodulates method :

$$r_s = \frac{N_s(T,l)/N_R - N_s(T_i,l)/N_R}{N_s(T_{i+1},l)/N_R - N_s(T_i,l)/N_R}$$
$$= \frac{R_s(T) - R_s(T_i)}{R_s(T_{i+1}) - R_s(T_i)} = \frac{R_a(T) - R_a(T_i)}{R_a(T_{i+1}) - R_a(T_i)} \qquad (18)$$

,Error formula as following :

$$\left| r_s - \frac{R_a(T) - R_a(T_i)}{R_a(T_{i+1}) - R_a(T_i)} \right| =$$
$$\left| \frac{e(T) - e(T_i) - r_s[e(T_{i+1}) - e(T_i)]}{\eta \chi_{SNR}[N_s(T_{i+1},l)/N_R - N_s(T_i,l)/N_R]} \right| \qquad (19)$$

$$e(T) - e(T_i) \approx e(T_{i+1}) - e(T_i), r_s < 1 \text{ ,formula}$$

(19) explains that the demodulation by formula (18) would restrain the system noise at a certain extent.

### 2.4 The relative sensitivity of system

according to formula (14) and (18):

$$\frac{dr_s}{r_s dT} / \frac{dr_a}{r_a dT} = \frac{1}{1 - R_a(T_i)/R_a(T)} \quad (20)$$

$$T_i < T, R_a(T_i)/R_a(T) < 1 \quad , \quad \text{then,} \quad \frac{dr_s}{r_s dT} >$$

$$\frac{dr_a}{r_a dT} = 1.065\%.$$

### 3  Experimental setup and measurements

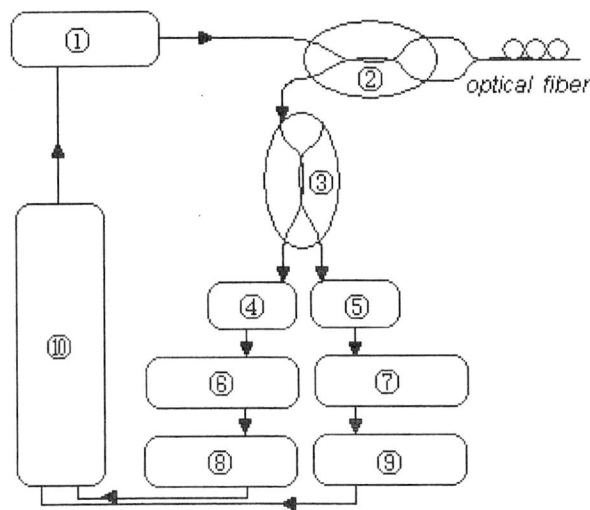

Fig.2 Construction of distributed temperature sensor,①pulse laser ②50:50coupling ③97:3coupling ④filter ⑤filter ⑥APD⑦ APD⑧A/D⑨A/D⑩computer

**Tab.1 demodulated temperature datum（℃）**

| No. | Formula（6） | Formula(9) |
|-----|-----------|-----------|
| 1 | 46.05 | 45.05 |
| 2 | 44.56 | 45.14 |
| 3 | 45.37 | 44.89 |
| 4 | 46.68 | 45.12 |
| 5 | 43.55 | 45.15 |
| 6 | 48.25 | 45.05 |
| 7 | 46.12 | 44.88 |
| 8 | 45.43 | 45.01 |
| 9 | 47.74 | 44.95 |
| 10 | 46.23 | 44.85 |
| average | 45.99 | 45.01 |
| uncertainty | 3.25 | 0.15 |

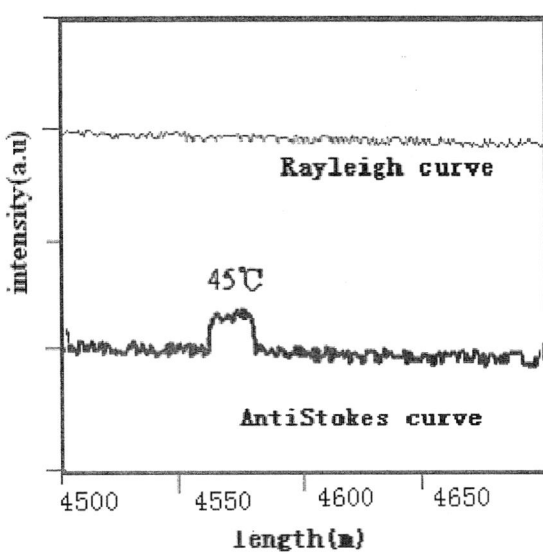

Fig.3 Raman temperature OTDR curves
from formula (14)

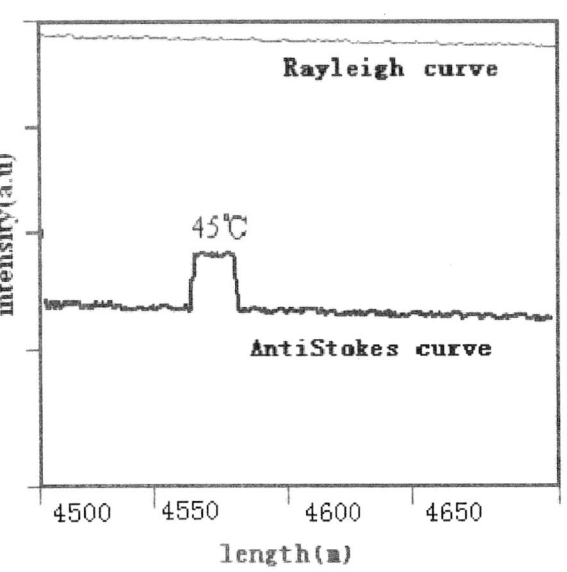

Fig.4 Raman temperature OTDR curves
from formula (18)

The experimental setup using a distributed temperature sensor system is shown on Fig.2.an average power 1W10ns pulse laser is launched into the sensor fiber. As it proceeds, it interacts with the silica and dopant matrix and causes some energy to be scattered in all direction. A proportion of this scattered light makes its way back to the same end of the fiber of 5km meters. There is a linear one-to-one correspondence between the time-varying output signal and the distance along

203

the fiber which lay into the homo-box of 5℃.

Adjust the box's temperature to 40℃,we obtained the down boundary temperature's Anti-Stokes and Rayleigh waveform; adjust the box's temperature to 50 ℃, we obtained the up boundary temperature's Anti-Stokes and Rayleigh waveform. when the fiber of 4.52km to 4.54m laid into the box ,adjust it's temperature to 45℃,we make ten times experiments at half hour's interval. The Rayleigh, anti-Stokes and Stokes waveform of the formula (14) is shown in Fig.2, the Rayleigh, anti-Stokes and Stokes waveform of the formula (18) is shown in Fig.3.

Fig.2 and Fig.3 show the detector can be different response to the anti-Stokes backscattering intensity. According to calculation of the Fig.2 and Fig.3 waveform ,Tab.1 lists a group of datum which are demodulated by formula (14) and (18).

### 4.CONCLUSIONS

The distributed optical fiber sensor (DOFS) system is a multipoint ,on line system of optical mechanic electric and computer integration developed recently to take real time measurement of spatial temperature field. We report on a 1.55 μ m 2×2 juxtaposed type optical fiber Raman temperature Sensor, the sensitivity of sensor is heightened.making up the experimental system ,we achieve 5km of length,1m of distance   resolution ,accuracy to ±0.1 ℃ .

### References

1 Dakin J P ,Pratt D J, Bibby G W.distributed optical fibre Raman temperature sensor using a semiconductor light source and detector,Electronics Letters, 1985,21:569-570

2 Hartog A H,Leach A P,Gold M P. distributed temperature sensing in solid-core fibres Electronics Letters,1985,21:1061-1062

3 Namkung J S,Aude C W,Lavarias A C.fiber optic distributed temperature sensor using Raman backscattering.SPIE, Vol.1918:82-87,1993

4 Zhang Z X, Wang J F, Liu H L. 30km Long Distance Distributed Optical Fiber Raman Temperature Sensor . Journal of Optoelectronics.Laser,2004,15(10):1174～11779

5 Zhang L X, Ou Z H, Liu Y Z. et al. A circulated demodulated method of distributed fiber Raman temperature sensor. ACTA PHOTONICA SINICA, 2005 34(8):1176~1178

6 Zhang L X, Liao Y, Dai Zh Y, et al. Demarcated method of distributed fiber Raman temperature sensor . Journal of Optoelectronics.Laser,2006,17(6):772～774

6 Glynn Williams, George Brown, William Hawthorne. Distributed temperature sensing (DTS)to characterize the performance of producing oil wells. Proceedings of SPIE Vol. 4202 (2000):39～54.

**Dr. Michael Titov**
**Technica-Pro**
**Moscow Russia**

## PORTABLE LASER THERAPEUTIC DEVICE WITH PULSE-MODULATED REGIMES MODE

The Firm "Technica-Pro" (Russia) creates new type portable laser therapeutic device &laquo;Creolka-M&raquo; on base of series device &laquo;Creolka&raquo; with taking into consideration of theoretical conceptions and experimental-clinical date. The one of peculiarities of new type of portable device is that fact that besides of ordinary regime with continuous radiation wavelength 635 or 650 nm we provide three additional regimes with various pulse-modulated radiation mode with modulation 1,2; 2,4 and 10 Hz. Many researches showed that these kind of modulation have favorable therapeutic action. The device is conducted with help only 2 control keys. There are 4 various electronic keys for each regime, and also system of light and audible indication. The construction of portable device provides creation of marked contrast between brightly lit part (place of effect of laser beam) and skin, the most part of which is closed by frame of device from external lighting; it is excluded re-reflection of light by tissue into space. As a result space distribution of absorbed energy in space becomes more uniform, that positively effects on the development of therapeutic effect Low Level Laser Therapy.

**M.R.Nasiri Avanaki**
**Electrooptics Department, MS.c Student of University of Semnan, Iran**


**Full Progress of Digital Signal Processing in Open Loop-IFOG**

The fiber optic gyroscope, which is solid state sensor, represents a highly successful impact of fiber optics on rotation sensing. It has advantages of small size and weight, and low cost. In the construction process, there are two things that cause to reduce accuracy; Quality of optical components and DSP program. In this paper we reviewed structure and operation of Interferometric Optical Fiber Gyro and in continue, described process of derivation of rotation rate from output current and simulated this approach.

## A Fast Convergence Self-Adaptive Bit Allocation Algorithm for OFDM Systems

**ZHENG Pei-chao1, SONG Han-tao1, LIU Bin1,2**

(1. School of Computer Science and Technology, Beijing Institute of Technology, Beijing 100081, China;
2. College of Economics and Management, Hebei University of Science and Technology, Shijiazhuang Hebei 050018, China)

Adaptive technology is proposed to enhance the capability for OFDM Systems. Dichotomy combining the Lagrange-multiplier method is the optimal bit allocation algorithm to estimate the value of . That algorithm has a good convergence in general, but it becomes very low when the target is far from the center of the estimating interval of . This paper presents a fast convergence self-adaptive bit allocation algorithm based on Lagrange-multiplier method for OFDM Systems. Simulation results show that the proposed algorithm has faster convergence than Dichotomy.

Key words: OFDM, Lagrange-multiplier, Bit Allocation

**Mr. Bin Liu**
**Beijing Institute of Technology, China**

**Dynamic Load Balancing Based on Restricted Multicast Tree in Homogeneous Multiprocessor Systems**

Multicast tree can effectively decrease the communication time delay in a network due to single source node multicasts messages to some other destination nodes. In the proposed dynamic load-balancing (DLB) algorithm, multicast tree is adopted to decrease the communication time delay of exchanging load information among processors. To avoid wrongly transferred or redundant load balancing messages that may be produced because of the overlapping of multicast trees, specified rules are proposed to restrict multicast tree construction. To improve the DLB efficiency in homogeneous multiprocessor systems, the proposed algorithm is designed as distributed controlled, sender initiated and can help heavily loaded processor with complete distribution of redundant loads with minimum number of executions. Experiments have been executed to compare the effects of the proposed DLB algorithm and other three ones, and the results prove the effect and practicability of the proposed algorithm in dealing great scale compute–intensive task.

9780978921705